Semiconductors: Physics, Materials and Properties

Semiconductors: Physics, Materials and Properties

Edited by **Jared Jones**

Willford Press

New York

Published by Willford Press,
118-35 Queens Blvd., Suite 400,
Forest Hills, NY 11375, USA
www.willfordpress.com

Semiconductors: Physics, Materials and Properties
Edited by Jared Jones

International Standard Book Number: 978-1-68285-058-9 (Hardback)

Printed in the United States of America.

Contents

Preface

This book has been an outcome of determined endeavour from a group of educationists in the field. The primary objective was to involve a broad spectrum of professionals from diverse cultural background involved in the field for developing new researches. The book not only targets students but also scholars pursuing higher research for further enhancement of the theoretical and practical applications of the subject.

This book aims to provide a comprehensive overview of semiconductor materials and devices. The objective of the text is to give an in-depth analysis of semiconductor materials, physical and electrical properties, applications of semiconductor devices in electronics, etc. The various studies that are constantly contributing towards advancing technologies and evolution of this field are examined in detail. The extensive content of this book provides the readers with a thorough understanding of the subject.

It was an honour to edit such a profound book and also a challenging task to compile and examine all the relevant data for accuracy and originality. I wish to acknowledge the efforts of the contributors for submitting such brilliant and diverse chapters in the field and for endlessly working for the completion of the book. Last, but not the least; I thank my family for being a constant source of support in all my research endeavours.

Editor

Enhanced superconducting properties of the Pb doped trilayer high-temperature ($Bi_2Sr_2Ca_{n-1}Cu_nO_{2n+4\delta}$) cuprate superconductor

Widad M. Faisal[1] and Salwan K. J. Al-Ani[2*]

[1]Department of Electronic and Communication, Faculty of Engineering and Petroleum, Hadramout University of Science and Technology, Yemen.
[2]Department of Physics, College of Science, University of Baghdad, Baghdad, Iraq.

A ceramic superconductor compounds with the composition $Bi_2Sr_2Ca_{n-1}Cu_nO_{2n+4\delta}$ were prepared by solid state reaction from the principle routs like Bi-2212,Ca_2CO_3 and CuO with high purity of 99.99%. Different measurement such as resistivity measurement, X-ray diffraction (XRD) and scanning electron microscope (SEM) were made to show the improvement in the superconductivity of the high phase. The lattice constants are obtained for undoped and Pb doped samples with different values of n. The critical temperatures (Tc) of the high temperature superconducting cuprate (HTSC) $Bi_2Sr_2Ca_{n-1}Cu_nO_{2n+4\delta}$ were obtained to be Tc =108K for the undoped sample and equal to 133, 115, 147 and 127K for n=3, n=3.5, n=4 and n=4.5 Pb doped systems, respectively.

Key words: $Bi_2Sr_2Ca_{n-1}Cu_nO_{2n+4\delta}$, HTSC, Pb doping, Electrical Resistivity, XRD, SEM.

INTRODUCTION

Since the discovery of the Bi-Sr-Ca-Cu-O superconducting compounds (Maeda et al., 1988), a great deal of work has been done to prepare the high-critical temperatures (Tc) phase. The most common method used to synthesize these superconducting oxides is the solid state reaction method (Maeda et al., 1988; Koyoma et al., 1988) which consists of the mixing nominal compositions of high purity compounds such as Bi_2O_3, $SrCO_3$, $CaCO_3$ and CuO, pelletization of calcined powder, under a pressure of 4 to 6 tons/cm^2 and heating the mixed powders for 10 to 20 h at 800 to 820°C. The formation of single high-Tc Bi-2223 phase however, is very difficult due to the appearance of other superconducting phases such as Bi-2212 or non superconducting phases. The most effective methods to enhance the volume fraction of the high-Tc phase were found to be:

1) By starting from composition with surplus Ca and Cu than the possible ideal composition of $Bi_2Sr_2Ca_2Cu_3O_x$ (Koyoma et al., 1988); by prolonging the sintering (Nobumasa et al., 1988) which is necessary to from extra layers of Cu-O_2 and Ca planes into the layer structure of the low -Tc phase.
2) The annealing in an atmosphere with low oxygen pressure (Nobumasa et al., 1988; Mizuno et al., 1988) which is effective in lowering the reaction temperature and enhancing the stability of the high - Tc.
3) The substitution of Pb for Bi (Takano et al., 1988; Usai et al., 1988) or the addition of Pb in the $Bi_2Sr_2Ca_2Cu_3O_x$ (Uehara et al., 1988) is very effective in increasing the high-Tc phase. It was found that Pb has a catalytic effect on the reaction to from the high-Tc phase, and a certain amount (about 0.3 mole) of Pb to the $Bi_2Sr_2Ca_2Cu_3O_x$ compound is necessary for the occurrence of this reaction (Usai et al., 1988). Furthermore, the Pb has a role in the stabilization of the high-Tc phase (Takano et al., 1988).

Uehara et al. (1988) studied the magnetic measurement of $BiSrCaCu_2O_x$ and they estimated the high-Tc volume fraction to be 21%. Kumakura et al. (1988) investigated the upper critical field of $BiSrCaCu_2O_x$,

*Corresponding author. E-mail: salwan_kamal@yahoo.com.

with Tc above 100K. They found that at zero fields, the majority of transition occurred above 103K a zero resistivity was attained below 80K, but as the magnetic field increased, the low temperature portion of transition curve shifts to low temperature, resulting in abroad transition. Matsuoka et al. (1990) investigated the effect on the superconducting behavior of Bi-Sr-Ca-Cu-Pb-O system, and found no effect on the Tc, but is effective in increasing the volume fraction of the $Bi_2Sr_2Ca_2Cu_3O_{10}$ compound.

Wu et al. (1992) investigated the effect of sintering temperature and duration on the formation of the 110K phase, in samples of nominal compositions of $(Bi_{3.2}Pb_{0.3})Sr_4Ca_xCu_yO_z$. They observed that the optimum compositional rang for obtaining large fraction of high -Tc phase (Tc=105K) occurs between PbBi-4457 and PbBi-4468.

The Bi:2223 phase has attracted considerable interest due to its higher Tc and the potential for applications. One of the reasons for the low critical current densities J is the granular nature of this sintered high temperature superconducting (HTS) compound. It is difficult to prepare single phase material for Bi:2223 phase. The Bi:2212 was frequently observed as a major impurity phase in Bi:2223 samples because of its greater thermodynamic stability with respect to the Bi:2223 phase. The partial substitution of Pb in Bi sites enhances chemical stability and promotes the formation of the 2223 phase (Pop et al., 2006).

Kovaleva et al. (2004) presented systematic study of the c-axis lattice dynamics in $Bi_2Sr_2Ca_{n-1}CunO_{4+2n}$ (n=1, 2, 3) cuprate superconductors (Bi2201, Bi2212 and Bi 2223) based on spectral ellipsometry investigation on single crystal and comparing them with theoretical shell model calculation.

The observation of the multilayer band splitting in the optimally doped trilayer cuprate $Bi_2Sr_2Ca_2Cu_3O_{10+\delta}$ (Bi2223) by angle-resolved photoemission spectroscopy is reported by Ideta et al (2010). They obtained values of energy gaps 43 (m eV) and 60 (m eV) larger than those for the same doping level of the double-layer cuprates, which leads to the large value of Tc in Bi2223. The aim of this work is to study the enhancement of the super-conducting properties of the Pb doped trilayer high-temperature $(Bi_2Sr_2Ca_{n-1}Cu_nO_{2n+4\delta})$ cuprate super-conductor.

EXPERIMENTAL

Materials

The samples were prepared by the solid state reaction method using highly pure 99.99% powders of Bi_2O_3,$SrCO_3$ and $CaCO_3$,CuO, with appropriate weights in proportion to their molecular weights through the following chemical reaction. Four type of superconductors systems were obtained, namely, $Bi_2Sr_2Ca_{n-1}Cu_nO_{2n+4\delta}$, $Bi_2Sr_2Ca_2Cu_3O_{10}$+Pb, $Bi_2Sr_2Ca_{2.5}Cu_{3.5}O_{11}$+Pb, $Bi_2Sr_2Ca_3Cu_4O_{12}$ +Pb and $Bi_2Sr_2Ca_{3.5}Cu_{4.5}O_{13}$+Pb. The preparation of ceramic superconductor by solid state reaction includes a series of stages as shown in the following, measuring the weight of each reactants with the required amount, using sensitive balance with (4-digits), type (STATON)462AL.

The chemical reactions

The systems were prepared by mixing appropriate amount of the sating_materials through the following chemical reactions:

1) The first system (n, 3)
$Bi_2O_3+2SrCO_3+2CaCO_3+3CuO \longrightarrow Bi_2Sr_2Ca_2Cu_3O_{10}+4CO_2$
2) The second system (n, 3.5)
$Bi_2O_3+2SrCO_3+2.5CaCO_3+3.5CuO \longrightarrow Bi_2Sr_2Ca_{2.5}Cu_{3.5}O_{11}+4.5CO_2$
3) The third system (n, 4)
$Bi_2O_3+2SrCO_3+3\,CaCO_3+4CuO \longrightarrow Bi_2Sr_2Ca_3Cu_4O_{12}+5CO_2$
4) The fourth system (n, 4.5)
$Bi_2O_3+2SrCO_3+3.5CaCO_3+4.5CuO \longrightarrow Bi_2Sr_2Ca_{3.5}Cu_{4.5}O_{13}+5.5CO_2$

Sample preparation

The mixture for each specimen was prepared by homogeneously mixing and grinding prescribed amounts of powders into a gate mortar. Appropriate amounts of these powders were mixed with alumna mortar and pestle for 2 h in 2- propanole and dried. The calcinations process performed at 810°C for 16 h by heating rate of 60°C/h, then cooled to room temperature with a rate of 60°C/h in order to remove the CO_2 from the mixture as gas and then crushed into fine powder. The calcinations and grinding procedure were repeated at least three times. This ensure the completely evolvement of CO_2 from the mixture. The resulting powder was pressed into pellets by using cylindrical die set with a stainless steel cylinder of 13 mm diameter and 1.5 to 1.8 mm thickness using manually hydraulic press PERKIN-ELMER, under a pressure of 0.5 GPa. The mixture with the final step of sample preparation was treated with high temperature and causes particles of the materials to join together, and gradually reduce the volume of pore space between them. The powder is compacted into a pellet shape with a certain pressure and then the powder particles was in contact with one another at numerous sites, with significant amount of pore spaces between them.

In order to reduce the boundary energy, atoms diffuse to the boundaries, permitting the particles to be bound together and eventually causing the pores to shrink. If sintering is carried out for a long time the pores may be eliminated and the material becomes dense. The programming data for this process include the rate of heating 60°C/h up to 860°C for 130 h with the flow of oxygen gas of about 1.25 L/min and then with slow rate of cooling 30°C/h down to room temperature. Resintering the pellets was done for 90 to 130 h using the same method as described earlier.

Doping process

The addition of Pb was done after the calcinations. At the end, the weight of the powder was measured. Then 6% wt of PbO was added as a doping. The mixture of the powder was grinded in agate mortar for 45 min with the present of a suitable amount of 2-propanol. It was oven-dried for 1 h at temperature of 60°C, after which the powder was pressed into pellets as explained above and sintered at a temperature of 850°C

Sample testing

The Tc of the superconducting sample has been measured by

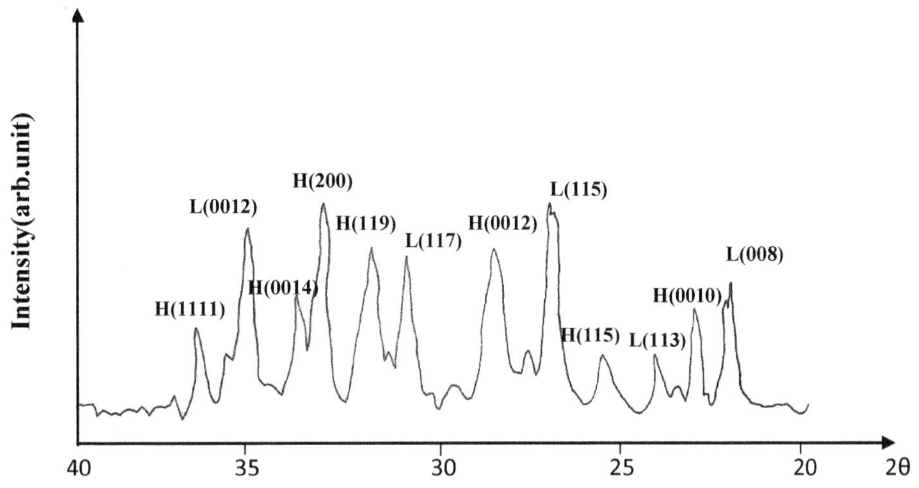

Figure 1. XRD patterns for n=3+Pb for components $Bi_2Sr_2Ca_{n-1}Cu_nO_{2n+4\delta}$.

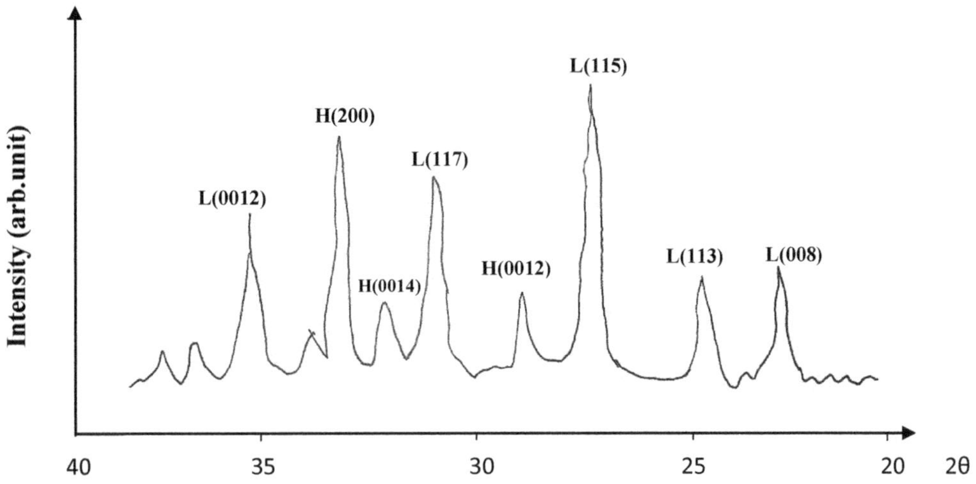

Figure 2. XRD patterns for n=3 for components $Bi_2Sr_2Ca_{n-1}Cu_nO_{2n+4\delta}$.

using the resistivity measurement carried out by four-probe technique, which is considered as a good method for studying the electrical behavior of superconducting materials and a good tool for determining the Tc, although the last sample was measured roughly by Meissner effect. In this method, a small current is passed through a sample and the voltage drop across it. The terminals distinct from those used for passing the main part of the current through the specimen and the electrical contacts of the sample were made of fine copper wires and adhered with silver paste. The cryostat system was used for the measurement of critical resistivity of the sample, with the presence of liquid nitrogen. The cryostat was joined to a rotary pump to get a pressure of $\sim10^{-2}$ mbar inside the cryostat.

Structure and surface morphology

The structure of the samples was studied by X-ray diffraction (XRD) type PHELIPS. A computational program has been used to find the lattice parameter of the unit cell from the pattern of XRD. Fine powder were obtained by grinding the pieces of the samples, then adhered to glass substrate and examined by the X-ray diffract meter. Scanning electron microscope (SEM) type JEOL JSM 6400, was used to study the surface morphology and grain size of the samples for the composition $Bi_2Sr_2Ca_{n-1}Cu_nO_{2n+4}$ with different values of n and Pb addition.

RESULTS AND DISCUSSION

XRD patterns of Pb doped $Bi_2Sr_2Ca_2Cu_3O_{10}$ and undoped high temperature superconducting cuprate (HTSC). These are presented in Figures 1 and 2, respectively. Figures 3 to 5 present the XRD patterns of Pb doped $Bi_2Sr_2Ca_{n-1}Cu_nO_{2n+4\delta}$ for n=3.5, n=4 and n=4.5, respectively.

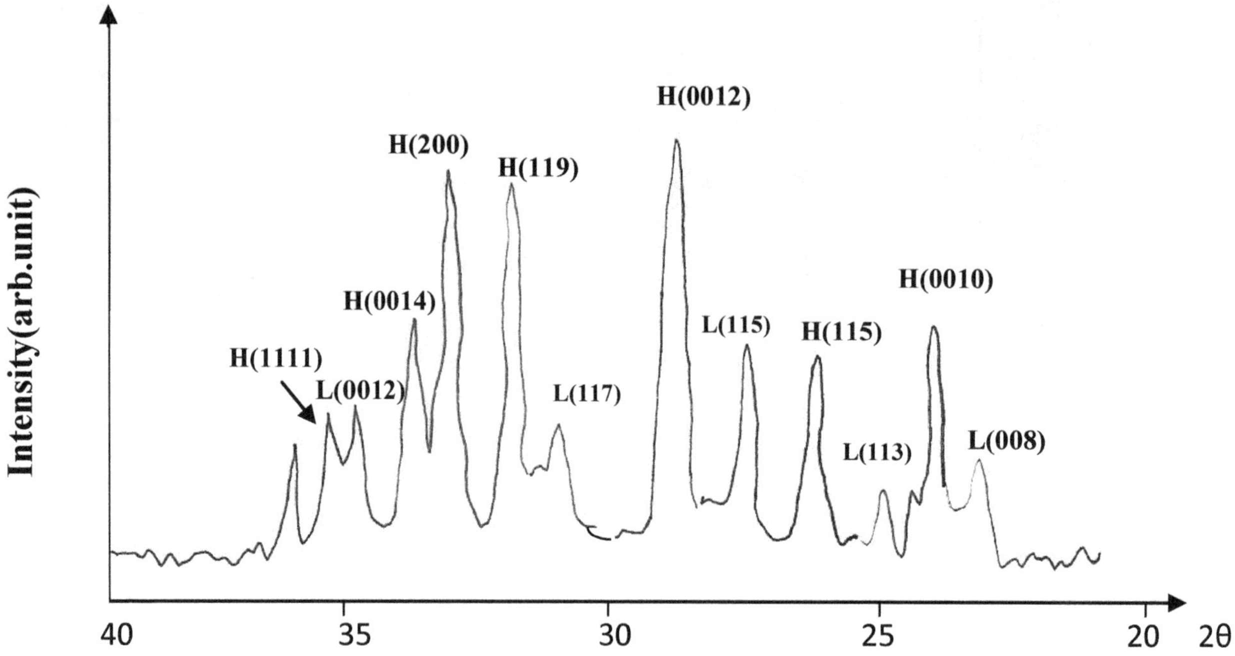

Figure 3. XRD patterns for n=3.5 + Pb for components $Bi_2Sr_2Ca_{n-1}Cu_nO_{2n+4\delta}$.

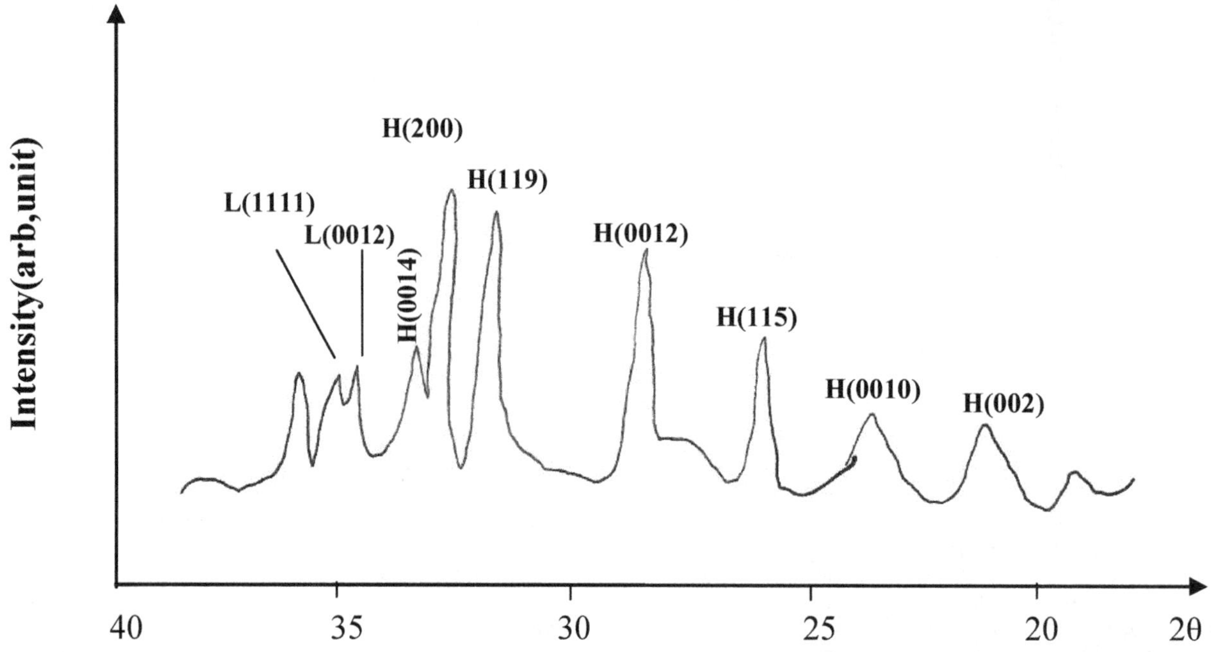

Figure 4. XRD patterns for n=4 + Pb for components $Bi_2Sr_2Ca_{n-1}Cu_nO_{2n+4\delta}$.

The lattice constants are calculated as [(a= 5.4102A°, b=5.4069A°, c=37.978A°); (a=5.4079A°, b=5.4109 A°, c=37.1893A°); (a=5.4094A°, b=5.411A°, c=37.1447A°) and (a=5.4126A°, b=5.4142A°, c=37.1747A°)] for (n=3, n=3.5, n=4 and n=4.5) Pb doped systems.

The presence of Pb in the structure of Bi-2223 compound has a direct influence on the increase of the high-Tc phase, which can be seen clearly in the XRD pattern in Figures 1 and 2. In the first one, the high-Tc phase reflection (0010, 115 and 1111) is created and the

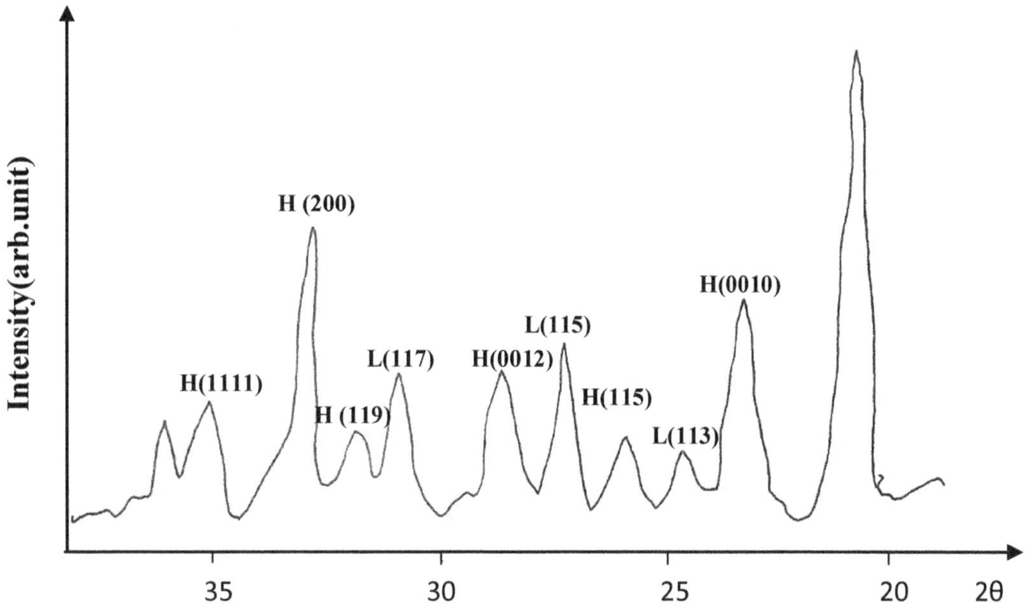

Figure 5. XRD patterns for n=4.5 + Pb for components $Bi_2Sr_2Ca_{n-1}Cu_nO_{2n+4\delta}$.

intensities of the reflection (0012, 119 and 0014) are increased by the addition of Pb. At the same time low- Tc phase reflection (008,113,115,117 and 0012) were reduced compared to the same reflections in Figure 2. We conclude that Pb^{2+} ions occur in a parallel plane between the Bi-O and Sr-O sheets on the c-axis at distance of 3.7, 3.09 and 2.065A°. The Pb^{2+} ions may occupy the Bi-positions (Vasumathi et al., 1990) or the Ca positions (Oota et al., 1988) and the presence of Pb in Bi positions enhances the structure stability of the superconducting phases (Mizuno et al., 2010; Cloots et al., 1993). Our experiments also revealed that the addition of Pb lowers the optimum sintering temperature required to form the high-Tc phase to about 848 to 850°C. This could be attributed to the presence of the $CaPbO_4$ phase in the system (Chavira et al., 1988), which changes the kinetic process of the formation of the Bi-2223 high-Tc phase. $CaPbO_4$ melts at 822°C and thus induces a liquid phase below 850°C and increases drastically, through a dissolution process, the diffusion of the reactive species (Ca^{2+} and Cu^{2+}). The Pb thus acts as a flux allowing the formation of the 2223 phase in a temperature range in which the Bi-2223 phase is thermodynamically stable (Mizuno et al., 2010; Cloots et al., 1993). The volume fraction of high-Tc phase present in the samples was measured using the following relation:

$$Volume\ fraction\ of\ high-Tc\ phase\% \frac{I_H(0012)}{[I_H(0012)+I_L(115)]} x\,100$$

A computer program was used as a helpful tool for the determination of the cell parameters. The XRD patterns for different values of n are shown in Figures 1 to 5. It can be noticed that the samples are mixture of a major high-Tc phase and a minor low-Tc phase; the intensity of the high-Tc phase reflections is greater and the peaks and are sharper than those of the low-Tc phase.

Mizuno et al. (2010) reported that effective methods to increase the volume fraction of high-Tc phase for Pb-Bi-Sr-Ca-Cu-O system are starting from nominal compositions with more Ca and Cu than in Bi-2223 and the addition of Pb to Bi-Sr-Ca-Cu-O system, and annealing at 870°C under higher oxygen pressure than 0.2 atm. Chavira et al. (1988) confirmed that the volume fraction of high-Tc phase is rapidly increased, when a small amount of Pb are incorporated in the Bi-Sr-Ca-Cu-O system. They showed that the XRD patterns reveal the possibility for Pb to occupy Ca or Bi sites and in agreement with our results.

The resistivity of the sample was measured as a function of temperature using standard four-probe technique. For the prepared $Bi_2Sr_2Ca_2Cu_3O_{10}$ sample, the onset Tc is at 108K and the resistivity drops to zero at 100K.

Figure 6 shows the resistivity measurements versus temperature and values of Tc =108K and Tc= 133K for the undoped and Pb-doped HTSC samples, respectively. Figures 7 to 10 show values of Tc=133K, 115K, 145K and 127K for samples (n=3, n=3.5, n=4 and n=4.5) Pb doped systems, respectively.

The SEM observing the surface fractures for all samples (figures 11-14) show the plate –like morphology which is characterized by thin elongated grains with the tendency to align parallel to each other. For n=3, the grains are elongated platelets and their size are between

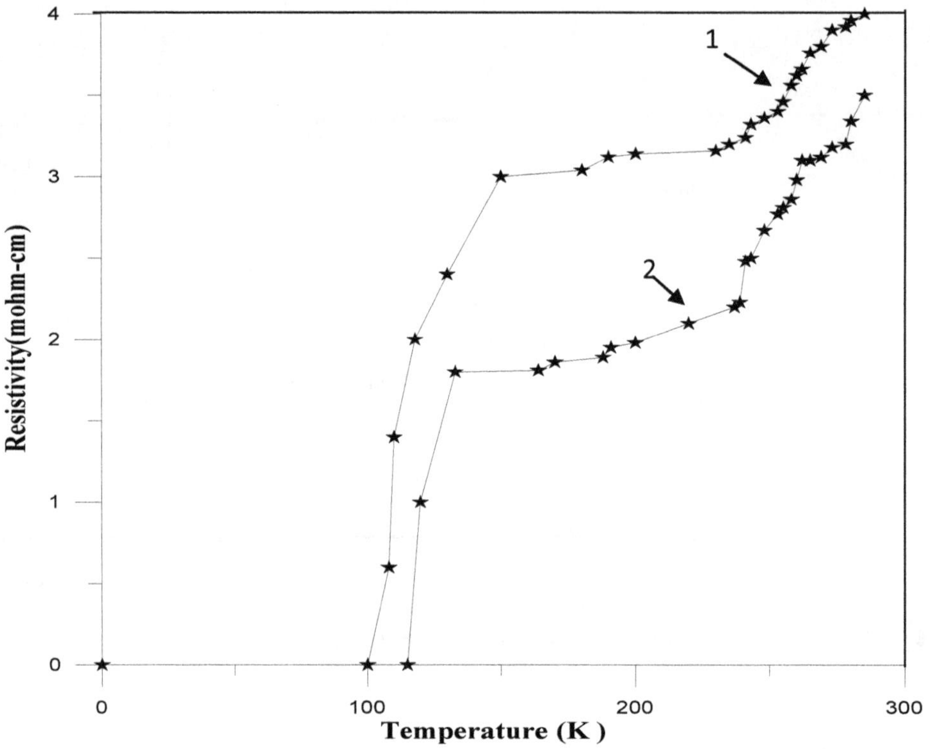

Figure 6. The resistivity versus temperature for HTSC (1) without Pb Tc=108K; (2) With Pb Tc=133K.

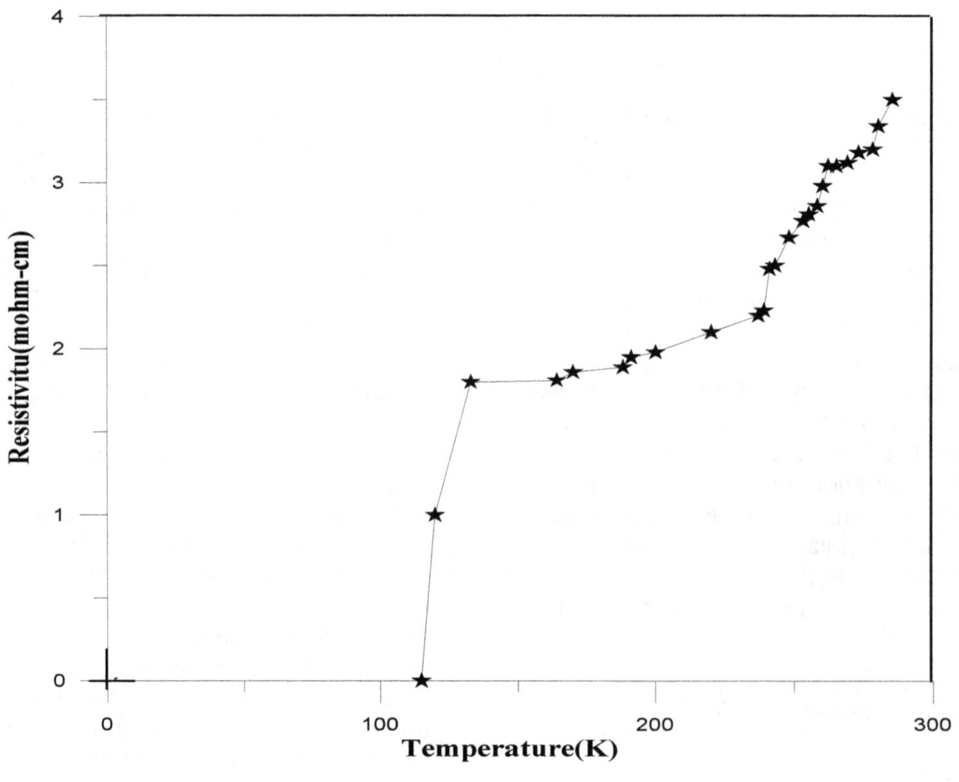

Figure 7. The resistivity versus temperature for HTSC $Bi_2Sr_2Ca_{n-1}Cu_nO_{2n+4\delta}$. Tc, 133K; n, 3+Pb.

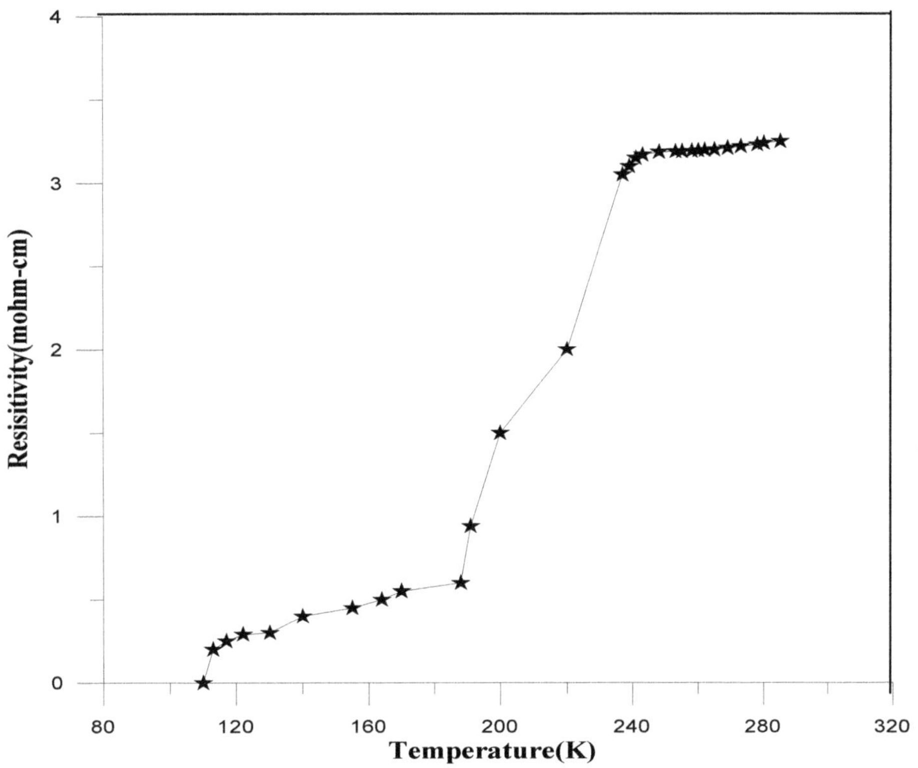

Figure 8. The resistivity versus temperature for HTSC $Bi_2Sr_2Ca_{n-1}Cu_nO_{2n+4\delta}$ + Pb. Tc, 115K; n, 3.5+Pb.

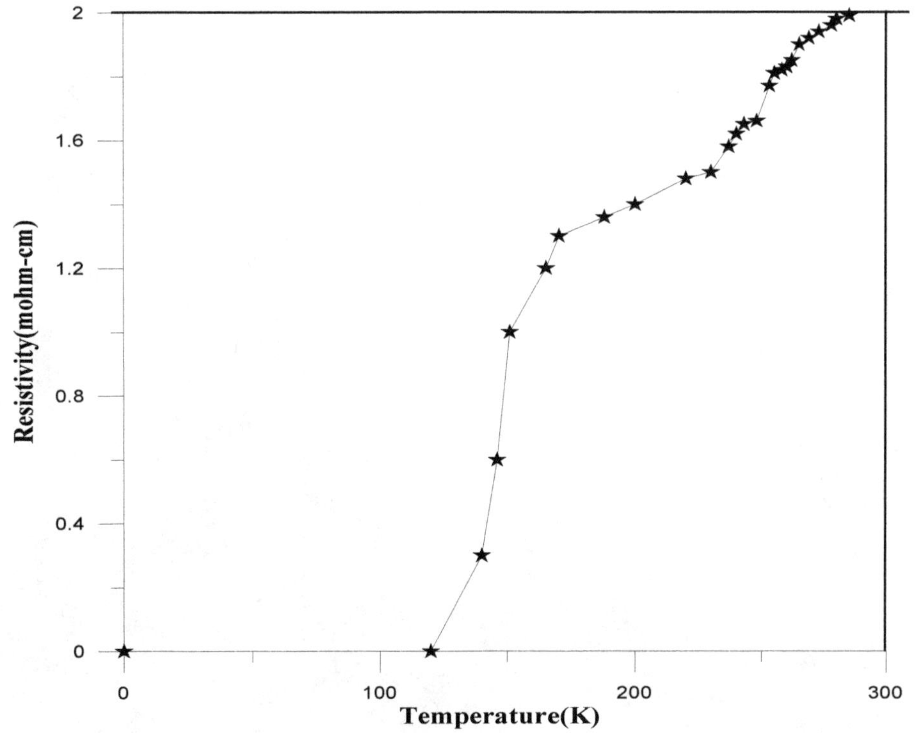

Figure 9. The resistivity versus temperature for HTSC $Bi_2Sr_2Ca_{n-1}Cu_nO_{2n+4\delta}$ + Pb. Tc, 147K; n, 4.

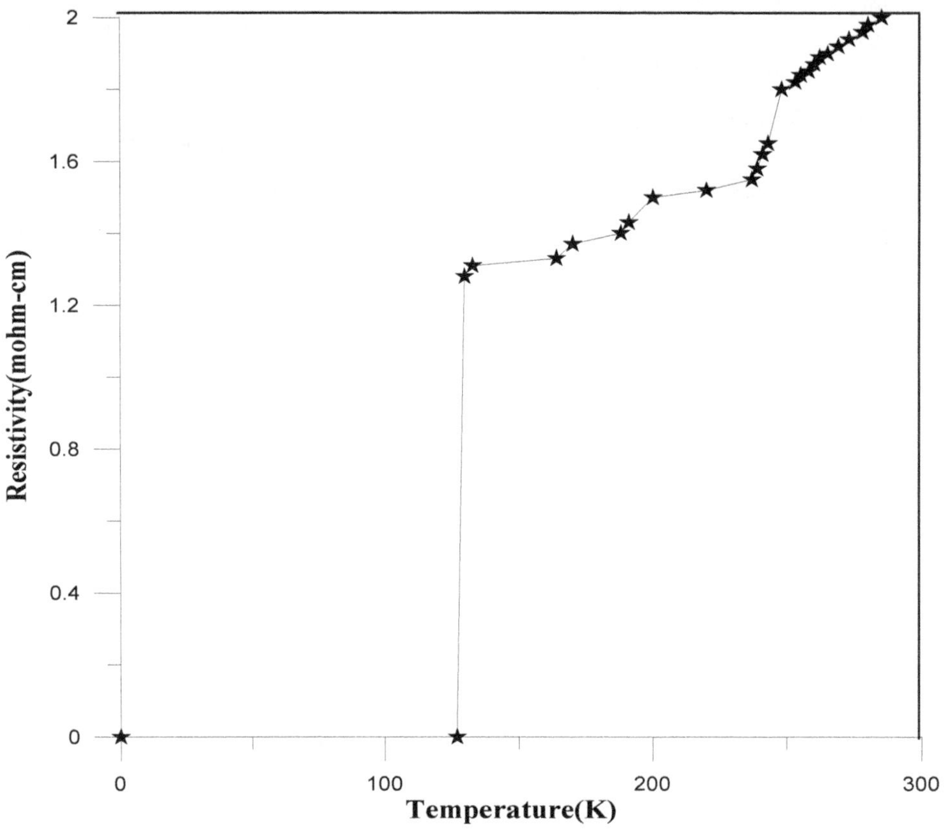

Figure 10. The resistivity versus temperature for HTSC $Bi_2Sr_2Ca_{n-1}Cu_nO_{2n+4\delta}$ + Pb. Tc, 127K; n, 4.5 + Pb.

Figure 11. Morphology of fracture surface of compound $Bi_2Sr_2Ca_{n-1}Cu_nO_{2n+4\delta}$. n, 3 + Pb.

Figure 12. Morphology of fracture surface of compound $Bi_2Sr_2Ca_{n-1}Cu_nO_{2n+4\delta}$. n, 3.5 + Pb.

Figure 13. Morphology of fracture surface of compound $Bi_2Sr_2Ca_{n-1}Cu_nO_{2n+4\delta}$. n, 4 + Pb.

5 to 9 μm in addition to these particles; they are small grains of 1 to 1.9 μm. According to Primo et al. (1992) these aggregates of small grains are mixture of the $Ca_2CuO_2$and CuO.

When n=3.5 and n=4, the size of the plate's shape become large than those when n=3+Pb the size of grains become as large as 48 μm or even larger. The composition with 3.5 ≤ n≤4 seems suitable to grow grains

Figure 14. Morphology of fracture surface of compound $Bi_2Sr_2Ca_{n-1}Cu_nO_{2n+4\delta}$. n, 4.5 + Pb.

of high-Tc phase. Many of these are needle like and big ones are bundle like and these may be responsible for the highest Tc that has been achieved in this study. The grains of samples n=4.5 have the same size. Syono et al. (1988) showed that the XRD patterns and SEM images of $Bi_3Sr_3Ca_2Cu_4O_y$ compound reveals a structure derived from the $Bi_4Ti_3O_{12}$ type in which double bismuth layers along the c-axis and the orthorhombic unit cell dimensions showed an approximate relation of a=√2ap, b=√2ap and c=8ap, where ap =3.82A the parameter of a cubic perovskite and their appearance are similar to samples of n=3 presented earlier.

REFERENCES

Chavira E, Escudero R, Rios D, Jara , Leon LM (1988).Influence of lead on the formation of the 110-K superconducting phase in the Bi-Sr-Ca-Cu-O compounds , Phys. Rev. B., 38(13): 9272-9275.

Cloots R, Romain AC, Rulmont A, Diko P, Duvigneaud PH, Hannay C, Gillet F, Godelain PA, Ausloos M (1993). Supercond study of the crystallization process in $Bi_{2-x}Pb_xSr_2Ca_2Cu_3O_{10-y}$glass systems: optical polarized light microscopy, electrical and magnetic properties, Superconductor Science and Technology, 6(12): 850-857.

Ideta S, Takashima K, Hashimoto M, Yoshida T, Fujimori A, Anzai H, Fujita T, Nakashima Y, Ino A, Arita M, Namatame H, Tanniguchi M, Ono K, Kubota M, Lud HY, Shen ZX, Kojima MK, Uchida S (2010). Phys.Rev.Lett., 104(22): 22700-22704

Kovaleva NN, Boris AV, Holden T, Ulrich C, Liang B, Lin CT, Keimer B, Bernhard C, Tallon JL, Munzar D, Stoneham AM (2004). *C*-axis lattice dynamics in Bi-based cuprate superconductors , Phys. Rev . B 69, 054511-1-054511 -16

Koyoma S, Endo U, Kawai T(1988).Formation of high Tc phase of Bi-Pb-Sr-Ca-Cu-O superconductors", Proceedings of the 1st international symposium on superconductivity, Tokyo Berlin, 88: 833-836.

Kumakura H, Shimizu H, Takashi K, Togano K, Maeda H (1988). Upper critical field of new oxide superconductor Bi-Sr-Ca-Cu-O),Jap. J. Appl. Phys., 27(4): L668- L669.

Maeda H, Tanaka Y, Fukutomi M, Asano T (1988). A new high-T_c oxide superconductor without a rare earth element, Jpn. J. Appl. Phys., 27: L209-L210.

Matsuoka D, Okada M, Murakami T, Cross K, Homma M, (1990).Superconducting behavior in the Bi-In-Sr-Ca-Cu-Pb-O system,Materials Transactions, JIM, 31(9): 755-758.

Mizuno M, Endo H, Tsuchiya J, Kijima N, Sumiyama A, Oguri Y (2010). Synthetic conditions and structural properties of the high - Tc phase in the superconducting Bi-Sr-Ca-Cu-Pb-O system, Phys. Rev. Lett., 104(22): 839-841.

Mizuno M, Endo H,Tsuchiya J, Kijima N, Sumiyama A, Oguri Y, (1988).Synthetic conditions and structural properties of high -Tc phase in the superconducting Bi-Sr-Ca-Cu-Pb-O, Proceedings of the 1st international symposium on superconductivity, Tokyo, Berlin 88:839-841.

Nobumasa H, Shimizu K, Kitano Y, Kawai T (1988). High T_c phase of Bi-Sr-Ca-Cu-O superconductor, Jpn. J. Appl. Phys., 27: L846-L848.

Oota A, Kirihigashi A, Sasaki Y, Ohba K (1988). The Effect of Pb Addition on Superconductivity in Bi-Sr-Ca-Cu-O, Jap, J. Appl. Phys. 27(12): L2289-L2292.

Pop AV, Marconi D, Pop V, Pop MJ (2006). Intergranular dissipation processes induced by Nano defects in (Bi,Pb): 2223 HTS superconductor, Optoelectronics Adv. Mate., 8(2): 476-479.

Primo V , Sapiá F, Sanchis MJ, IbÜez R, BeltrJn A, BeltrJn D (1992).A new improved synthesis of the 110 K bismuth superconducting phase: freeze-drying of acetic solutions, ,Mate. Letters, 15(3): 149-155.

Syono Y, Hiraga K, Kobayashi N, Kikuchi M, Kusaba K, Kajitani T, Shindo D, Hosoya S, Tokiwa A, Terada S , Muto Y (1988). An x-ray diffraction and electron microscopic study of a new high-T_c

superconductor based on the Bi-Ca-Sr-Cu-O system, Jap, J. Appl. Phys., 27(4): L569-L572.

Takano M, Takada J, Oda K, Kitaguchi H, Miura Y, Ikeda Y, Tomii Y, Mazaki H (1988). High-T_c phase promoted and stabilized in the Bi, Pb-Sr-Ca-Cu-O System, Jap, J. Appl. Phys., 27(6): L1041-L1043.

Uehara M, Asada Y, Maeda H, Ogawa K (1988). Magnetic properties of $BiSrCaCu_2O_x$ superconductors,Jap. J. Appl. Phys., 27(4): L665-L667.

Usai T, Sadakata N, Ikeno Y, Kohno O, Osanai H (1988). Preparation of high Tc Bi-Sr-Ca-Cu-O and T1- Ba-Ca-Cu-O superconductor, International symposium on superconductivity (Iss.88), August 28-31 Nagoya Springier -Verlag ,Tokyo, Berlin, Heidelberg.

Vasumathi D, Sundar CS, Bharathi A, Sood AK, Hariharan Y (1990). A positron annihilation study of the decomposition of $Y_1Ba_2Cu_3O_{7-x}$ Physica, C167: 149-156.

Wu Q, Fu Z, Zhang A, Huang J, Tang D, Yao P, Chu S, Yi S, Rong X, Zhang A, Cheng X (1992). Synthesis and crystal growth of the high Tc phase in Bi–Pb–Sr–Ca–Cu–O system with variation of excess concentrations of CaO and CuO, J. Appl. Phys., 71: 2772-2776

2

Structural properties of binary semiconductors

D. S. Yadav[1] and Chakresh Kumar[2]

[1]Department of Physics, Ch. Charan Singh P G College, Heonra, Etawah-206001 (U.P.) India.
[2]Department of Electronics and Communication Engineering Tezpur University, Napam-784001, India.

Using the plasma oscillations theory of solids, 2 empirical relations have been proposed for the calculation of the structural properties such as bond-stretching central force constant (*α*) and bond-bending non-central force constant (*β*) for II-VI and III-V group binary semiconductors. We find that $\alpha = D (\hbar\omega_p)^2$ and $\beta = S (\hbar\omega_p)^2$, where D and S are constants. The numerical values of *D* and *S* are respectively, 0.151 and 0.016 for II-VI and 0.177 and 0.031 for III-V group binary semiconductors. The structural properties of binary semiconductors exhibit a linear relationship when plotted on a log-log scale against the plasmon energy $\hbar\omega_p$ (in eV), which lies on the straight lines. We have applied the proposed empirical relations on these binary semiconductors and found a better agreement with the experimental data as compared to the values evaluated by earlier researchers.

Key words: Plasmon energy, structural properties, III-V and II-VI Semiconductors.

INTRODUCTION

In recent years (Kamran et al., 2008; Kumar et al., 2010; Hasan et al., 2009, 2010; Breidi et al., 2010), II-VI and III-V group binary tetrahedral semiconductors have been extensively studied because of their technical and scientific importance and because they have zinc-blende and wurtzite structure. Using the valence-force-field model of Keating (Keating, 1966), the elastic properties of II-VI and III-V group semiconductors with a sphalerite-structure have been analyzed by Martin (Martin, 1970) and several other workers (Phillips, 1970; Yogurtcu et al., 1981).

A considerable amount of discrepancies have been obtained between theory and experiment in determining vibrational modes on the basis of the model parameters derived from elastic constant data. Nowadays more reliable elastic constant data are available which differ partially from those obtained by Martin (Martin, 1970). In the Martin analysis, the contribution of Coulomb force to the elastic constants has been described in terms of macroscopic effective charge which is responsible for the splitting of transverse and longitudinal optical modes. Lucovsky et al. (1971) has pointed out that the Martin approach (Martin, 1970) is incorrect. The ratio of $\beta/\alpha = 0.3\,(1-f_i)$ measures the importance of covalent bonds in determining the stability of the tetrahedral structures. The overall trend of (β/α) tends to zero for purely ionic crystal that is, for which ionicity (f_i) is unity.

Neumann (1985, 1989) has extended the Keating model (Keating, 1966) considering localized effective charge to account for long range Coulomb force and dipole-dipole interaction in analysing the vibrational properties of binary and ternary compounds with a sphalerite-structure. Neumann (1985) has taken experimental values of bond length d (in nm) and spectroscopic bond iconicity (f_i) (Phillips, 1970) to determine the constant associated with the above theory. Reddy et al. (2003) have reported a simple correlation between lattice energy, bond stretching and bond bending force constants, ionicity and micro-hardness for $A^{II}B^{VI}$ and $A^{III}B^{V}$ semiconductors.

Recently, Yadav and Singh (2012) have calculated the static and dynamical properties of II-VI and III-V group binary solids with the help of plasma oscillations theory. This is based on the fact that the plasmon energy, $\hbar\omega_p = \hbar\ (4\pi ne^2/m)$, is related to the effective number of valence electrons (n) in a compound. In many cases empirical relations do not give highly accurate results for each specific material, but they still can be very useful.

In particular, the simplicity of empirical relation allows a broader class of researchers to calculate useful properties and often trend become more evident. Therefore, we thought it would be of interest to give an alternative explanation for bond stretching central force constant (α in N/m) and bond-bending non-central force constant (β in N/m) of II-VI and III-V group binary semiconductors.

The purpose of this work is to obtain structural properties of the II-VI and III-V group binary semiconductors using the plasma oscillations theory of solids. The theoretical concept is given. We also presented and discussed the simulation results for structural properties of binary semiconductors. Finally, the conclusion is given.

THEORETICAL CONCEPTS

The nearest-neighbour bond-stretching central forces have been characterized by the parameter *α*, and next neighbor bond-bending non-central forces by the parameter *β*. These parameters depend on inter-atomic distance obtained from lattice vibration data. The lattice vibration data have been further obtained from various types of two-body inter-atomic potential given in literature. Such potentials have the advantage of keeping the repulsive and attractive forces in the same mathematical form.

The simplest form of inter-atomic potential has been described by Neumann (1989), Harrison (1983) and Harrison and Freeman (1989) in which it has been assumed that both the repulsive and attractive parts of inter-atomic potential are described by the power law of inter-atomic distance (*d*). This form of potential for the total energy or cohesive energy per pair of atom can be written as (Neumann, 1989):

$$V_1(d) = C/d^m - D/d^n \qquad (1)$$

Where *C, D, m* and *n* are numerical constants and *d* is the nearest neighbour distance.

These parameters have been estimated for equilibrium condition when the repulsion is half of the attraction that is m = 2n (Neumann,1989) and the following equation has been obtained:

$$\alpha = \alpha_0\, d^{-x} \qquad (2)$$

where α_0 and *x* are the constants. The other form of potential is based on Morse potential. In this type of potential both the repulsive and attractive terms are described by exponential functions of inter-atomic distance.

The general form of Morse potential is given by Neumann (1989):

$$V_2(d) = A \exp(-ad) - B \exp(-bd) \qquad (3)$$

Where *A, B, a* and *b* are constants. The above equation has been further used to describe the two-body interaction in total energy calculation of Si (Sobotta et al., 1986). Neumann (1985, 1989) has also extended it to ternary chalcopyrites. Solving above equation (3) for equilibrium condition a = 2b, the following relation has been obtained Neumann (1989).

$$\alpha = \alpha_1 \exp(-bd) \qquad (4)$$

where α_1 and *b* are the numerical constants. Presently, Reddy et al. (2003) has described the correlation between lattice energy, bond stretching force constant (*α*) and bond banding force constant (*β*) for $A^{II}B^{VI}$ and $A^{III}B^{V}$ compound semiconductors as follows:

$$\alpha = m_\alpha U - b_\alpha \qquad (5)$$

$$\beta = m_\beta U - b_\beta \qquad (6)$$

where m_α, m_β, b_α, and b_β are constants.

Recently, Verma (2009) has derived an empirical relations for the bond-stretching and bond-bending force constants of $A^{II}B^{VI}$ and $A^{III}B^{V}$ group binary solids in term of the product of ionic charges of cation and anion with nearest-neighbor distance *d* (in A^0) by the following expressions as:

$$\alpha = A\,(Z_1 Z_2)^S\, d^{-3} \qquad (7)$$

$$\beta = V\,(Z_1 Z_2)^B\, d^{-3} \qquad (8)$$

Where *A, S, V* and *B* are the numerical constants depending upon the group of materials. In a previous work (Yadav and Singh, 2012) we proposed the simple expressions for static and dynamical properties such as bulk modulus B (in GPa) and cohesive energy E_{coh} (in Kcal/mole) of II-VI and III-V group binary solids in term of the plasmon energy $\hbar\omega_p$ (in eV) by the following relations:

$$B = D\,(\hbar\omega_p)^S \qquad (9)$$

$$E_{Coh} = D^*\,(\hbar\omega_p)^{S^*} \qquad (10)$$

Where *D, S D** and *S** are numerical constants. Using this idea to get better agreement with experimental and theoretical data for the bond-stretching and bond-bending force constants of binary semiconductors, equations (7) and (8) may be extended as:

$$\alpha = D\,(\hbar\omega_p)^2 \qquad (11)$$

$$\beta = S\,(\hbar\omega_p)^2 \qquad (12)$$

Where D and S are numerical constants, depends upon the structure of the compounds, semiconductor and plasmon energy ($\hbar\omega_p$) (Yadav and Singh, 2012; Kumar et al., 1996). The values of D and S turn out to be equal to 0.151 and 0.016 for II-VI group and 0.177 and 0.031 for III-V group binary solids, respectively. A detailed study of structural properties of these materials has been given elsewhere (Keating, 1966; Martin, 1970; Phillips, 1970; Yogurtcu et al., 1981; Lucovsky et al., 1971; Van-Vechten, 1969) and will not be presented here.

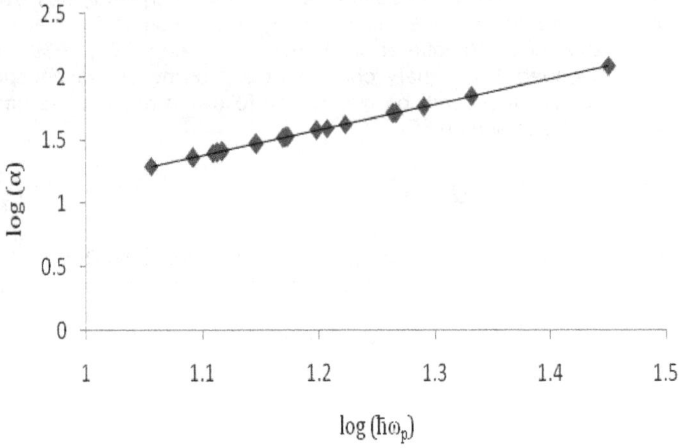

Figure 1. Plot of bond-stretching force constant versus plasmon energy of II-VI group binary semiconductors.

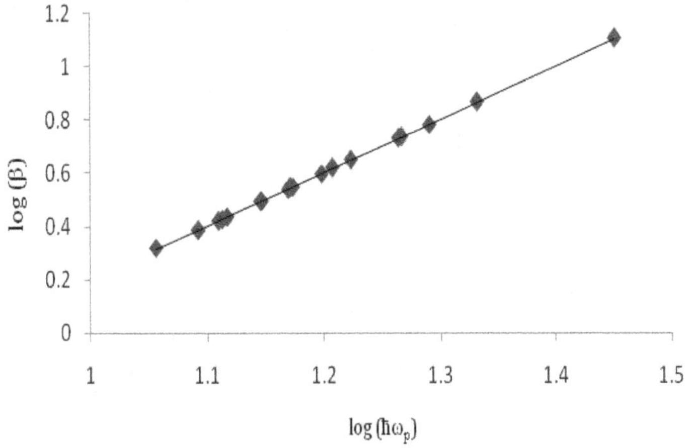

Figure 2. Plot of bond-bending force constant versus plasmon energy of II-VI group binary semiconductors.

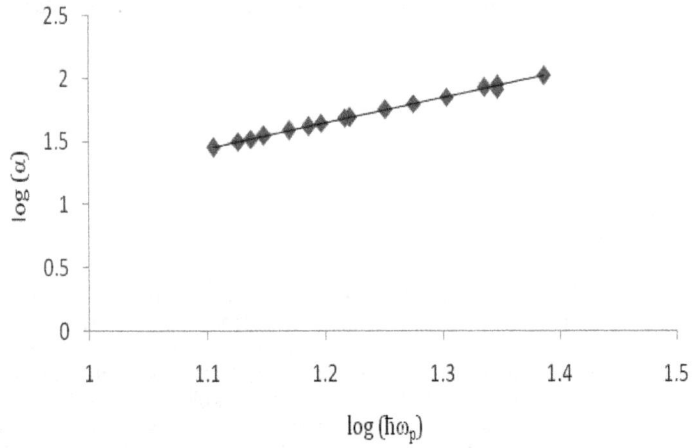

Figure 3. Plot of bond-stretching force constant versus plasmon energy of III-V group binary semiconductors.

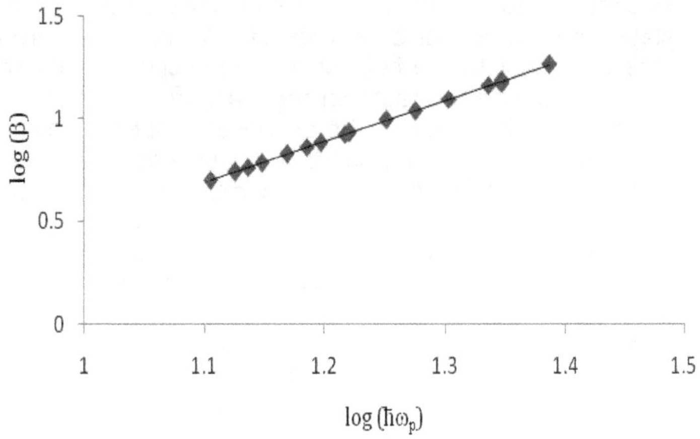

Figure 4. Plot of bond-bending force constant versus plasmon energy of III-V group binary semiconductors.

RESULTS AND DISCUSSION

The present paper reports 2 expressions between the bond stretching force constant α (in N/m) and bond-bending force constant β (in N/m) with their plasmon energy of binary semiconductors. This can be successfully employed to estimate the bond stretching and bond-bending force constant from their plasmon energies. We have plotted log-log curve for II-VI and III-V group binary semiconductors, which are presented in Figures 1 to 4, we observed that in the plot of inter-atomic force constants and $\hbar\omega_p$ of these materials, the data points lies on the straight lines.

From this Figure 1 to 4, it is quite obvious that the bond stretching force constant (α) and bond-bending force constant (β) trends in these compounds increase with increasing their plasmon energy and lies on the straight line. The proposed empirical relations (11) and (12) have been applied to calculate the structural properties of binary semiconductors. The calculated values are presented in Tables 1 and 2, respectively, along with their literature values. We note that the values of the bond stretching and bond-bending force constant calculated from our proposed empirical relations are in fairly good agreement with the values reported by earlier researchers (Neumann, 1985, 1989; Reddy, 2003; Verma, 2009).

These results shows that our current approach is quite reasonable and can give us a useful guide in calculating and predicting the inter-atomic force constants of these materials.

Conclusion

From the above results and discussion obtained by using the proposed empirical relation based on plasma oscillations theory of solids, it is quite obvious that the

Table 1. Structural properties of II-VI group binary semiconductors.

$A^{II}B^{VI}$	$\hbar\omega_p$ (eV)	α (in N/m)				β (in N/m)			
	[16, 21]	Calc.	[20]	[12-14]	[15]	Calc.	[20]	[12-14]	[15]
ZnO	21.48	69.67	---	---	71.31	7.38	---	---	6.973
ZnS	16.71	42.16	42.22	44.73	48.67	4.46	4.46	4.36	4.561
ZnSe	15.78	37.60	36.34	38.61	44.18	3.98	3.84	4.65	4.083
ZnTe	14.76	32.89	29.40	32.04	39.02	3.48	3.11	4.47	3.534
ZnPo	12.36	23.07	---	---	---	2.44	---	---	---
CdO	18.46	51.45	---	---	---	5.45	---	---	---
CdS	14.88	33.43	33.81	---	39.70	3.54	3.57	---	3.606
CdSe	14.01	29.64	30.08	---	35.21	3.14	3.18	---	3.128
CdTe	13.09	25.87	24.38	29.44	30.21	2.74	2.58	2.48	2.602
CdPo	11.43	19.73	--	---	---	2.09	---	---	---
BeO	28.25	120.50	---	---	---	12.76	---	---	--
BeS	19.52	57.53	---	---	62.64	6.08	----	---	6.018
BeSe	18.39	51.06	---	---	56.96	5.41	---	---	5.445
BeTe	16.12	39.24	---	---	45.75	4.16	--	---	4.250
BePo	14.91	33.56	---	---	---	3.55	---	---	---
HgS	14.84	33.25	33.41	---	---	3.52	3.53	---	---
HgSe	13.99	29.55	29.74	37.43	---	3.13	3.14	2.37	---
HgTe	12.85	24.93	24.65	29.32	---	2.64	2.61	2.54	---
MgTe	12.97	25.40	---	---	29.83	2.69	---	---	2.555

Table 2. Structural properties of III-V group binary semiconductors.

$A^{III}B^{V}$	$\hbar\omega_p$ (eV)	α (in N/m)				β (in N/m)			
	[12, 30*]	Calc.	[15]	[20]	[12- 14]	Calc.	[15]	[20]	[12- 14]
BN	22.75	91.60	88.357	---	---	16.04	17.179	---	---
BP	21.71	83.71	75.037	87.14	---	14.61	14.447	15.60	---
BAs	20.12	70.80	67.044	74.95	---	12.54	12.855	13.42	---
BSb	17.85	56.39	---	56.61	---	9.87	---	10.14	---
AlN	22.27	87.78	81.091	97.30	---	15.37	15.705	17.42	---
AlP	16.65	49.06	49.122	48.41	---	8.59	9.219	8.67	---
AlAs	15.75	43.90	44.278	44.34	---	7.68	8.236	7.94	---
AlSb	13.72	33.31	33.137	33.81	35.74	5.83	5.975	6.05	6.63
GaN	20.46	74.10	76.248	95.76	---	14.97	14.722	17.15	---
GaP	16.50	48.18	48.395	48.41	48.57	8.44	9.071	8.67	10.40
GaAs	15.35	41.70	42.098	43.27	43.34	7.30	7.794	7.75	8.88
GaSb	13.38	31.68	31.200	34.19	34.42	5.55	5.582	6.12	7.16
InN	18.82	62.69	60.505	70.70	---	10.97	11.528	12.66	---
InP	14.76	38.56	38.950	38.83	44.29	6.75	7.155	6.95	6.26
InAs	14.07	35.04	35.075	35.79	37.18	6.14	6.369	6.41	5.47
InSb	12.73	28.68	27.567	28.68	30.44	5.02	4.845	5.14	4.73

bond stretching and bond-bending force constant reflecting the structural properties can be expressed in term of plasmon energy of these materials; this is a surprising phenomenon. The calculated values of these parameters are presented in Table 1 and 2, respectively. We come to the conclusion that plasmon energy of any compound is the input parameter for calculating the structural properties. The inter-atomic force constants of binary semiconductors exhibit a linear relationship when plotted on a log-log scale against the plasmon energy $\hbar\omega_p$ (in eV), which lies on the straight lines.

In the present study, we find that both the inter-atomic force constants α and β directly depends upon the plasmon energy of these compounds semiconductor. Thus, this theory can be easily extended to binary semiconductors. A fairly good agreement between our calculated values of inter-atomic force constants with the values reported by earlier researchers has been found. It is also note worthy that the proposed empirical relations are simple and widely applicable.

REFERENCES

Breidi A, Hasan Haj El F, Nouet G, Drablia S, Meradji H, Pages O, Ghemid S (2010). Theoretical study **of** InAs, InSb and their alloys InAsxSb1-x. J. Alloys Compounds 49380.

Harrison WA (1983). A theory of force constant models is derived for tetrahedral semiconductors from Harrison's Phys. Rev. B 27:3592.

Harrison WA, Freeman WH (1989). Electronic Structure and the Properties of Solids (Dover, New York, 1989).

Hasan El Haj F, Annane F, Meradji H, Ghemid S (2010). First principle investigation of AlAs and AlP compounds and ordered $AlAs_{1-x}P_x$ alloys, Comput. Mater. Sci., 50: 274-278.

Hasan El Haj F, Drablia S, Meradji H, Ghemid S, Nouet G (2009). First principles investigation of barium chalcogenide ternary alloys Computational Mater. Sci. 46:376–382.

Kamran S, Chen K, Chen L (2008). Semi-empirical formulae for elastic moduli and brittleness of covalent crystals", Phys. Rev. B 77:094109.

Keating PN (1966). Effect of Invariance Requirements on the Elastic Strain Energy of Crystals with Application to the Diamond Structure, Phys. Rev. 145:637–645.

Kumar V, Prasad GM, Chetal AR, Chandra D (1996). Microhardness and bulk modulus of binary tetrahedral semiconductors, 57(4):503–506.

Kumar V, Shrivastava AK, Jha V (2010). Bulk modulus and microhardness of tetrahedral semiconductors, J. Phys. Chem. Solids, 71:1513.

Lucovsky G, Martin RM, Burstein E (1971). Localized Effective Charges in Diatomic Crystals Phys. Rev. B 4:1367–1374.

Martin RM (1970). Elastic Properties of ZnS Structure Semiconductors. Phys. Rev. B1:4005.

Neumann H (1985). Interatomic force constants and localized effective charges in sphalerite-structure compounds. Cryst. Res. Technol. 20(6):773-780.

Neumann H (1989). Interatomic force constants in AIIBIVCV2 chalcopyrite compounds. Cryst. Res. Technol. 24(6):619-624.

Phillips JC (1970). Bonds and Bands in Semiconductors ,Rev. Mod. Phys. 42:317.

Reddy RR, Nazeer Ahammed Y, Abdul Azeem P, Gopal KR, Sasikala D, Rao TVR, Behere SH (2003). Interatomic force constants, ionicity and microhardness of binary tetrahedral semiconductors. Ind. J. Phys. 77(3):237-241.

Sobotta H, Neumann H, Reide V, Kuhn G (1986). Lattice vibrations and interatomic forces in LiInS2. Cryst. Res. Technol. 21:1367-1371

Van Vechien JA (1969). Quantum dielectric theory of electronegativity in covalent systems. I. Electronic dielectric constant. Phys. Rev. 182:891-905.

Verma AS (2009). Electronic and Optical Properties of Rare-earth Chalcogenides and Pnictides, Afr. Phys. Rev. 3:0003.

Yadav DS, Singh DV (2012). Static and dynamical properties of II–VI and III–V group binary Solids,Phys.Scr. 85:015701.

Yogurtcu YK, Miller AJ, Saunders GA (1981). Pressure dependence of elastic behaviour and force constants of GaP. J. Phys. Chem. Solids 42(1):49-56.

Electrical characterization of conducting poly(2-ethanolaniline) under electric field

Seyed Hossein Hosseini[1] , Ghasem Asadi[2] and S. Jamal Gohari[3]

[1]Department of Chemistry, Faculty of Science, IslamShahr Branch Islamic Azad University, Tehran-Iran.
[2]Department of Chemistry, Faculty of Science and Engineering, Shar-e-Rey Branch, Islamic Azad University, Tehran-Ghom Express Way, Tehran-Iran.
[3]Department of Chemistry, Faculty of Science, Imam Hossein University, Babaee Express Way, Tehran – Iran.

We have studied conductivity and molecular weight obtained poly(2-ethanolaniline), P2EANi. P2EANi synthesized according to the best ratio of obtained molar of initiator to monomer at different reaction times. Then we measured the mass and conductivity of the obtained polymers in the best time of polymerization. Next, we repeated these reactions under different electric fields in the most appropriate time and measured the mass and conductivity of obtained polymers. As a result, intensity of the required electric field for polymerization was determined. After which the polymerization was carried out at the best electric field at different times. Finally, the best time and amount of the electric field for polymerization were determined. Moreover, we studied the doping of polymerization in the presence of an electric field by applying different dopants and other initiators. Then compared the obtained results to the results of a similar condition; but without the electric field. As a result we found the best condition for the reaction was determined as follows; the P2EANi with a high molecular weight was synthesized under the electric field, Mw=193749 g/mol, with Mw/Mn=2.3. The conductivity of the black films oxidized by ammonium peroxydisulfate and doped with dodecylbenzene sulfonic acid cast from NMP was higher than 0.118 S/cm under 10 KV/Cm^2 electric field and showed an enhanced resistance to aging. It can be concluded that polymers synthesized under electric field probably have better physical properties as a result of less branching and high electrical conductivity.

Key words: Polymers, chemical synthesis, electrical conductivities, electric field.

INTRODUCTION

Conducting polymers constitute an emerging class of materials. Among the organic conducting polymers, polyaniline is the only conducting polymer whose properties depend on the oxidation state, its protonation state/doping level as well as the nature of dopants. Among conducting polymers, polyainilne has received greater attention due to its advantages over other conducting polymers. Simplicity of its preparation from cheap materials, superior resistance to air oxidation, and controllable electrical conductivity by doping and de-doping (Pron and Ranou, 2002), makes it very useful in preparing light-weight batteries (Senadeera and Pathirathne, 2004), liquid crystalline polymers (Gato et.al., 2001; Hosseini and Mohammadi, 2009), optical activities (Li et al., 2004), ion exchange materials (Hosseini and Noor, 2005) and sensors (Hosseini et al., 2005, 2006; Hosseini, 2006).

A number of mechanisms for the electric field effects in the chemical reactions have been suggested, well documented, and undergone a proper theoretical analysis (Buchachenko, 2000). Any polar molecule is composed of at least two atoms. Furthermore, a molecule, has an inherent dipole moment with random direction, at room temperature, dipoles and hence the bulk materials are not polarized. An external electric field, applies two equal and opposite forces of any dipole and tries to adjust them with itself. If the field is sufficiently intense and there is no other effective factor, like thermal energy, polarization reaches its saturation value, $P_s=NP_m$. In which P_m is the molecular dipole moment and N is the number of molecules per unit volume. Polarization is between zero and saturation value that can be calculated using a statistical mechanism. From this point of view, the polarizability is (Reits et al., 1979): $\alpha = P_0^2/3kT$ and we have: $P=\alpha E_m$. If we show a monomer molecular dipole moment, by P_0, for a dimmer $P_m=2P_0$, and for a polymer string that is composed of n monomers, $P_m=nP_0$. So as the polymer grows, its polarizability increases as the following: $\alpha= n^2P_0^2/3kT$. This order will increase the electrical conductivity of conductive polymers. Therefore, effects of electric field on doping of conductive polymers can be investigated (MacDiarmid, 2001; Manoha et al., 2002).

Polyaniline and poly2-ethanol aniline are the most of the promising conducting polymers because of its chemical stability and high conductivity. The addition of substituent into the side chain of polyaniline enhances its solubility and processability. On the other hand, there has been no report on polymerization of poly2-ethanol aniline under electric or magnetic fields. Therefore, we think polymerization of poly2-ethanol aniline under electric field can be modified in terms of all its properties. Considering molecular weight, orientation and future applications such as liquid crystalline and optical activities, modified poly2-ethanol aniline can be used. Here, we present the syntheses and properties of poly2-ethanol aniline with a side chain group under an electric field. Therefore, the effect of applying the electric field on polymerization with specific amount of electrical conductivity and molecular weight is rather limited.

EXPERIMENTAL

Chemicals used in this study were American Chemical Society (ACS) grade. 2-Ethanol aniline (Merck) was dried with NaOH, fractionally distillated under reduced pressure from CaH_2. Other chemicals such as, ammonium peroxodisulfate (APS), dodecylbenzene sulfonic acid (DBSA), champhor sulfonic acid (CSA), methane sulfonic acid (MSA), p-toluene sulfonic acid (PTSA), acetic acid (AA), $FeCl_3.6H_2O$, KI and other reagents were purified as per standard procedure before use.

Instrumentals

Conductivity changes were measured with a four probe device (made to the, ASTM Standards, F 43-93). An Electric field device was applied by Hipotronics S.O. No. 004390-00, HV power supply, model 830.50 made in USA. A fourier-transform infrared spectrometer, FT-IR, (Bruker) was used in the spectral measurements of the polymer and graft copolymer and reported (sh=sharp, w=weak, m=medium, b=broad). Proton and carbon nuclear magnetic resonance (FT-^{1}H and ^{13}C NMR) spectra were recorded at 250 MHz on a BruKer WP 200 SY spectrometer. NMR data are reported in the following order: chemical shift (ppm), spin multiplicity (as singlet, doublet, triplet, quartet, multiplet, and broad peak), integration UV-Visible spectra were obtained by Perkin Elmer Lambda 15 spectrophotometer. Molecular weight was measured at 30°C with a gel permeation chromatography (GPC), (Waters Associates, model 150-C). Three styragel packed columns with different pore sizes ($10^4 - 10^6$ A°) were used. The mobile phase was m-cresol with flow rate of 1.5 ml/min. The thermal properties (thermo gravimetric analysis (TGA)) and differential scanning calorimetry (DSC)) of polymer were performed by STA 625-PL Thermal Science and heating rate of 10 °C min^{-1}. Scanning electron microscopy (SEM) was used to study the type of surface morphology of polymer. A Cambridge S-360 SEM was used for this purpose.

Preparation of P2EANi in the absence of electric field

Project participants followed the same instructions to oxidize 0.001 mol 2-ethanol aniline hydrochloride with 0.001 mol ammonium peroxydisulfate, $(NH_4)_2S_2O_8$, in aqueous medium. 2-Ethanol aniline hydrochloride (purum; 0.1 g) was dissolved in 10 ml HCl 1 M in a 50 ml volumetric flask. Ammonium peroxydisulfate (purum; 0.228 g) was dissolved in 10 ml HCl 1 M, and then added. Both solutions were kept for different times (5, 10, 15, 20, 30, 40, 50, 60 and 120 min) at room temperature (~18 to 24°C). The P2EANi precipitate was collected on a filter, and washed with three 10 ml portions of 0.1 M HCl, and acetone. P2EANi hydrochloride powders were dried air and then in a vacuum at 60°C. Weighed and also their electrical conductivities were measured by four probe method (Hosseini et al., 2005, 2006).

The P2EANi powders with $FeSO_4$, KI, $FeCl_3.6H_2O$ as initiators and $HClO_4$, H_2SO_4, AA, PTSA, MSA, DBSA and CSA as dopants were synthesized in similar manner. UV (DMSO); $\square_{max}$= 320 nm (2.45 intensity). FT-IR (KBr): 3550-3100(b), 3015(w), 2930(w), 1615(m), 1450(w), 1515(sh), 1349(m), 1250(m), 1130(sh), 830(m) cm^{-1}. ^{1}H-NMR (DMSO); δ 2.45 (2H, s), 3.50 (2H, t), 3.70 (2H, t), 4.50 (broad), 7.20 (1H, s), 7.35 (1H, s), 7.55 (1H, d), 7.70 (2H, s) ppm. ^{13}CNMR (DMSO); δ 51.5, 54.5, 119.8, 126.5, 127.7, 128.7, 135.1 ppm.

Preparation of P2EANi under electric field

P2EANi was prepared under the same chemical circumstances but with different electric fields (5, 10, 15 and 20 KV/Cm^2), in 60 min, at room temperature. Then, P2EANi was prepared by the same reaction but, in various times (30, 45, 60, 90 and 120 min) intervals at the best amount of electric field. UV (DMSO); $\square_{max}$ = 340 nm (2.80 intensity), 450 nm (0.5 intensity), 590 nm (0.3 intensity). FT-IR (KBr): 3550-3100(b), 3010(w), 2923(w), 1604(m), 1503(sh), 1341(m), 1247(m), 1156(sh), 822(m) cm^{-1}. ^{1}H-NMR (DMSO); δ 2.51 (2H, s), 3.34 (2H, t), 3.67 (2H, t), 4.61 (broad), 7.10 (1H, s), 7.30 (1H, s), 7.34 (1H, d), 7.51 (2H, s) ppm. ^{13}CNMR (DMSO); δ 52.0, 55.4, 121.9, 127.4, 128.9, 129.1, 136.2 ppm.

RESULTS AND DISCUSSION

Spectrometric studies

Figure 1 shows the FT-IR spectrum of P2EANi under the

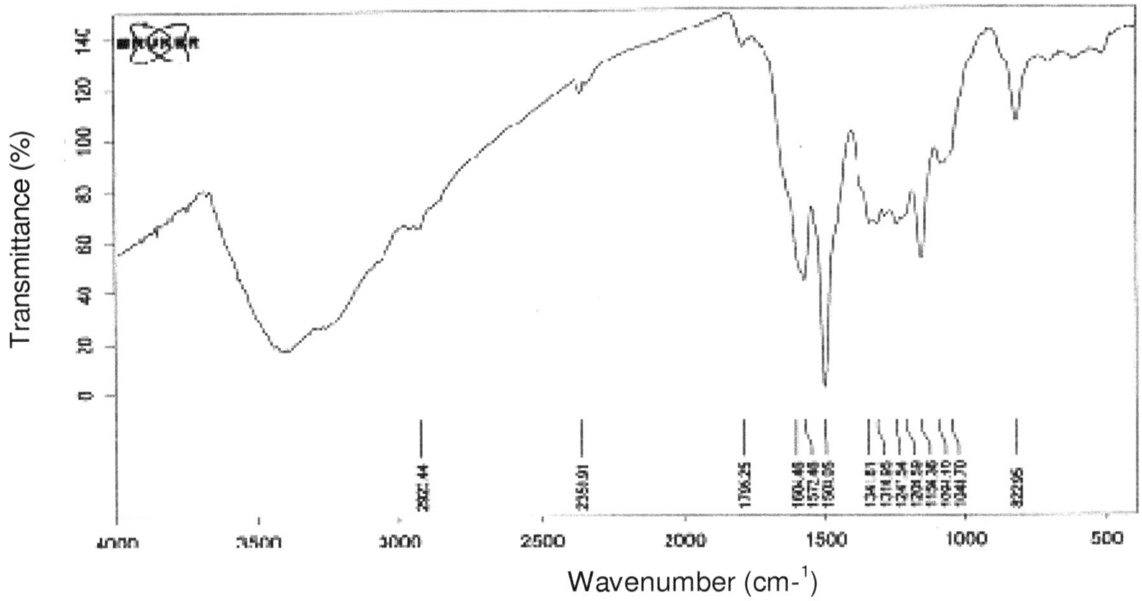

Figure 1. FT-IR spectrum of P2EANi after electric field.

electric field. The special peaks of this spectrum are 3330 cm^{-1} (O-H Stretching H-bonding), 2923 cm^{-1} (C-H stretching aliphatic), 1604 and 1503 cm^{-1}(C=C stretching aromatic) and 1156 cm^{-1} (C-O stretching) and you can see C-H stretching aromatic overlapped with OH stretching.

Figure 2a shows the UV-Vis spectrum of P2EANi prepared without an electric field at the DMSO solvent. In this spectrum the 298 nm peak relates to benzenoid diamine forms. At this spectrum the quinoid diimine from the peak cannot be seen because the polymer is lattice matrix. The UV-visible spectrum of P2EANi under electric field is shown in Figure 2b. This figure shows three peaks, 340, 450 and 590 nm, of which two last ones are related to quinoic form of P2EANi. The two last peaks are not in Figure 2a which relates to P2EANi prepared without an electric field. As we know, by increasing the length of chain and subsequently rising the number of conjugated double bonds, a decrease in energy difference $\pi \rightarrow \pi^*$ occurs, so that it causes an increase in the wavelength. The spectrum of 340 nm region is for the second peak of aniline groups and the $\pi \rightarrow \pi^*$ is a conjugated couple system for the benzoic states and wide peak in 590 nm comes from the transfer of $\pi \rightarrow \pi^*$ quinion of aniline groups (for the polaron and bipolaron transfers).

The ^{1}HNMR spectrum of P2EANi under an electric field in d^6-DMSO as solvent is shown in Figure 3. The peaks observed in 7.1 to 7.5 ppm related to different aromatic protons in polymer. The peak of 4.5ppm is related to OH group of ethanol on the polymeric string. The 3.3 and 3.6 ppm regions are related to ethylenic protons and 2.5 ppm signal as well as protons of DMSO impurity.

Figure 4 shows the ^{13}CNMR spectrum of P2EANi. At this spectrum, peaks of 52.3 and 55.4 regions are related to aliphatic carbons of the ethanol group. The signals of 121.9, 127.4, 128.9, 129.1 and 136.2 ppm regions are related to aromatic carbons of ring.

Thermal properties studies

Figure 5a and b shows the TGA thermograms of P2EANi before and after electric field, respectively. In comparison of figures shows that P2EANi produced under electric field is more stable than P2EANi produced in the absence of an electric field. In Figure 5b, we see that the polymer is stable up to 170°Cb, and loses it's weight at region 170 to 280°C slightly which relates to evaporation of solvent and probable water , and ethanol groups of P2EANi begins to be destroyed and reaches its maximum value at 280°C. The 33.8% loss of molecular weight is the special characteristic of this temperature. At 390°C the lost weight is 59.5% and the polymer will be destroyed completely at 700°C, and 27.7% of initial weight remains to form the ash. Also, the peaks at 250 and 350°C, are related to exothermic states and the peaks at 280 and 390°C, are related to endothermic states of thermal treatment of the polymer that is due to weight loss of the polymer.

Scanning electron microscopic studies

Figure 6a and b shows SEM images of P2EANi synthesized in absence and presence of an electric field,

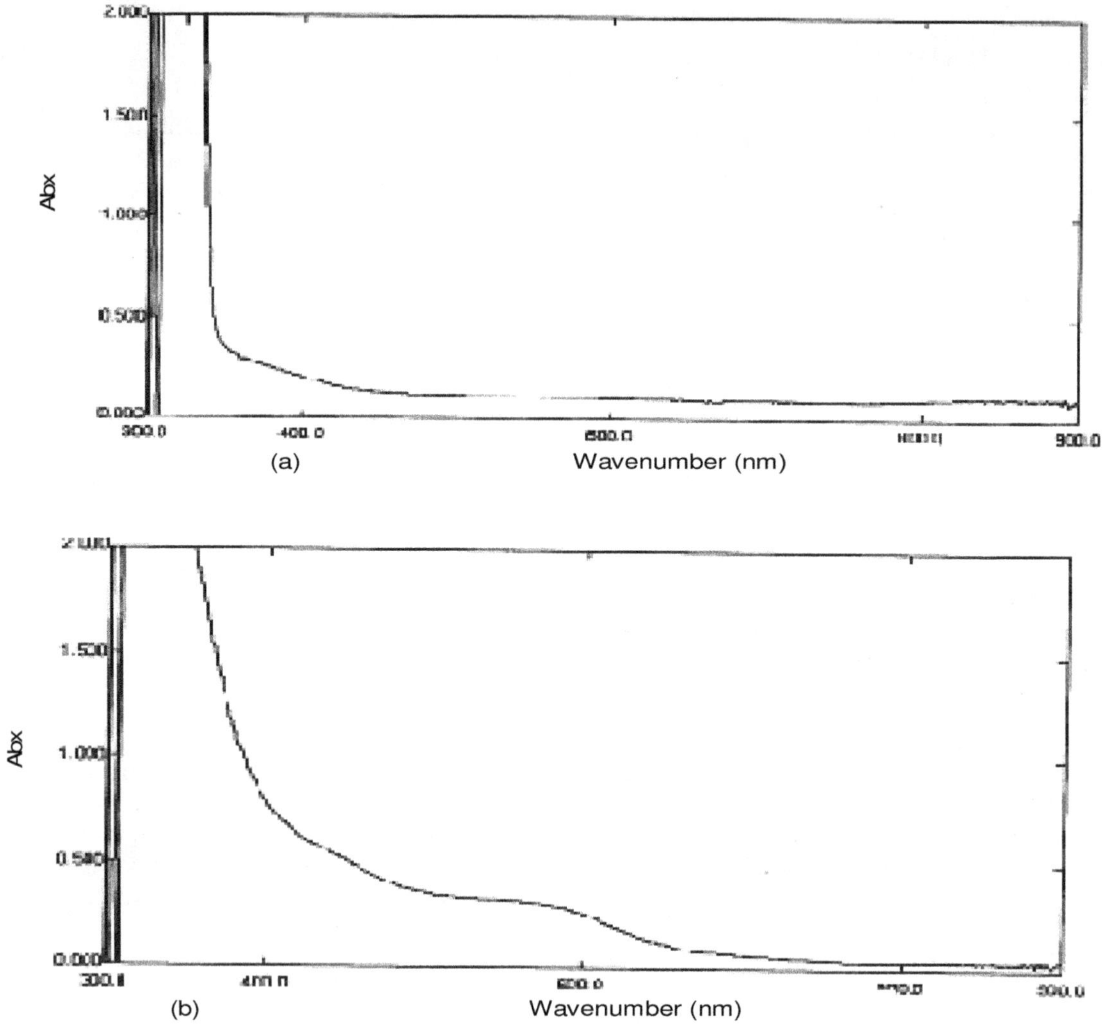

Figure 2. UV-visible of P2EANi prepared in a) absence and b) presence of electrical field in DMSO.

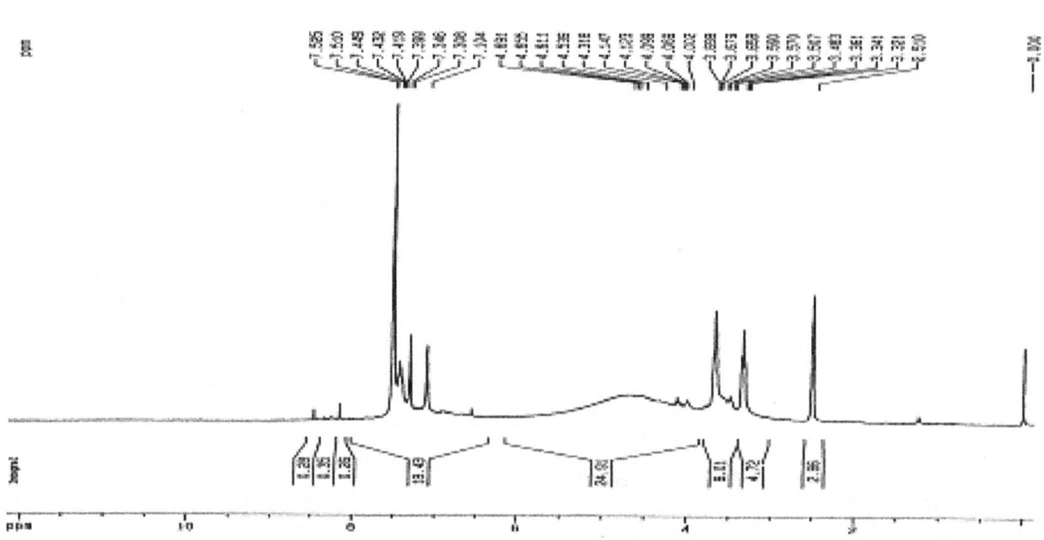

Figure 3. ^{1}HNMR spectrum of P2EANi after electric field.

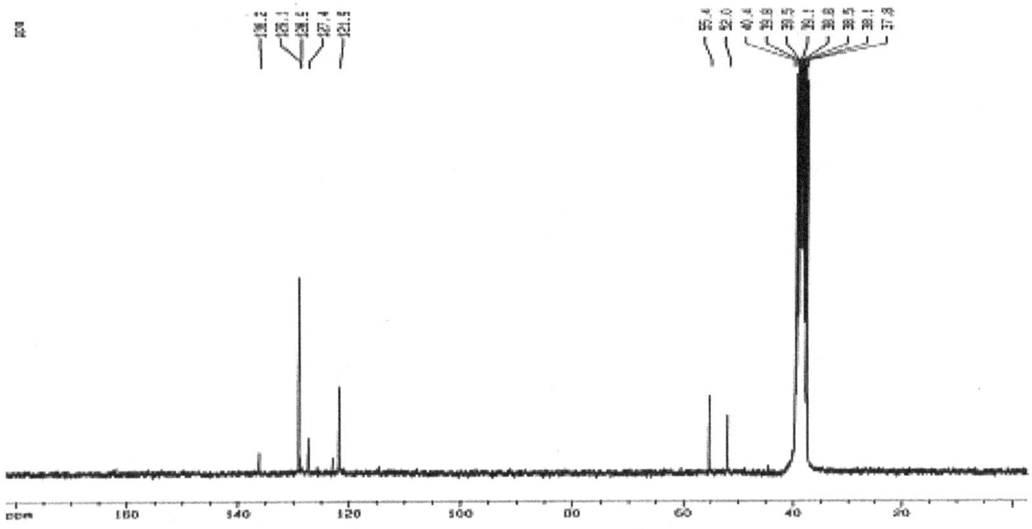

Figure 4. ^{13}CNMR spectrum of P2EANi after the electric field applied.

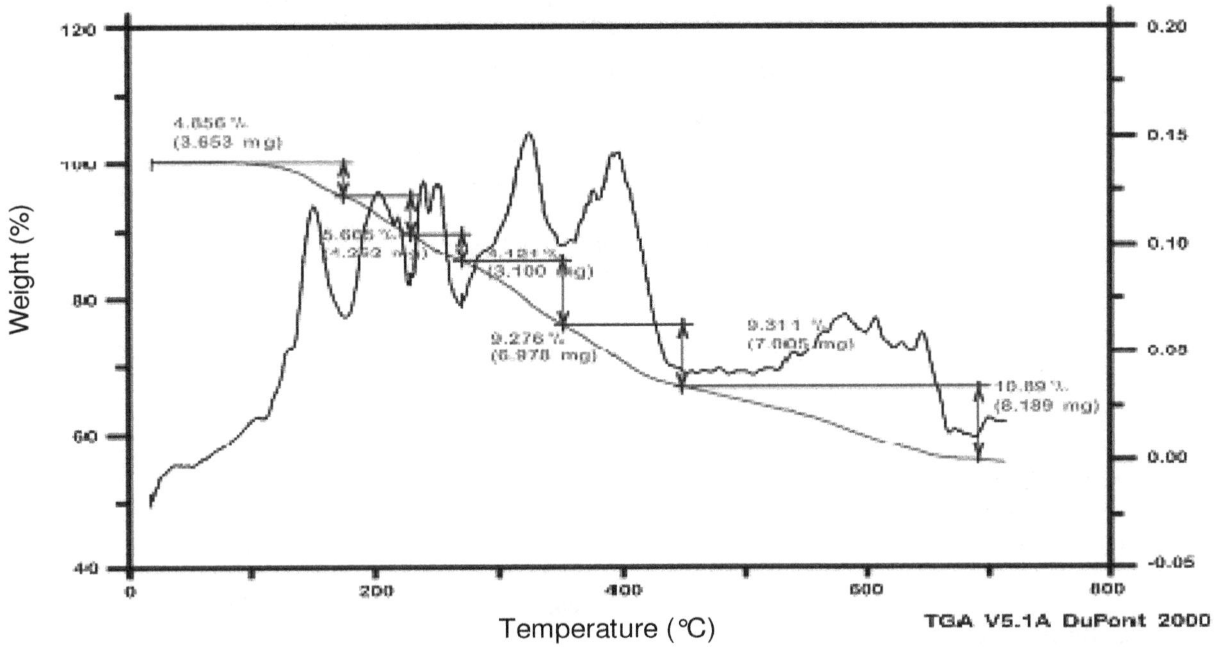

Figure 5. TGA thermograms of P2EANi prepared in a) absence of an electrical field and b) the presence of an electrical field.

respectively. The Figure 6a shows spherical masses of P2EANi chains that are grown on the form of bulk polymerization. But Figure 6b shows the SEM of synthesized P2EANi in the presence of an electric field. Figure 6b shows that P2EANi has a smooth form and relatively homogeneous surface. This is caused by growing more polymer chains with gathering and developing an order in the polymer morphology under an electric field.

Molecular weight distribution studies

To obtain the molecular weight distribution, we used GPC method. In this way, we found that for the prepared polymer in absence of the electric field state, the mean molecular weight is 35440 g/mol, and the scattering is 3.7. Also, the mean weight of molecular mass is 131128 g/mol while for the polymer prepared at the electric field, the mean molecular weight is 84239 g/mol, the scattering

(a) Absent of an electrical field

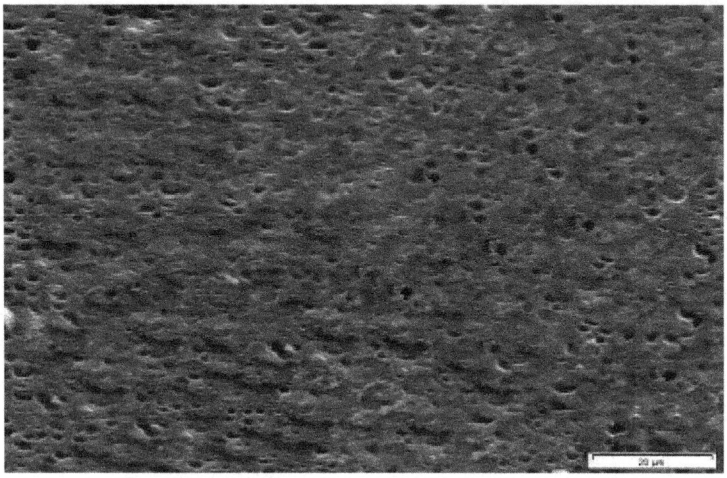

(b) Electrical field present

Figure 6. SEM images of P2EANi prepared in (a) absence and (b) presence electrical field.

is 2.3 and the mean weight of molecular mass is 193749 g/mol. This shows that the electric field has increased the molecular mass by a factor 2.3.

Electrical conductivity studies

The obtained polymer has a high purity because the excess initiator was readily washed with acetone. The purity was confirmed by elemental analysis. Conductivity of all samples were measured at room temperature by a four-probe method on pellets compressed at 700 MPa, 13 mm in diameter and 1 to 1.5 mm thick (Hosseini and Entezami, 2003; Hosseini et al., 2005, 2006). The 2-ethanol aniline was polymerized by using the four different initiators, such as APS, $FeCl_3.6H_2O$, KI and $FeSO_4$. For each initiator, the different dopants like HCl, $HClO_4$, H_2SO_4, AA, PTSA, MSA, DBSA and CSA were studied. Then the polymerization time for any initiator-dopant combination was changed from 30 to 120 min. In the next step, all experiments were repeated at different electric fields. The best molar ratio (monomer/initiator) experimentally was defined as 1/1. We used this ratio in all experiments. In all experiments, we focused on the obtained mass of polymer, m, electrical conduction of the obtained polymer in the absence of an electric field, σ, electrical conduction of the obtained polymer in the presence of an electric field, σ_E, and their ratio $r=\sigma_E/\sigma$. To identify the best molar ratio of APS initiator to monomer, values 1/0.5, 1/1 and 1/2 were examined by using HCl solution as dopant. The best electrical conduction was $\sigma=6\times10^{-3}$ S/cm for 1/1 ratio. Then to identify the best time of experiment, the polymerization was done for 1/1 molar ratio at different times such as 30, 45, 60, 90 and 120 min, and the best electrical conduction was $\sigma=9.5\times10^{-3}$ S/cm in 60 min. In this condition, the most obtained product mass was m=0.24 g too. Increasing the time does not affect the product mass, but it decreases the electrical conduction (Table 1 and Figure 7).

Table 1. Electrical conductivity of P2EANi produced by APS (HCl 1M) as initiator (1/1 mol ratio) in different times and absence of electric field.

Sample	Time of polymerization (min)	Color change time (min)	Weight of produced polymer (g)	Electrical conductivity ($\delta \times 10^{-3}$ S/cm)
1	30	5	0.21	2.4
2	45	5	0.23	5.9
3	60	5	0.24	9.5
4	90	5	0.24	9.1
5	120	5	0.25	8.8

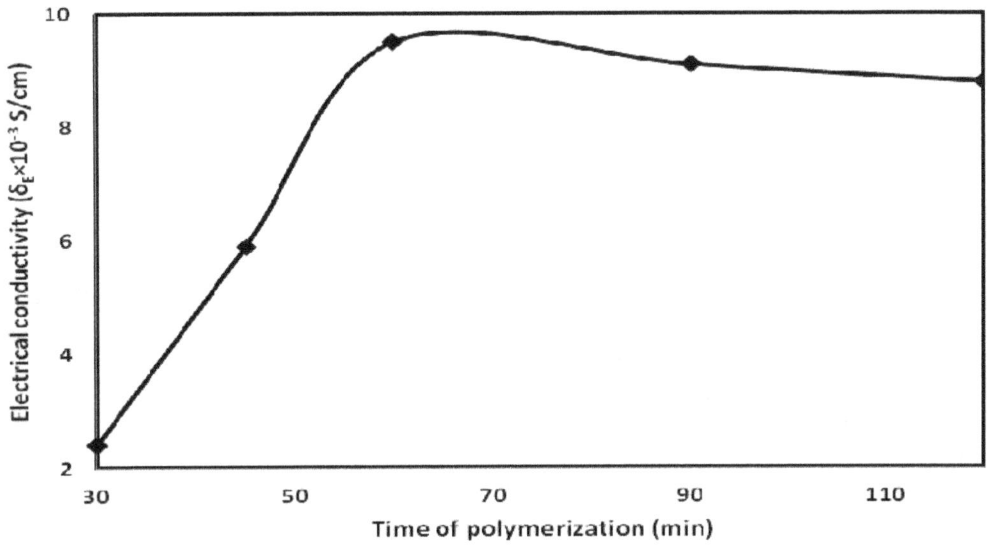

Figure 7. Electrical conductivity of P2EANi in different times and absence of electric field.

Table 2. Electrical conductivity of P2EANi produced by APS (HCl 1M) as initiator (1/1 mol ratio) in different electric field after 60 min.

Sample	Electric field (KN/C)	Color change time (min)	Weight of produced polymer (g)	Electrical conductivity ($\delta_E \times 10^{-3}$ S/cm)
1	35	6	0.21	22
2	69	7	0.28	87
3	104	7	0.29	85
4	138	8	0.28	74

In other experiments APS initiator and different dopants were used done with molar ratio of initiator to monomer, 1/1, in the 60 min, and the optimized condition under a uniform electric field with different intensities such as 5, 10, 15 and 20 KV/Cm^2. The most effective electric field intensity was 10 KV/Cm^2 in which we gained 0.28 g of product with σ_E=87×10^{-3} S/cm at Table 2 and Figure 8. On the other hand, we repeated the electrical conductivity of P2EANi produced by APS as an initiator and different dopants in the absence of the electric field and under 10 KV/Cm^2 one after 60 min. These results were summarized in Table 3.

In Table 3, we see that for all dopants without an electric field, the electrical conductivities are low while they are increased under electric field conductivities. Therefore, we also see that for APS initiator, the most effective dopants is DBSA with m=0.35 g, σ_E= 0.118 S/cm and r=7.87. So imposing the polymerization by a uniform electric field causes the electrical conduction improve by a factor 7.87. Then polymerization of 2-ethanol aniline at the optimized time (60 min) and with the best molar ratio initiator to monomer, (1/1), by using $FeCl_3.6H_2O$ as initiator, was done with different dopants in different electric field intensities. These results were compared with the ones obtained from incorporating polymer without applying electric field conditions. The results are

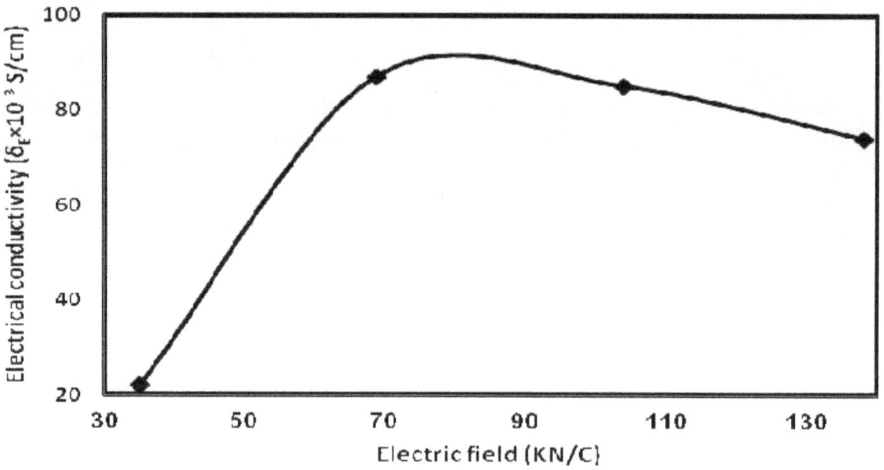

Figure 8. Electrical conductivity of P2EANi in different electric field after 60 min.

Table 3. Electrical conductivity of P2EANi produced by APS (HCl 1M) as initiator (1/1 mol ratio) with different dopants in absence and under 69 KN/C electric field after 60 min.

	Absence electric field		Under E=69 N/C		
Dopant (1M)	Weight of produced polymer (g)	Electrical conductivity (δ×10^{-3} S/cm)	Weight of produced polymer (g)	Electrical conductivity (δ_E×10^{-3} S/cm)	r= δ_E/δ
HCl	0.24	9.5	0.28	87	9.16
$HClO_4$	0.24	11	0.26	89	8.09
H_2SO_4	0.20	8.5	0.22	85	10.0
AA	0.18	2.1	0.21	15	7.14
MSA	0.24	4.5	0.24	42	9.33
PTSA	0.26	5.5	0.31	53	9.64
DBSA	0.29	15	0.35	118	7.87
CSA	0.27	11	0.31	105	9.54

Table 4. Electrical conductivity of P2EANi produced by $FeCl_3.6H_2O$ as initiator (1/1 mol ratio) different dopants in absence and under 69 KN/C electric field after 60 min.

	Absence electric field		Under E=69 N/C		
Dopant (1M)	Weight of produced polymer (g)	Electrical conductivity (δ×10^{-3} S/cm)	Weight of produced polymer (g)	Electrical conductivity (δ_E×10^{-3} S/cm)	r= δ_E/δ
HCl	0.22	4.5	0.23	22	4.88
$HClO_4$	0.22	5.5	0.24	25	4.54
H_2SO_4	0.21	3.0	0.23	22	7.33
AA	0.17	0.3	0.19	1.5	5.0
MSA	0.22	0.9	0.22	12	13.3
PTSA	0.23	6.0	0.23	21	3.50
DBSA	0.26	7.5	0.28	35	4.67
CSA	0.24	6.5	0.26	30	4.62

summarized in Table 4. Here we see that for $FeCl_3.6H_2O$ initiator, the best dopant is DBSA of which the best results are m=0.28g, σ_E= 30×10^{-3} S/cm and r=4.62. In addition, all described polymerizations were repeated by using KI as an initiator. The findings can be seen at Table 5. This table shows that for KI initiator, the best

Table 5. Electrical conductivity of P2EANi produced by KI as initiator (1/1 mol ratio) different dopants in absence and under 69 KN/C electric field after 60 min.

	Absence electric field		Under E=69 N/C		
Dopant (1M)	**Weight of produced polymer (g)**	**Electrical conductivity ($\delta\times10^{-3}$ S/cm)**	**Weight of produced polymer (g)**	**Electrical conductivity ($\delta_E\times10^{-3}$ S/cm)**	**r= δ_E/δ**
HCl	0.23	6.5	0.24	34	5.23
$HClO_4$	0.24	12	0.24	38	3.17
H_2SO_4	0.22	9.0	0.24	55	6.11
AA	0.19	2.5	0.21	3.0	1.20
MSA	0.23	6.5	0.24	18	2.77
PTSA	0.23	8.8	0.25	27	3.07
DBSA	0.27	19	0.30	55	2.89
CSA	0.27	25	0.31	60	2.4

Table 6. Electrical conductivity of P2EANi produced by $FeSO_4$ as initiator (1/1 mol ratio) different dopants in absence and under 69 KN/C electric field after 60 min.

	Absence electric field		Under E=69 N/C		
Dopant (1M)	**Weight of produced polymer (g)**	**Electrical conductivity ($\delta\times10^{-3}$ S/cm)**	**Weight of produced polymer (g)**	**Electrical conductivity ($\delta_E\times10^{-3}$ S/cm)**	**r= δ_E/δ**
HCl	0.21	2.5	0.22	15	6.0
$HClO_4$	0.22	5.5	0.24	25	4.54
H_2SO_4	0.23	5.5	0.24	25	4.54
AA	0.17	0.3	0.18	1.5	5.0
MSA	0.21	0.8	0.22	10	12.5
PTSA	0.22	5.5	0.23	13	2.36
DBSA	0.23	6.5	0.24	23	3.54
CSA	0.22	6.5	0.24	23	3.54

dopants are $HClO_4$, DBSA and CSA. The results of CSA are m=0.31 g, $\sigma_E=60\times10^{-3}$ S/cm and r=2.4. Finally, all above-described experiments were repeated by using $FeSO_4$ as an initiator. The results can be seen in Table 6. Table 6 shows $FeSO_4$ as initiator, and the best dopants are H_2SO_4, $HClO_4$ and DBSA. The results of DBSA are m=0.24g, $\sigma_E=23\times10^{-3}$ S/cm and r=3.54. In the experiments, we had 24 initiator–dopant combinations with widely different results. To compare the results, we collected all the best results (for initiator and dopant combination) in Table 7. This table shows all combinations in which conduction occurs, imposing electric field causes the electrical conduction to be improved. This also, shows that the best combination is APS-DBSA in which σ_E= 0.118 S/cm.

Conclusion

A molecule of 2-ethanol aniline is an electroactive monomer. It can be well polymerized through connecting the head to tail, and formed conductive polymer. They are head to head, tail to tail, or secondary connections. In this work, the electroactive monomers are arranged by using the electric field, and the head to tail connections are increased. Therefore, obtained polymer have a larger length chain, higher electrical conductivity, better space order and more appropriate physical resistance in comparison with the obtained one in the absence of electric field. The results of the experiments show that the electrical conductivity of produced P2EANi in different electric fields and the most constant time (60 min), increased with increasing intensity of electric field. But after 10 KV/Cm^2, the electrical conductivity has been decreased steadily.

Furthermore, the findings shows that the electrical conductivity of produced P2EANi in constant intensity of electric field – bout 10 KV/Cm^2– increases with increasing the amount of time. But after about 60 min, the electrical conductivity has been decreased steadily too.

On the other hand, considering the results of applying the various types of dopants and different initiators, the best initiator was APS, and the best dopant was dodecylbenzene sulfonic acid. It seems that by increasing

Table 7. Final results for any initiator–dopant combination.

Initiator	APS		$FeCl_3.6H_2O$		KI		$FeSO_4$	
Dopant	$\delta\times10^{-3}$	$\delta_E\times10^{-3}$	$\delta\times10^{-3}$	$\delta_E\times10^{-3}$	$\delta\times10^{-3}$	$\delta_E\times10^{-3}$	$\delta\times10^{-3}$	$\delta_E\times10^{-3}$
HCl	9.5	87	4.5	22	6.5	34	2.5	15
$HClO_4$	11	89	5.5	25	12	38	5.5	25
H_2SO_4	8.5	85	3.0	22	9.0	55	5.5	25
AA	2.1	15	0.3	1.5	2.5	3.0	0.3	1.5
MSA	4.5	42	0.9	12	6.5	18	0.8	10
PTSA	5.5	53	6.0	21	8.8	27	5.5	13
DBSA	15	118	7.5	35	19	55	6.5	23
CSA	11	105	6.5	30	25	60	6.5	23

the amount of time and intensity of the electric field after passing a special amount, the destructive processes in the polymer chain began as a result of increasing the length of polymer chain and dipolar moment forces. And also, due to polymeric confusion in the reaction solution, polymers involve defect reactions such as producing secondary branches on the polymer chain. Finally, mass and conductivity measurements of polymers showed that molecular weight and electric conductivity of the polymers were increased under electric field about 2 and 100 times, respectively.

REFERENCES

Buchachenko AL (2000). Mechanism of $Fe(OH)_2$ (aq) photolysis in aqueous solution, Pure Appl. Chem., 72:2243-2248.

Gato H, Akagi K, Itoh K (2001). Synthesis of liquid crystalline polyaniline derivatives and their orientational behaviors under magnetic field, Synth. Met., 117:91-93.

Hosseini SH (2006). Investigation of sensing effects of polystyrene-graft-polyaniline for cyanide compounds, J. Appl. Polym. Sci. 101(6):3920-3926.

Hosseini SH, Abdi Oskooe SH, Entezami AA (2005). Toxic Gas and Vapour Detection with Polyaniline Gas Sensors. Iranian Polym. J. 14(4):333-344.

Hosseini SH, Dabiri M, Ashrafi M (2006). Chemical and electrochemical synthesis of conducting graft copolymer of acrylonitrile with aniline, Polym. Int. 55(9):1081-1089.

Hosseini SH, Entezami AA (2003). Chemical and electrochemical synthesis of conducting poly di-heteroaromatics from pyrrole, indole, carbazole and their mixed containing hydroxamic acid groups and studies of its metal complexes, J. Appl. Polym. Sci., 90:63-71.

Hosseini SH, Mohammadi M (2009). Preparation and characterization of new poly-pyrrole having side chain liquid crystalline moieties, Mater. Sci. Eng. C, 29:1503-1509.

Hosseini SH, Noor P (2005). Ion exchange properties and kinetic behaviour of polyaniline-coated silica gel for p-toluenesulphonic acid and methanesulphonic acid, Iranian Polym. J. 14(1):55-60.

Li G, Zheng P, Wang NL, Long YZ, Chen ZJ, Li JC, Wan MX (2004). Optical study on doped polyaniline composite films, J. Phys. Condens. Matter., 16:6195–6204.

MacDiarmid AG (2001). A Conducting Composite of Polyaniline and Wood, Angew. Chem. Inst. Ed., 40:2581-2587.

Manoha SK, MacDiarmid AG, Epstein AJ (2002). Dependency of conductivity of selected doped conducting polymers on an unusual "Through space" electric field effect, Polymeric Materials: Sci. Eng. 86:1-5.

Pron A, Ranou P (2002). Processible conjugated polymers: from organic semiconductors organic metals and superconductors, Prog. Polym. Sci., 27:135-190.

Reits JR, Milford J, Christy RW (1979). Foundations of Electro Magnetic Theory, Third Edition Addison, Wesley.

Senadeera GKR, Pathirathne WMTC (2004). Utilization of conducting polymer as a sensitizer in solid-state photocells, Res. Communications-Current Sci. 87(3):339-342.

4

Synthesis, structural and dielectric properties of zinc sulfide nanoparticles

Sagadevan Suresh

Crystal Growth Centre, Anna University, Chennai-600 025, India.

Zinc sulfide (ZnS) nanoparticles were synthesized by the wet chemical method. The crystal structure and grain size of the particles were determined, using X-ray diffraction (XRD). The optical properties were studied by the Ultraviolet-Visible (UV-Vis) absorption spectrum. The dielectric properties of ZnS nanoparticles were examined using a HIOKI 3532-50 LCR HITESTER over the frequency range of 50 Hz – 5 MHz at different temperatures. The variation of the dielectric constant and dielectric loss were studied. The dielectric constants of the ZnS nanoparticles are high at low frequencies, and decrease rapidly when the frequency is increased. Further, electronic properties like valence electron plasma energy, Penn gap, Fermi energy and electronic polarizability of the ZnS nanoparticles, were estimated.

Key words: Nanoparticles, zinc sulfide (ZnS), x-ray diffraction (XRD), ultraviolet (UV) analysis, dielectric constant and dielectric loss.

INTRODUCTION

In recent years, many efforts have been devoted to the synthesis and study of the physicochemical characterrization of nanometer-scale semiconductors. Nanoparticle synthesis has opened up alternative paths in the design of materials with new properties. The interest in semiconductor nanoparticles is justified by the fact that their fundamental physical and chemical properties can be very different from those of the bulk materials. The II - VI nanostructures with their distinct properties have become potential candidates for applications in electronics and optoelectronics. Nanosized particles of semi conducting compounds in particular display grain size dependent optoelectronic properties, due to the size quantization effects (Bangal et al., 2005). The photo emission wavelengths, the band gap and lattice parameter are strongly dependent on the grain size rendering the tailarabilty of these properties as functions of grain size possible. Such unique tunability cannot be achieved in bulk semiconductors. The biggest hurdle in nanotechnology seems to be production of uniform sized nanoparticles and the control of grain size in a few nanometer ranges with considerable reproducibility.

There has been considerable interest recently in semiconductors of nanometer dimensions due to the quantum size effect they exhibit (Pathak et al., 2012, 2013). Nanocrystalline semiconductors have electronic properties intermediate between those of molecular entities and macro crystalline solids, and are at present the subject of intense research (Weller, 1993; Henglein 1989; Dounghong et al., 1982). ZnS nanostructures have gained a lot of attention that can be attributed to the properties arising from their size in the nanometer range (Gupta et al., 1997). Nanobelts (Jiang et al., 2003; Meng et al., 2003), nanowires (Zhu et al., 2003; Verna et al., 1995), nanocables, nanorods, nanocable- aligned tetra pods, nanoparticles (Sugimoto et al., 1998) and

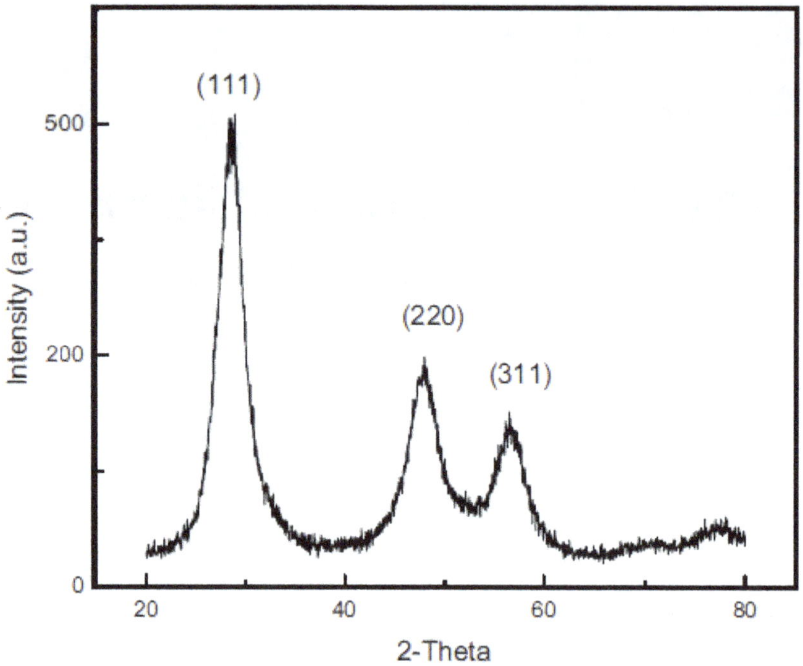

Figure 1. The XRD pattern of the ZnS nanoparticles

nanotubes were synthesized. These modify electronic, mechanical, luminescent and optical properties, and are used in nanoelectronics, photonics and as tools in biomedical applications. This paper deals with the preparation of ZnS nanoparticles using the wet chemical method. The prepared nanoparticles were characterized structurally and optically, using the powder XRD and the UV-Vis absorption spectrum. The dielectric studies have been carried out in the frequency range of 50Hz to 5MHz at different temperatures. Some of the electronic properties like plasma energy, Penn gap, Fermi energy and electronic polarizability of the ZnS nanoparticles were also determined.

EXPERIMENTAL PROCEDURE

ZnS nanoparticles were synthesized by the wet chemical method, using zinc chloride and sodium sulfide as the starting materials and were kept stirred using magnetic stirrer. The white precipitate of the ZnS nanoparticles is formed slowly in the solution. The nanoparticles were collected by centrifugation for 30 min. The obtained precipitate was then filtered and dried at 110°C for 2 h. The prepared ZnS was washed and dried. After drying, nanoparticles were grinded to obtain a fine powder for characterization. The XRD patterns of the synthesized samples were obtained using CuK_α radiation (λ = 1.5481nm), the 2θ range used was from 10 to 70° at a scanning rate of 0.02 deg/sec. The optical absorption spectrum of the ZnS nanoparticles has been taken by using the VARIAN CARY MODEL 5000 spectrophotometer in the wavelength range of 300 to 700 nm. The dielectric constant and the dielectric loss of the pellets of ZnS nanoparticles in disk form were studied using a HIOKI 3532-50 LCR HITESTER in the frequency range of 50 Hz to 5 MHz.

RESULTS AND DISCUSSION

XRD studies

The structural properties of the prepared nanoparticles were studied using X-ray diffraction. Figure 1 shows the XRD pattern of the ZnS nanoparticles.The XRD pattern exhibits prominent broad peaks at 2θ values of 28.90°, 48° and 56.50°. This shows that the ZnS nanoparticles have a zinc blended structure, and the peaks correspond to diffraction at (111), (220) and (311) planes, respectively (Mahamuni et al., 1993). The broad peak indicates the nanocrystalline behavior of the particles. The synthesized nanoparticles have good crystallinity, and the average particle size obtained using the diffraction pattern was 2.6 nm. These values are in good agreement with the reported value of the particle size 2.8 nm.

Optical absorption studies

Figure 2 shows the variation of the optical absorbance with the wavelength of the as-prepared ZnS nanoparticles. The optical absorption coefficient has been calculated in the wavelength range of 300 to 700 nm. The absorption edge has been obtained at a shorter wavelength. The optical absorption edge of the ZnS nanoparticles at 320 nm (3.80 eV), is slightly blue shifted from that of the bulk ZnS (340 nm, E_g = 3.65 eV). This nearness of the absorption peak to the bulk ZnS crystals is attributed to the near-band-edge free exactions (Mingwen et al., 2000). The broadening of the absorption

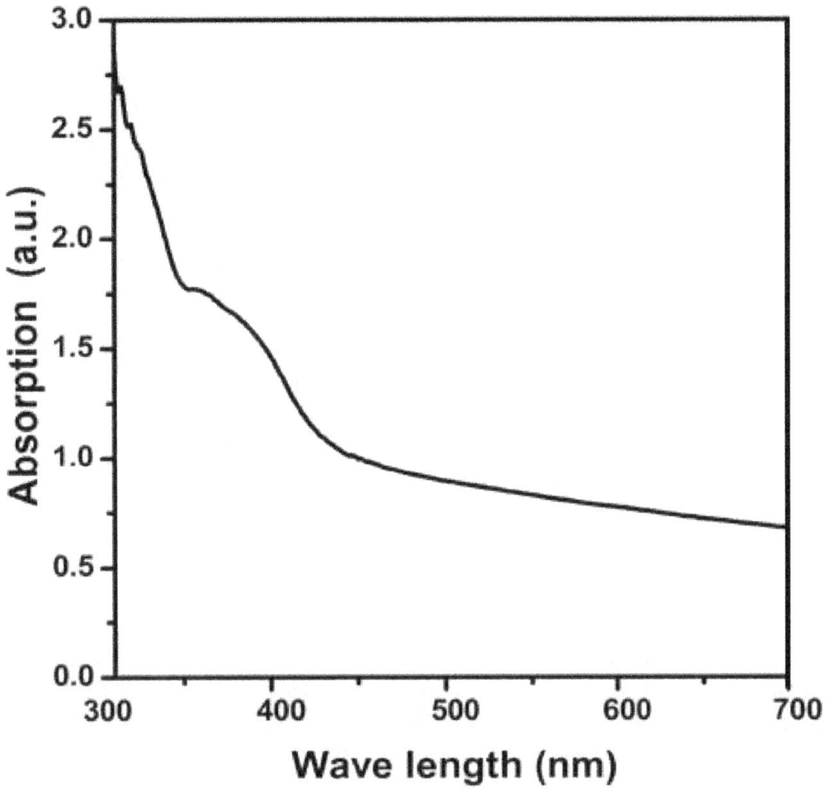

Figure 2. Optical absorption spectrum of ZnS nanoparticles.

spectrum could be due to the quantum confinement of the nanoparticles. From the absorption peak the optical energy bandgap of the ZnS nanoparticles has been calculated using the formula:

$$E_{gn} = h\nu = hc/\lambda \quad (1)$$

Where E_g, is the bandgap, *h,* is Planck's constant, *c*, is the velocity of light, and λ is the maximum absorption of the ZnS nanoparticles. The value of the bandgap is found to be 3.80 eV. The absorption band edge was shifted to 480 nm and the corresponding bandgap is 3.80 eV, which is higher compared to the bulk ZnS bandgap (3.65 eV). The measured transmittance (T) was used to calculate the absorption coefficient (α) using the relation,

$$\alpha = \frac{2.3026\log\left(\frac{1}{T}\right)}{t} \quad (2)$$

where *T* is the transmittance and *t* is the thickness of the crystal. Optical band gap (E_g) was evaluated from the absorption spectrum and optical absorption coefficient (α) near the absorption edge is given by,

$$\alpha h\nu = A(h\nu - E_g)^{1/2} \quad (3)$$

where E_g is the optical band gap of the crystal and *A* is a constant. The Tauc's plot of $(\alpha h\nu)^2$ against the photon energy ($h\nu$) at room temperature (Figure 3) shows a linear behaviour, (α-absorption coefficient and *h*-Planck's constant) which can be considered as an evidence of the indirect transition. Hence, assuming indirect transition between valence band and conduction band, the bandgap (E_g) is estimated by extrapolation of the linear portion of the curve to the point $(\alpha h\nu)^2 = 0$ .Using this method, the band gap of the ZnS nanoparticles was found to be 3.80 eV.

Dielectric properties of ZnS nanoparticles

Dielectric studies shows the effects of temperature and frequency on the conduction phenomenon in nano-structured materials. Dielectric behavior can effectively be used to study the electrical properties of the grain boundaries. The dielectric properties of materials are mainly due to contributions from the electronic, ionic, dipolar and space charge polarizations. Among these, the most important contribution to the polycrystalline materials

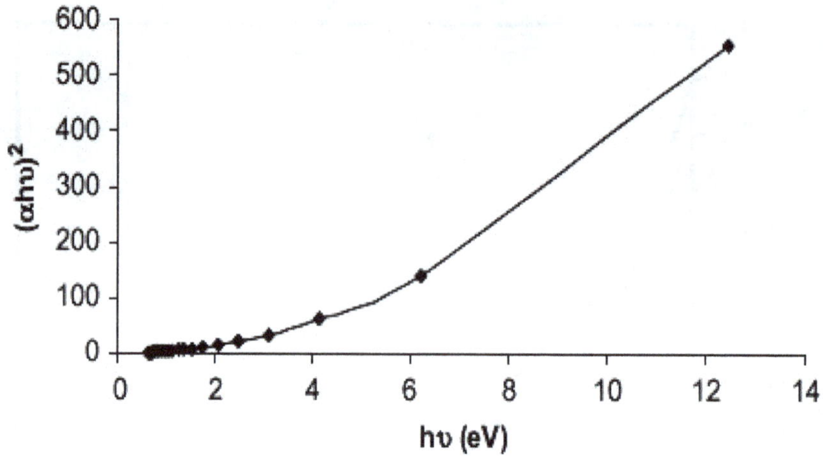

Figure 3. Plot of a vs. photon energy for ZnS nanoparticles.

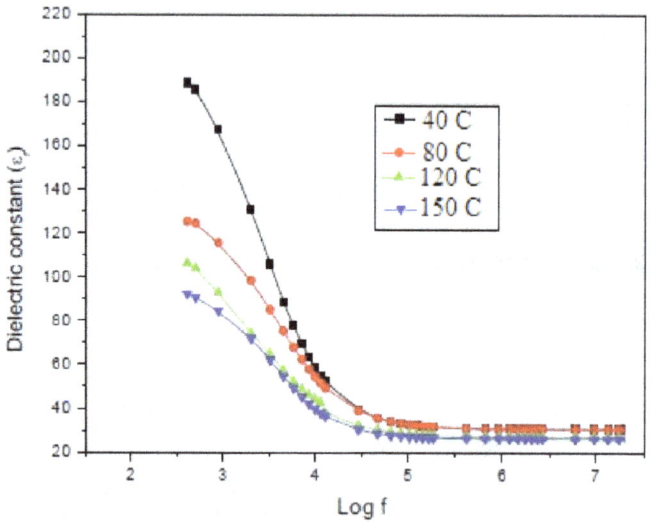

Figure 4. Dielectric constant of ZnS nanoparticles, as a function of frequencyat different temperatures.

in bulk form is from the electronic polarization, present in the optical range of frequencies. The next contribution is from ionic polarization, which arises due to the relative displacement of the positive and negative ions. Dipolar or orientation polarization arises from molecules having a permanent electric dipole moment that can change its orientation when an electric field is applied. Space charge polarization arises from molecules having a permanent electric dipole moment that can change its orientation when an electric field is applied. The dielectric parameters, like the dielectric constant (ε_r) and dielectric loss (tanδ) are the basic electrical properties of the ZnS nanoparticles. The measurement of the dielectric constant and loss as a function of frequency and different temperatures reveals the electrical processes that take place in ZnS nanoparticles and these parameters have been measured. The variations of the dielectric constant and dielectric loss of the ZnS nanoparticles at frequencies of 50 Hz to 5 MHz and at different temperatures of 40 to 150℃ are displayed in Figures 4 and 5. The dielectric constant is evaluated using the relation:

$$\varepsilon_r = \frac{Cd}{\varepsilon_0 A} \tag{4}$$

where d is the thickness of the sample and A, is the area of the sample. The results suggest that the dielectric constant and dielectric loss strongly depend on the frequency of the a.c. signal and the different temperatures of the ZnS nanoparticles. The dielectric

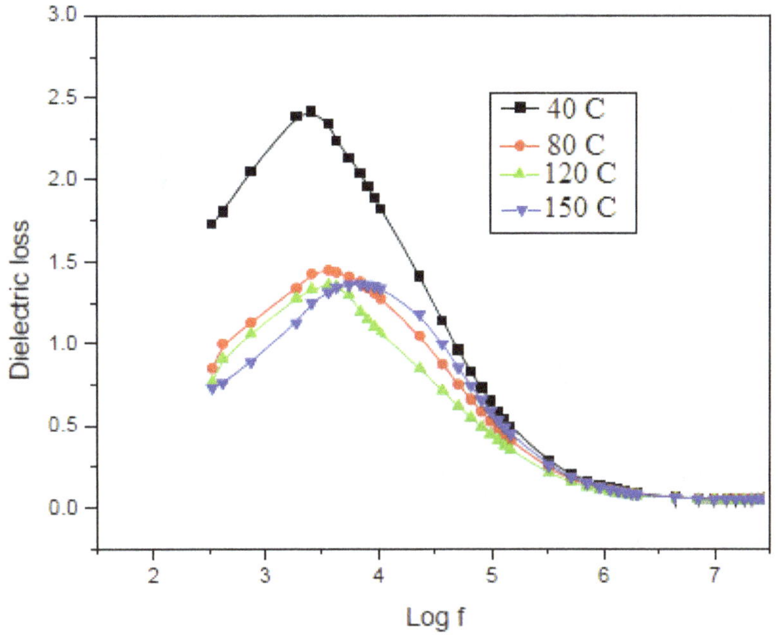

Figure 5. Dielectric loss' of ZnS nanoparticles, as a function of frequencyat different temperatures.

constant has higher values in the lower-frequency (50 Hz) and then it decreases up to the high frequency (5 MHz). The dielectric constant of ZnS nanoparticles is high at lower frequencies due to the contribution of the electronic, ionic, dipolar and space charge polarizations, which depend on the frequencies (Xue and Kitamura, 2002). Space charge polarization is generally active at lower frequencies and indicates the purity and perfection of the nanoparticles. Its influence is strong at higher temperature and is noticeable in the low frequency range (Smyth, 1956).

Most of the atoms in the nanocrystalline materials reside in the grain boundaries, which become electrically active as a result of charge trapping. The dipole moment can easily follow the changes in the electric field, especially at low frequencies. Hence, the contributions to the dielectric constant increase through space charge and rotation polarizations, which occur mainly in the interfaces. Therefore, the dielectric constant of nanostructures materials should be larger than that of the conventional materials (Meera et al., 2010). One of the reasons for the large dielectric constant of nanocrystalline materials at sufficiently high temperature is the increased space charge polarization due to the structure of their grain boundary interfaces. Also, at sufficiently high temperature the dielectric loss is dominated by the reason for the sharp increase of the dielectric constant at low frequencies and at lower temperatures. As the temperature increases, the space charge and ion jump polarization decrease, resulting in a decrease in the dielectric constant.

In dielectric materials, dielectric losses usually occur due to the absorption current. The orientation of the molecules along the direction of the applied electric field in polar dielectrics requires a part of the electric energy to overcome the forces of internal friction. Another part of the electric energy is utilized for rotations of the dipolar molecules and other kinds of molecular transfer from one position to another, which also involve energy losses. In nanophase materials, the grain boundaries have an amorphous or glassy structure. All the in homogenities, defects, space charge formation etc. together produce an absorption current, which results in dielectric losses. Due to the presence of the dangling bonds on the surface layers, the nanoparticles will be highly reactive, and there is a chance of adsorption of gases like oxygen or nitrogen. These adsorbed gases can also cause an increase in the dielectric loss. In nanophase materials, in homogeneities like defects and space charge formation in the inter phase layers produce an absorption current resulting in a dielectric loss. Figure 5 shows the variation of the dielectric loss with respect to the logarithm of frequency at different temperatures of 40, 80, 120 and 150°C. Dielectric loss also shows a trend similar to the one shown by the dielectric constant. The decrease in the dielectric loss with the increase in frequency for all the temperatures suggests that the dielectric loss is strongly dependent on the frequency of the applied field. The high values of dielectric loss at low frequencies could be related to the charge lattice defect of the space charge polarization (Xue and Kitamura, 2002).

In the proposed relation only one parameter, such as the high frequency dielectric constant is required as the input, to evaluate the electronic properties, like valence

Table 1. Electronic properties of the ZnS nanoparticles

Parameter	Value
Plasma energy ($h\omega_p$)	16.69eV
Penn gap (E_p)	3.56 eV
Fermi energy (E_F)	12.46eV
Electronic polarizability (using Penn analysis)	5.39 × 10^{-24} cm^3
Electronic polarizability (using Clausius-Mossotti relation)	5.40 × 10^{-24} cm^3
Electronic polarizability (using bandgap)	4.91 × 10^{-24} cm^3

electron plasma energy, average energy gap or Penn gap, Fermi energy and electronic polarizability of the ZnS nanoparticles. The theoretical calculations show that the high frequency dielectric constant is explicitly dependent on the valence electron Plasmon energy, an average energy gap referred to as the Penn gap and Fermi energy. The Penn gap is determined by fitting the dielectric constant with the Plasmon energy (Ravindra et al., 1981). The valence electron plasma energy, $\hbar\omega_p$, is calculated using the relation (Kumar and Sastry, 2005),

$$\hbar\omega_P = 28.8\left(\frac{Z\rho}{M}\right)^{1/2} \tag{5}$$

According to the Penn model (Penn, 1962), the average energy gap for the ZnS nanoparticles is given by:

$$E_P = \frac{\hbar\omega_P}{(\varepsilon_\infty - 1)^{1/2}} \tag{6}$$

where $\hbar\omega_P$ is the valence electron plasmon energy and the Fermi energy (Ravindra et al., 1981) given by:

$$E_F = 0.2948(\hbar\omega_P)^{4/3} \tag{7}$$

Then we obtained the electronic polarizability α, using a relation (Ravindra and Srivastava, 1980; Penn, 1962),

$$\alpha = \left[\frac{(\hbar\omega_P)^2 S_0}{(\hbar\omega_P)^2 S_0 + 3E_P^2}\right] \times \frac{M}{\rho} \times 0.396 \times 10^{-24}\, cm^3 \tag{8}$$

Where S_0 is a constant given by

$$S_0 = 1 - \left[\frac{E_P}{4E_F}\right] + \frac{1}{3}\left[\frac{E_P}{4E_F}\right]^2 \tag{9}$$

The value of α obtained from equation (6) closely matches with that obtained using the Clausius-Mossotti relation,

$$\alpha = \frac{3}{4}\frac{M}{\pi N_a \rho}\left[\frac{\varepsilon_\infty - 1}{\varepsilon_\infty + 2}\right] \tag{10}$$

Considering that the polarizability is highly sensitive to the bandgap (Reddy et al., 1995), the following empirical relationship is also used to calculate α,

$$\alpha = \left[1 - \frac{\sqrt{E_g}}{4.06}\right] \times \frac{M}{\rho} \times 0.396 \times 10^{-24}\, cm^3 \tag{11}$$

Where E_g is the bandgap value determined through the UV absorption spectrum. The high frequency dielectric constant of a material is a very important parameter for calculating its physical or electronic properties. All the above parameters as estimated are shown in Table 1.

Conclusion

Nanoparticles of ZnS are synthesized using the wet chemical method. The crystal structure and grain size of the particles are determined using XRD studies, and the particle size of the ZnS nanoparticle is found to be 2.6 nm. From the optical absorption spectrum, the blue shift of 320 nm with respect to its bulk counterpart is contributed by the quantum confinement effect. The value of the bandgap is found to be 3.80 eV. The dielectric constant and dielectric loss of the ZnS nanoparticles are measured in the frequency range of 50 Hz to 5 MHz at different temperatures. The dielectric studies reveal that both the dielectric constant and dielectric loss decrease with an increase in the frequency. The dielectric characterization shows the low value of the dielectric constant at higher frequencies. The dielectric constant of the ZnS nanoparticles is found to be much higher than that of the bulk ZnS. Some of the electronic properties like the plasma energy, Penn gap, Fermi energy and electronic polarizability of the ZnS nanoparticles have been calculated.

REFERENCES

Bangal M, Ashtaputre S, Marathe S, Ethiraj A, Hebalkar N, Gosavi SW, Urban J, Kulkarni SK (2005). Semiconductor Nanoparticles. Hyperfine Interact. 160:81-94.

Dounghong D, Ramsden J, Gratzel M (1982). Dynamics of interfacial electron-transfer processes in colloidal semiconductor systems. J. Am. Chem. Soc.104:2977-2985.

Gupta S, McClure JS, Singh VP (1997). Phosphor Efficiency and Deposition Temperature in ZnS: Mn A.C. Thin Film Electroluminescence Display Devices. Thin Solid Films. 33:299.

Henglein A (1989). Small-particle research: physicochemical properties of extremely small colloidal metal and semiconductor particles. Chem. Rev. 89:1861-1873.

Jiang Y, Meng XM, Liu J, Xie ZY, Lee CS, Lee ST (2003). Hydrogen-Assisted Thermal Evaporation Synthesis of ZnS Nanoribbons on a Large Scale. Adv. Mater. 15:323-327.

Kumar V, Sastry BSR (2005). Heat of formation of ternary chalcopyrite semiconductors. J. Phys. Chem. Solids. 66:99-102.

Mahamuni S, Khosravi AA, Kundu M, Kshirsagar A, Bedekar A, Avasare DB, Singh P, Kulkarni SK (1993). Thiophenol-capped ZnS quantum dots. J. Appl. Phys. 73:5237-5241.

Meera J, Sumithra V, Seethu R, Prajeshkumar JM (2010). Dielectric Properties of Nanocrystalline ZnS. Acad. Rev. 1:93-100.

Meng XM, Jiang Y, Liu J, Lee CS, Bello I, Lee ST (2003). Gallium nitride nanowires doped with silicon. Appl. Phys. Lett. 83:4241-4244.

Mingwen W, Lingdong S, Xuefeng F, Chunsheng L, Chunhua Y (2000). Synthesis and optical properties of ZnS:Cu(II) nanoparticles. Solid State Comm. 115:493-496.

Pathak CS, Mandal MK, Agarwala V (2013). Optical properties of undoped and cobalt doped ZnS nanophosphor. Mater. Sci. Semicond. Process. 16:467-471.

Pathak CS, Mandal MK, Agarwala V (2013). Synthesis and Characterization of Zinc Sulphide Nanoparticles Prepared by Mechanochemical Route. Super lattices Microst. 58:135-143.

Pathak CS, Mishra DD, Agarwala V, Mandal MK (2012). Blue Light Emission from Barium Doped Zinc Sulphide Nanoparticles" Ceramics Int. 38:5497-5500.

Pathak CS, Mishra DD, V. Agarwala V, Mandal MK (2013). Optical Properties of ZnS Nanoparticles Prepared by High Energy Ball Milling. Mater. Sci. Semicond. Process. 16:525-529.

Penn DR (1962). Wave-Number-Dependent Dielectric Function of Semiconductors.Phys. Rev. 128:2093-2097.

Ravindra NM, Bharadwaj RP, Sunil Kumar K, Srivastava VK (1981). Model based studies of some optical and electronic properties of narrow and wide gap materials. Infrared Phys. 21:369-381.

Ravindra NM, Srivastava VK (1980). Properties of liquid lead monosulfiede, lead selenide and lead telluride. Infrared Phys. 20:399-418.

Reddy RR, NazeerAhammed Y, Ravi Kumar M (1995). Variation of magnetic susceptibility with electronic polarizability in compound semiconductors and alkali halides. J. Phys. Chem. Solids. 56:825-829.

Smyth CP (1956). Dielectric Behavior and Structure. Acta. Cryst. 9:838-839.

Sugimoto T, Chen S, Muramatsu A (1998). Synthesis of uniform particles of CdS, ZnS, PbS and CuS from concentrated solutions of the metal chelates. Colloids Surf. 135:207-226.

Verna AK, Ranchfuss TB, Wilson SR (1995). Donor Solvent Mediated Reactions of Elemental Zinc and Sulfur, sans Explosion. Inorg. Chem. 34:3072-3078.

Weller H (1993). Quantized Semiconductor Particles: A novel state of matter for materials science. Adv. Mater. 5:88-95.

Xue D, Kitamura K (2002). Dielectric characterization of the defect concentration in lithium niobate single crystals. Solid State Commun.122:537-541.

Zhu YC, Bando Y, Uemura Y (2003). ZnS-Zn nanocables and ZnS nanotubes. Chem. Commun. 7:836-837.

5

Determination of the crust-mantle electrical conductivity-depth structure of Niger Delta using solar quiet day (Sq) current

Obiora D. N.[1]*, Okeke F. N.[1] and Yumoto K.[2]

[1]Department of Physics and Astronomy, University of Nigeria, Nsukka, Nigeria.
[2]Space Environment Research Centre, Kyushu University, Fukuoka, Japan.

Innovative attempt has been made to study the crust-mantle electrical conductivity-depth structure of the Niger Delta region in Nigeria by applying solar quiet (Sq) day current. The study involved the use of geomagnetic data obtained from Lagos by magnetic data set (MAGDAS), Japan in 2010. Gauss spherical harmonic analysis (SHA) technique was employed which separated the internal and external field contributions to Sq current system. The result revealed that the conductivity-depth structures could be determined using Sq current systems. The conductivity-depth profile started from 0.004 S/m at a crustal depth of 300 m and increased sharply to 0.178 S/m at 1 km depth. It then started decreasing until it got to 0.021 S/m at sub-crustal depth of 44 km. The profile then rose steadily from 0.041 S/m at 100 km to 0.09 S/m at 221 km. It continued increasing downward until it got to 0.211 S/m at 888 km depth and reached 0.243 S/m at a depth of 1179 km at the lower mantle. The downward increase in conductivity agrees with the global models which depicted a steep rise in conductivity from 300 to 700 km.

Key words: Crust-mantle, conductivity-depth structure, electrical conductivity, Niger Delta, solar quiet (Sq) day current, spherical harmonic analysis (SHA).

INTRODUCTION

The systematic flow of varying electric current in the part of the Earth's upper atmosphere (ionosphere) gives rise to magnetic field which in turn induces electric current in the Earth's crust and mantle. The composite of both the external field and the internal field is measurable at the Earth's surface in magnetic observatories. The depth of penetration of this induced current depends on the characteristics of the source current and upon the distribution of electrically conducting materials in the Earth.

The measured field at the Earth's surface is a mixture of the external (source) and internal (induced) components from the current. Separating these currents into their external and internal components using spherical harmonic analysis (SHA) will be used to determine the electrical conductivity of the Earth.

Several geophysical research works had been carried out on the subsurface structure of Niger Delta due to the presence of minerals, especially hydrocarbon. Such research works include the works of Short and Stauble (1967), Weber and Daukoru (1975), Avbovbo (1978), Doust and Omatsola (1990), Obiora (2006), and Magbagbeola and Willis (2007). These geophysical methods failed to resolve the subsurface structure of Niger Delta region beyond 13 km from the crust as shown in Table 1. They are also cumbersome and costly, hence the need to use Sq current system. Equally, Vassal et al. (1998) employed magnetotelluric and geomagnetic deep sounding method for mapping the conductivity of the upper mantle at latitudes along the dip equator, yet they found no results.

*Corresponding author. E-mail: daniel.obiora@unn.edu.ng.

Table 1. Formations and thicknesses of the Niger Delta by various researchers (Obiora, 2006).

Benin (m)	Agbada (m)	Akata (m)	Researchers
1,200	2,300	3,100	Short and Stauble (1967)
2,000	5,000	8,000	Weber and Daukoru (1973)
1,087	3,097	-	Obiora (2006)

The penetration of an electromagnetic wave into the conducting Earth depends upon the wavelength of the source field and the conductivity profile of the region into which the wave travels. Due to hemispherical extent of the ionospheric Sq source currents and the 24, 12, 8 and 6 h periodicity of the variations; it is possible to examine the electrical conductivity of the Earth's upper mantle region for depths of about 50 to 600 km (Jacobs, 1989). Chapman (1919), and Chapman and Bartels (1940) made the first significant Earth conductivity determinations with the separated external and internal fields.

Schmucker (1970) devised a method of profiling the Earth's conductivity with a transfer function that uses the external and internal spherical harmonic coefficients at a given site. This function allowed the determination of a depth to equivalent substitute conductors that would produce the observed fields at the Earth surface. Weidelt et al. (1980), Jones (1983) and Schmucker (1987) demonstrated the equivalence of the Schmucker (1970) method to other depth-conductivity profile transformation. Campbell and Andersen (1983), and Campbell (1998) expanded the technique for application to the continental solar quiet (Sq) system.

In this study, we made use of the Sq day ionospheric current to determine the crust-mantle conductivity-depth structure of the Niger Delta. The results of this work will be compared with the results of previous workers in the region who used geophysical methods and workers in other regions who used Sq method.

METHODOLOGY

Source of data

The hourly averaged geomagnetic data used in this study was obtained from Lagos (latitude: 6.813 and longitude: 3.460) by magnetic data set (MAGDAS), Japan in 2010.

Method of analysis

Campbell and Anderssen (1983) generalized the form of Schmucker's transfer function C_n^m in the following way:

$$C_n^m = z - ip \tag{1}$$

Which is a complex number in which the real (z) and imaginary (-p) parts are given by

$$z = \frac{R}{n(n+1)}\left\{\frac{a_n^m\left[na_n^{me} - (n+1)a_n^{mi}\right] + b_n^m\left[nb_n^{me} - (n+1)b_n^{mi}\right]}{\left(a_n^m\right)^2 + \left(b_n^m\right)^2}\right\} \tag{2}$$

$$p = \frac{R}{n(n+1)}\left\{\frac{a_n^m\left[nb_n^{me} - (n+1)b_n^{mi}\right] - b_n^m\left[na_n^{me} - (n+1)a_n^{mi}\right]}{\left(a_n^m\right)^2 + \left(b_n^m\right)^2}\right\} \tag{3}$$

Where R is the Earth's radius in km, z and p are given in kilometers and the coefficient sums are given by:

$$a_n^{me} + a_n^{mi} = a_n^m \tag{4a}$$

$$b_n^{me} + b_n^{mi} = b_n^m \tag{4b}$$

For each n, m set of coefficients, the depth (in km) to the uniform substitute layer is given by

$$d_n^m = z - p \tag{5}$$

The substitute layer conductivity (S/m) σ_n^m, is given by

$$\sigma_n^m = \frac{5.4\times10^4}{m(\pi p)^2}\,Siemens/meter \tag{6}$$

a_n^{me} and b_n^{me} are the external cosine and sine coefficients and a_n^{mi} and b_n^{mi} are the internal cosine and sine coefficients from the SHA of the Sq field.

The magnetically quiet days selected from the five internationally quiet days in each month, which was averaged were first obtained. The Fourier analyses of the two components of the magnetic field (H, the horizontal and Z, the vertical components) were computed, followed by the computation of Fourier analyzed values of H- and Z-components. Schmidt functions were then calculated for the co-latitude, followed by the determination of intermediate coefficients. The SHA coefficients were computed to degree 12 and order 4. Using Equation 5, the depths to conductive layers were computed and from Equation 6, the associated conductivities were calculated.

RESULTS AND DISCUSSION

The electrical conductivity of Earth's materials varies over many orders of magnitude. It depends upon many factors, including: rock type, porosity, connectivity of pores, nature of the fluid, metallic content of the solid matrix, temperature, pressure, partial melt, presence of volatile materials and oxygen fugacity. Electrical

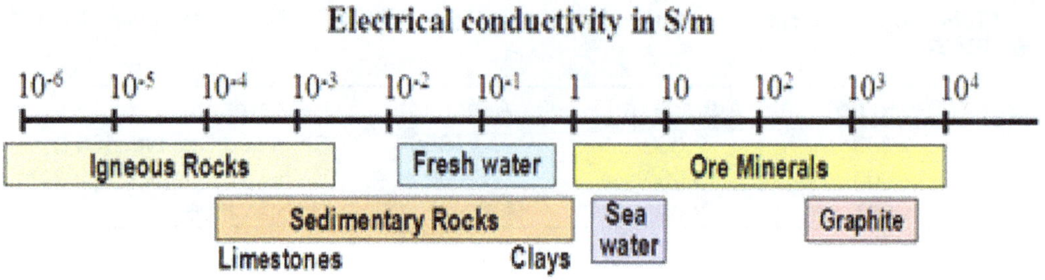

Figure 1. Electrical conductivity of earth materials (http://www.eos.ubc.ac/ubcgif/iag/foundations).

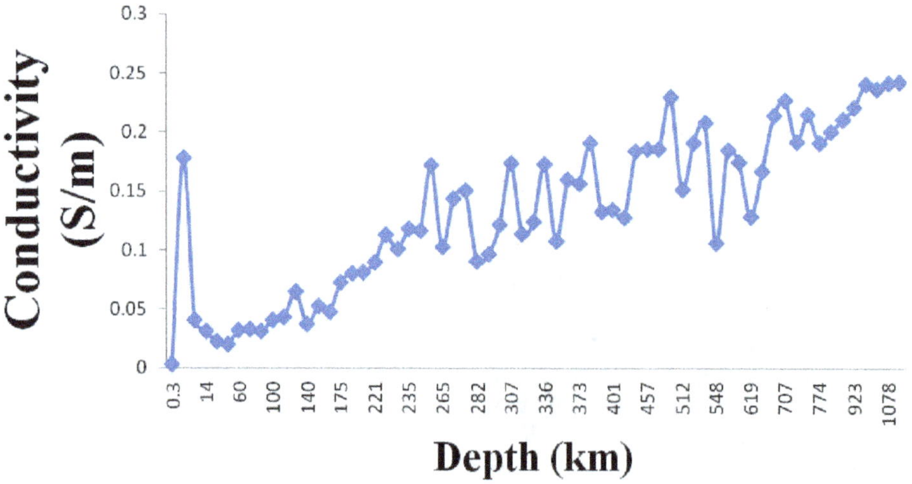

Figure 2. Electrical conductivity-depth structure of crust-mantle of Niger-Delta region.

conductivity for earth materials may be as low as 10^{-5} S/m for igneous rocks and as high as 10^{5} S/m for graphite.

A very rough indication of the range of conductivity for rocks and minerals is shown in Figure 1. A very small change in fluid saturation can lead to a big change in measured electrical conductivity, which was observed within the crustal layer. The crust-mantle electrical conductivity-depth structure of Niger delta region in Nigeria is shown in Figure 2. The scatter observed in the conductivity-depth values could be as a result of variability of source current location, error from the SHA fitting, magnetic field contributions produced by other than quiet time field conditions and error from field measurements.

Within the crust, the conductivity observed at a depth of 300 m was 0.004 S/m which rose sharply to 0.178 S/m at 1 km depth. It suddenly decreased to a value of 0.041 S/m at 2 km depth and continued fluctuating until it got to 0.021 S/m at a crustal depth of 44 km. The mineral discovered within the crust of Niger delta region was mainly hydrocarbon with conductivity values ranging from about 0.011 to 0.2 S/m starting from a depth of about 600 m to 6 km which is within the range of our conductivity-depth profile result.

The conductivity at a depth of about of 60 km was 0.033 S/m which rose steadily to a value of 0.041 S/m at 100 km and 0.09 S/m at 221 km which corresponds with the global seismic low velocity region, the asthenosphere (Dziewonski and Anderson, 1981; Kennett and Engdahl, 1991; Tarits, 1992). The profile then rose to a value of 0.186 S/m at 483 km, 0.191 S/m at 741 km and 0.243 S/m at a depth of 1179 km within the lower mantle.

The important physical parameters that control the electrical conductivity at upper mantle depths are phase changes, temperature and water content. Olivine is considered to be the main constituent of mantle rocks. It is not stable at high temperature and pressure conditions (Katsura and Ito, 1989), implying that the mantle mineralogy changes with depth. Laboratory studies on olivine have shown that the electrical conductivity increases with increasing temperature. Intrinsic semi-conduction is the most dominant mode of conduction throughout the mantle. The conductivity-temperature relation is given by Xu et al. (1998) as $\sigma = \sigma_0 \exp(-\Delta H/KT)$, Where σ_0 is the pre-exponential term, T is the absolute temperature, K is the Boltzman constant and $\Delta H (= \Delta U + P\Delta V)$ is the activation enthalpy (where ΔU and ΔV designate the activation energy and activation volume respectively and P is the pressure).

Comparing our results with the results of previous workers in other regions of the world, we therefore infer as follows:

(i) The general correspondence observed in this work between high conductivity zone and low velocity zone, the asthenosphere, is in agreement with the global results of Tarits (1992), Dziewonski and Anderson (1981), and Kennett and Engdahl (1991).
(ii) The conductivity-depth profiles show that the uppermost mantle could be viewed as a stack of inhomogeneous layers in variance with the global model obtained under the assumption of a spherically symmetric earth (Banks, 1972; Berdichevski et al., 1979) in which the upper mantle down to 400 to 500 km depth is seen as a single homogeneous layer (Arora et al., 1995).
(iii) The rapid increase in conductivity observed below 400 km depth is in conformity with the works of Campbell and Schiffmacher (1988) that the upper mantle under Africa and Asia regions is highly conductive.
(iii) The downward increase in conductivity agrees with the global models which depicted a steep rise in conductivity from 300 to 700 km (Campbell and Schiffmacher, 1988; Arora et al., 1995; Campbell et al., 1998).

Conclusion

The crust-mantle electrical conductivity-depth structure of Niger Delta region has been determined by the application of Sq day ionospheric current. This is the first time this method is employed in the area understudy. The conductivity-depth structure was found to fluctuate mostly in the crust from 0.004 S/m at a depth of 300 m to 0.178 S/m at 1 km and suddenly it started decreasing until it got to 0.021 S/m at crustal depth of 44 km. Within the mantle, the conductivity rose from 0.041 S/m at 100 km depth to 0.09 S/m at 221 km depth. It continued rising until it got to a value of 0.243 S/m at a depth of 1179 km in the lower mantle. The global mantle seismic discontinuity at around 370 and 720 km was not evident in this study. Our results compared with regional seismic velocity model, show correspondence between high conductivity and low velocity layers. It should be noted that Vassal et al. (1998), employed magnetotelluric (MT) and geomagnetic deep sounding methods of probing the conductivity of the upper mantle, but could not get any result. Unlike geophysical methods, MT and geomagnetic deep sounding methods, Sq method is not very costly and cumbersome. It can resolve the subsurface structure of the solid Earth beyond 1000 km depth which the other methods could not reach.

We are planning in future to extend this study to other regions in Nigeria since this is the first time this type of work is being done in Niger Delta region.

ACKNOWLEDGEMENTS

Authors are grateful to MAGDAS, Japan, for the data used in this work and Dan Okoh of Centre for Basic Space Science, University of Nigeria, Nsukka for his contributions.

REFERENCES

Arora BR, Campbell WH, Schiffmacher ER (1995).Upper mantle electrical conductivity in the Himalayan region. J. Geomag. Geoelectr. 47:653-665.

Avbovbo AA (1978). Tertiary lithostratigraphy of Niger Delta. Am. Assoc. Petrol. Geol. Bull. 62:295-300.

Banks RJ (1972). The overall conductivity distribution of the Earth. J. Geomag. Geoelectr. 24:337-351.

Berdichevski MN, Vanyan LL, Lagutinskaya LP, Rotanov NM, Fainber EB (1979). The experience of the Earth frequency sounding by the results of the spherical harmonic analysis of geomagnetic field variations. Geomag. Aeron. 10:374-377.

Campbell WH, Anderssen RS (1983). Conductivity of the subcontinental upper mantle: An analysis using quiet-day geomagnetic record of North America. J. Geomag. Geoelectr. 35:367-382.

Campbell WH, Barton CE, Chamalaun FH, Welsh W (1998). Quiet-day ionospheric currents and their application to upper mantle conductivity in Australia. Earth Planets Space 50:347-360.

Campbell WH, Schiffmacher ER (1988). Upper mantle electrical conductivity for seven subcontinental regions of the earth. J. Geomag. Geoelectr. 40:1387-1406.

Chapman S (1919). The solar and lunar diurnal variation of the Earth's magnetism. Phil. Trans. Roy. Soc. London A218:1-118.

Chapman S, Bartels J (1940). Geomagnetism. Oxford University Press. p. 1049.

Doust H, Omatsola E (1990). Niger Delta in Edwards, J. D. and Santogrossi, P. A., Eds: Divergent/Passive Margin Basins, AAPG Memoir 48, Tulsa pp. 239-248.

Dziewonski AM, Anderson DL (1981). Preliminary reference Earth model. Phys. Earth Planet Int. 25:297-356.

Jacobs JA (1989). Geomagnetism - volume 3. Academic Press, London.

Jones AG (1983). On the equivalence of the 'Niblett' and 'Bostick' transformations in magnetotelluric method. J. Geophys. Res. 53:72-73.

Katsura T, Ito E (1989). A temperature profile of the mantle transition zone. Geophys. Res. Lett. 16:425-428.

Kennett BLN, Engdahl ER (1991). Traveltimes for global Earthquake location and phase identification. Geophys. J. Int. 105:429-465.

Magbagbeola AO, Willis JB (2007). Sequence stratigraphy and syndepositional deformation of the Agbada formation, Robertkiri field, Niger Delta, Nigeria. AAPG Bull. 91(7):945-958.

Obiora DN (2006). Determination of depth to subsurface structure in the area of Niger Delta. Niger. J. Space Res. 1(1):131-141.

Schmucker U (1970). An introduction to induction anomalies. J. Geomag. Geoelectr. 22:9-33.

Schmucker U (1987). Substitute conductors for electromagnetic response estimates. Pure Appl. Geophys. 125:341-367.

Short KC, Stauble AJ (1967). Outline of geology of Niger Delta. AAPG Bull. 51:761-779.

Tarits P (1992). Electromagnetic studies of global geodynamic processes. 11th Workshop on Electromagnetic Induction in the Earth, Victoria University of Wellington, 26 August-2 September, 1992. 7.1-7.21.

Vassal J, Menvielle M, Cohen Y, Dukhan M, Doumouya V, Boka K, Fambitakoye O (1998). A study of transient variations in the earth's electromagnetic field at equatorial electrojet latitudes in West Africa (Mali and Ivory Coast). Ann. Geophysicae 16:677-697.

Weber KJ, Daukoru EM (1975). Petroleum geological aspects of the Niger delta. Niger. J. Miner. Geol. 12:9-32.

Weidelt P, Muller W, Losecke W, Knodel K (1980). Die Bostick transformation. In Protokoll Uber das Kolloquium Elektromagnetische Tief-enforschung (ed. V. Haak and Homilius). pp. 227-230. Berlin-Hannover.

Xu Y, Poe BT, Shankland TJ, Rubie DC (1998). Electrical conductivity of olivine, wadsleyite and rinwoodite under upper-mantle conditions. Science 280: 1415, doi.10, 1126/Science 280.5368.1415.

6

Structural and DC Ionic conductivity studies of carboxy methylcellulose doped with ammonium nitrate as solid polymer electrolytes

K. H. Kamarudin and M. I. N. Isa

Advanced Materials Research Group, Renewable Energy Research Interest Group, Department of Physical Sciences, Faculty of Science and Technology, Universiti Malaysia Terengganu, 21030 Kuala Terengganu, Terengganu, Malaysia.

Solid polymer electrolytes have been prepared by solution casting technique. Carboxy methylcellulose (CMC)-ammonium nitrate (AN) films were studied with varied AN salt concentration from 5-50 wt.% at ambient temperature. X-ray diffraction (XRD) pattern shows the amorphous nature of polymer electrolyte samples. IR-spectra confirm the polymer-salt complexes in the range of 1633 - 829 cm^{-1}. Impedance analysis reveals that polymer electrolyte film containing 45wt.% AN exhibits the highest ionic conductivity of (7.71 ± 0.04) x10^{-3} Scm^{-1}, while pure CMC film gives the lowest ionic conductivity of (1.86 ± 0.03) x10^{-8} Scm^{-1}. It was evident from this study that the increase of ionic conductivity depends on the AN salt concentration. The present polymer-salt system has potential application in electrochemical devices based on the results obtained.

Key words: Carboxy methylcellulose, ammonium nitrate, solid polymer electrolytes, ionic conductivity.

INTRODUCTION

Since the earliest breakthrough of polymer-salt complexes by Wright in 1975, there was a plethora of research focusing on the study and development of solvent free polymer electrolytes (PEs). Extensive research on PEs were driven by their advantages such as ease of preparation, light weight, leakage free, mechanically stable and flexibility for packaging design over gel/liquid counterparts (Ramesh et al., 2002; Quartarone et al., 1998). PEs have become promising materials for electrochemical device applications, namely, high energy density rechargeable batteries, fuel cells, supercapacitors, sensors and electrochromic displays (Bhargav et al., 2009; Ma et al., 2013). Poor ionic conductivity as a result of the low segmental mobility of the polymer chain at ambient temperature is the major drawback possessed by solid Pes (Armand, 1994; Tarascon and Armand, 2001). To overcome the challenge, several approaches have been employed to enhance the room temperature conductivity as well as to improve the mechanical stability and interfacial activity of PEs including using chitin/chitosan, starch and cellulose derivates as organic/biodegradable polymer matrix incorporating with inorganic salts such as sodium salt and ammonium salt (Kumar et al., 2011; Hassan et al., 2010). Typically, ionic conduction in PEs are governed by the degree of crystallinity (Kumar et al., 2011), salt/acid concentration (Sit et al., 2012; Idris et al., 2009) as well as the nature of polymer and salt/acid.

Carboxy methylcellulose (CMC) is a natural anionic polysaccharide which is widely used in many industrial sectors including food, textiles, paper, adhesives, paints, pharmaceutics, cosmetics and mineral processing. CMC

Figure 1. The highly translucent and flexible CMC-AN solid polymer electrolyte films.

is a cellulose derivative prepared through etherification of the hydroxyl groups with sodium monochloroacetate in the presence of aqueous alkali (Pushpamalar et al., 2006). It is a natural organic polymer that is non-toxic, renewable, available in abundance, biocompatible and biodegradable (Adinugraha et al., 2005). Due to its highly hydrophilic properties, CMC can easily dissolve in cold/hot water.

Inorganic AN salt is widely utilized in the manufacture of nitrogen-rich inorganic fertilizers, as a major component in explosive mixtures (AN-fuel oil for mining) and in gas generator propellant formulations as well as rocket propellant oxidizer (Lang and Vyazovkin, 2008). AN exhibits five polymorphic crystal phases at atmospheric pressure from cryogenic temperatures up to its melting temperature of 442 K. The IV-phase shows an orthorhombic crystal structure at room temperature (Chellappa et al., 2012).

In this paper, we report the effect of ammonium salt concentration on the structural and ionic conductivity of CMC-AN polymer electrolytes at ambient temperature in preparation of solid-state rechargeable proton battery.

EXPERIMENTAL PROCEDURE

Polymer electrolytes

The sodium salt of CMC was obtained from Acros Organics (purity >99.9%; average MW = 90,000 and DS = 0.7). AN was purchased from Sigma Aldrich (purity 99%) and both materials were directly used without further treatment. Solution casting method was employed to obtain film samples with varied amount of ammonium salt concentration (5 – 50 wt.%). Pure CMC film without ammonium salt was also prepared as a control. In a clean beaker, weighted amounts of CMC powder and AN crystals were dissolved in 100 ml distilled water at room temperature. Complete dissolution was achieved after several hours stirring at room temperature using magnetic stirrer. The final clear solution was then poured into separate Petri dishes and left to dry at room temperature to form highly translucent and flexible thin films. Polymer electrolyte films were transferred to a dessicator for further drying prior to sample characterization.

Characterization techniques

In this work, X-ray diffraction (XRD) analysis was carried out using a MiniFlex II Rigaku with CuK_α radiation (λ = 1.5418 Å) at room temperature. The XRD patterns were recorded at Bragg angle (2θ) in the range of 10 - 60° with a scan speed of 2°/min. This technique was employed to determine the crystallinity of polymer electrolyte films.

The occurance of complexation and the presence of functional groups of polymer electrolyte films were analyzed using IR technique. The Fourier Transform InfraRed (FTIR) spectra were recorded using Thermo Nicolet 380 FTIR spectrometer equipped with an attenuated total reflection (ATR) accessory with a germanium crystal. The sample was pressed on a germanium crystal and infrared light was passed through the sample with the frequency ranging from 700 - 4000 cm^{-1} at spectra resolution of 4 cm^{-1}. The FTIR data was recorded in the transmittance mode.

Impedance spectroscopy measurements were performed to determine the ionic conductivity of polymer electrolyte films over a wide range of frequency. The measurements were carried out with an electrical impedance spectroscopy (EIS) HIOKI 3532–50 LCR Hi Tester interfaced to a personal computer in 50 – 1M Hz frequency range. A π cm^2 round piece of electrolyte film was pressed between two steel electrodes of sample holder in a temperature controlled MEMMERT oven.

RESULTS AND DISCUSSION

XRD analysis

Figure 1 shows a highly translucent and flexible thin film of CMC-AN solid polymer electrolytes obtained from solution casting technique. Typical XRD patterns obtained from CMC-AN polymer electrolyte films with varied AN concentrations are shown in Figure 2i. Figure 2ii and iii represent XRD patterns of pure AN crystals and

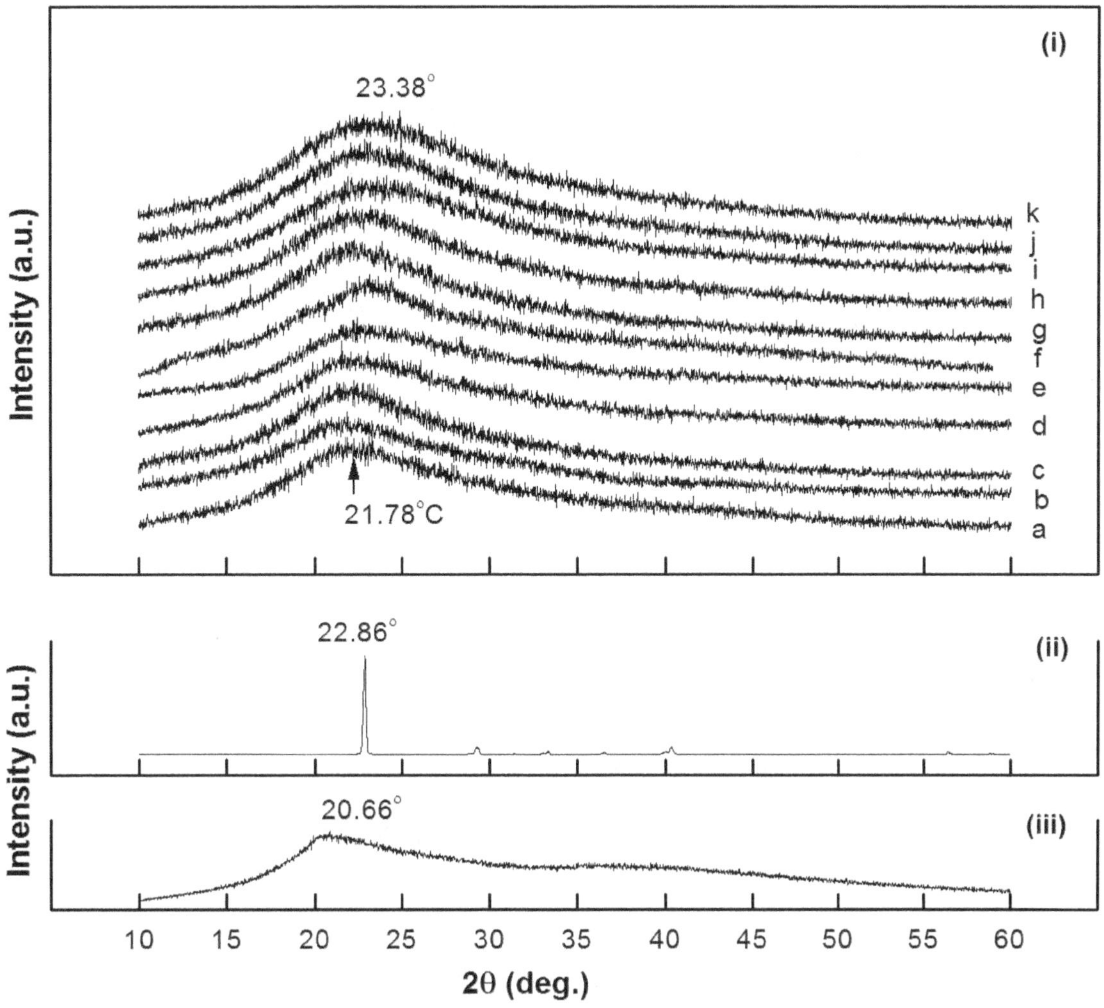

Figure 2. The XRD pattern of (i): Thin film samples containing CMC with varied amount of AN salt concentration, (a) Pure CMC, (b) CMC-5wt.% AN, (c) CMC-10wt.% AN,(d) CMC-15wt.% AN,(e) CMC-20wt.% AN,(f) CMC-25wt.% AN,(g) CMC-30wt.% AN,(h) CMC-35wt.% AN,(i) CMC-40wt.% AN,(j) CMC-45wt.% AN 5 and (k) CMC-50wt.% AN, (ii): Pure AN salt crystals, and (iii): Pure CMC powder.

CMC powder as references.

The XRD pattern for pure CMC powder (Figure 2iii) display a broad diffused hump centered at 2θ = 20.66°C and a small shoulder between 35°C and 45°C corresponding to the amorphous nature of pure CMC. The pattern for pure AN crystals show four polycrystalline peaks with the strongest intensity peak located at 2θ = 22.86°C (Figure 2ii). Upon addition of AN salt, the broad peaks tend to broaden and shift to a higher Bragg angle from 2θ = 21.78°C (pure CMC film) to 23.38°C (CMC-50wt.% AN) as can be seen in Figure 2i. The increase in broadness implies the amorphous nature of polymer electrolyte films with some localized ordering of the complex system (Selvasekarapandian et al., 2010). No additional peaks have been observed in the complex system indicating a complete dissociation of AN salt in the CMC polymer matrix. XRD patterns confirmed the amorphous nature of all polymer electrolyte films (Baril et al., 1997).

FTIR analysis

The occurance of complexation and the presence of functional groups in polymer electrolyte films were confirmed using IR Spectrophotometer. Different vibration modes of various functional groups of CMC and AN correspond to the observed IR characteristic peaks of polymer-salt system.

CMC and AN backbones

Pure CMC exhibits absorption band in the range of

Table 1. FTIR vibration modes of CMC and AN.

Band Assignment	CMC Wavenumber (cm^{-1})	AN Wavenumber (cm^{-1})
Bending mode of ether(glycosidic) linkage, δ(C-O-C)	1040, 1110	-
Bending mode of C-H band, δ(CH)	1377	-
Stretching mode of O-H in plane	1410	-
Stretching mode in carboxylic group, υ(C=O)	1605	-
Stretching mode of C-H band, υ(CH)	2920	-
Intra-molecular hydrogen bond, υ(OH)	3439	-
Symmetric bending band, δ(NO_3^-)	-	828
Symmetric stretching mode, υ(NO_3^-)	-	1043
Asymmetric stretching mode, υ(NO_3^-)	-	1363
Stretching mode of N-H band, υ(NH)	-	3031
Intra-molecular hydrogen bond, υ(OH)	-	3251

1072 - 3439 cm^{-1} (Lii et al., 2002; Jiang et al., 2009). The main backbones in CMC observed at 1605 cm^{-1} and 1377 cm^{-1} correspond to C=O band and C-H bonding, respectively. In the case of AN, vibration bands have been found in the range of 828 - 3251cm^{-1} (Kadir et al., 2011; Majid and Arof, 2005). Typically, the O-H bonding can be found in the range of 3000 - 3800 cm^{-1} for both backbones. The vibration modes of CMC and AN are summarized in Table 1.

CMC-AN complexes

The occurrence of complexation between the polymer host and salt has been correlated to the spectral changes including shift of band, emergence of new bands or/and disappearance of bands. The complexation between CMC and inorganic salts/acids was reported to occur in the band range of 1600 - 1040 cm^{-1} (Chai and Isa, 2013; Samsudin and Isa, 2012). Therefore, detailed observation on the frequency region of 1800 - 800 cm^{-1} is required in this study. Outside of this region, insignificant or very weak peaks have been discovered.

Figure 3 depicts that adding a range of AN concentrations to CMC polymer electrolyte films resulted in the emergence of complexation bands in the range of 1633 - 829 cm^{-1}. The COO^- stretching band of CMC shifts to the lower wavenumber from 1591 - 1583 cm^{-1} and a new peak emerges at 1633 cm^{-1} as the AN concentration increases to 50wt.%. The new peak is associated to the H^+ ion originating from AN. Strong absorption peaks at 1417 and 1321 cm^{-1} assigned to O-H stretching in plane and symmetrical stretching of C-H bands of CMC respectively have shifted to 1431 and 1331 cm^{-1}. The shifting proved the complexation of CMC with the nitrate ions of AN. Similar behavior has been detected in the

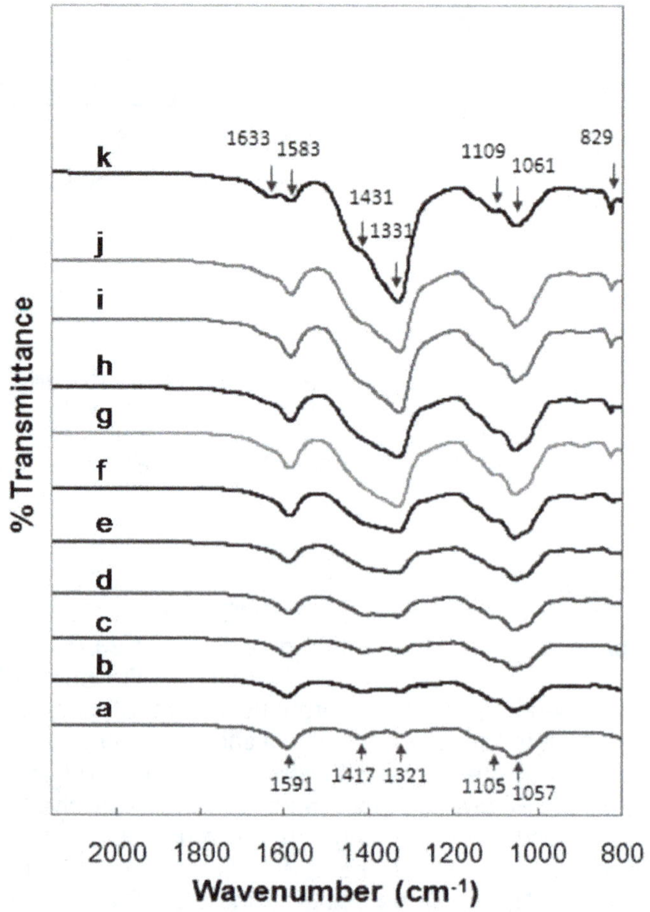

Figure 3. FTIR spectra of polymer electrolyte films containing CMC with varied amount of AN salt concentration, (a) Pure CMC film, (b) CMC-5wt.% AN, (c) CMC-10wt.% AN, (d) CMC-15wt.% AN, (e) CMC-20wt.% AN, (f) CMC-25wt.% AN, (g) CMC-30wt.% AN, (h) CMC-35wt.% AN, (i) CMC-40wt.% AN, (j) CMC-45wt.% AN 5 and (k) CMC-50wt.% AN.

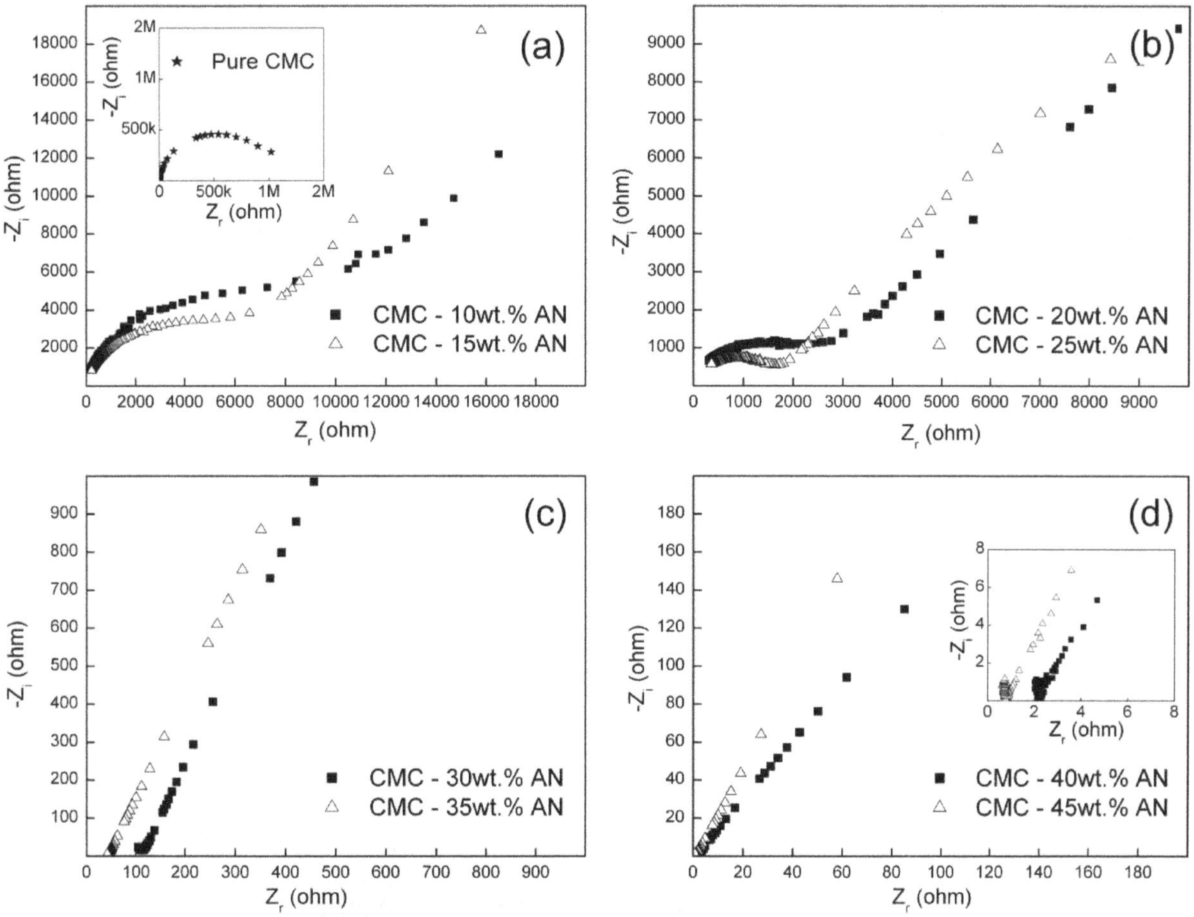

Figure 4. The Cole-Cole (Nyquist) plots of the real impedance, Z_r versus imaginary impedance, $-Z_i$ polymer electrolytes with varied amount of AN concentration at ambient temperature; (a) CMC-10wt.% AN and CMC-15wt.% AN (Inset: Pure CMC film), (b) CMC-20wt.% AN and CMC-25wt.% AN, (c) CMC-30wt.% AN and CMC-35wt.% AN, and (d) CMC-40wt.% AN and CMC-45wt.% AN (Inset: Enlargement of (d)).

region of 1105 - 1057 cm^{-1} which signifies the characteristic of polysaccharide skeleton. No obvious changes of intensity have been observed except shifting of the bands to the higher wavenumbers. A new peak at 829 cm^{-1} corresponds to the nitrate ion of AN. It can be concluded from the analysis that the protonation has occurred in the present polymer-salt complexes and the interactions between CMC and AN has been established.

Impedance analysis

Cole-cole plots

Figure 4 shows the complex impedance plots of pure CMC film and CMC-AN doped with different concentration of AN at ambient temperature. The figure shows a part of a depressed semicircle for pure CMC film. The high frequency semicircle is associated to the parallel combination of the bulk resistance and bulk capacitance as a result of protons migration and immobile polymer chains, respectively (Selvasekarapandian et al., 2010; Subban et al., 2005). As the salt concentration begins to increase, the semicircle in the plots become to lessen. The depressed semicircle and inclined spikes (Figure 4a-b) implies that the ions have different relaxation times (Subban et al., 2005). Beyond 25wt.% AN, the appearance of low frequency spikes indicate that only the resistive component of the polymer electrolyte predominate (Selvasekarapandian et al., 2010) as illustrated in Figure 4c-d. The bulk resistance, R_b can be calculated from the intercept of high frequency semicircle or the low frequency spike on the Z_r-axis.

dc ionic conductivity

The high ionic conductivity of solid polymer electrolytes is mainly governed by two important factors; that is, ionic

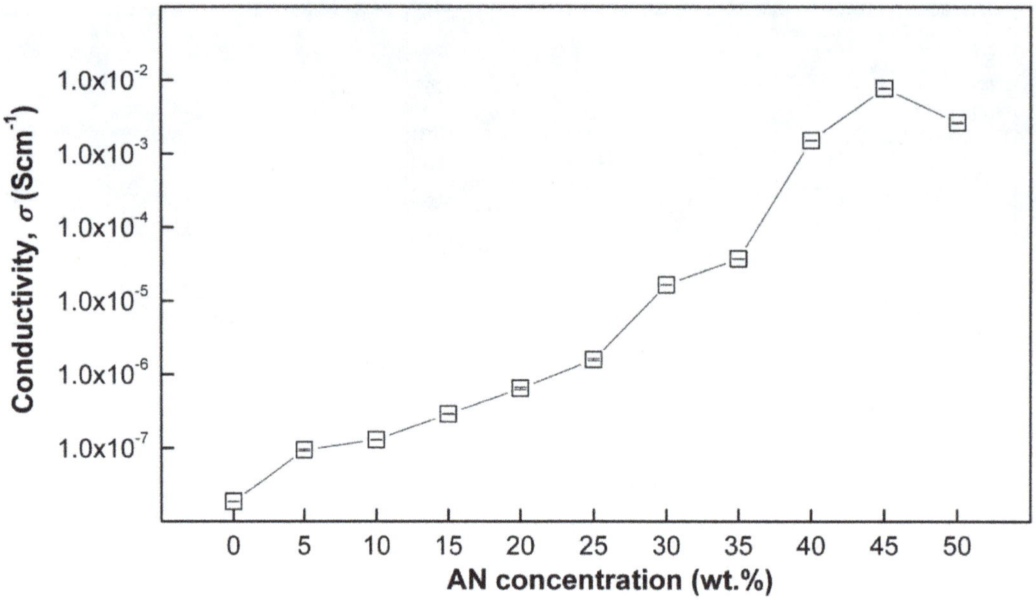

Figure 5. Graph of ionic conductivity, σ versus AN salt concentration of polymer electrolyte films at ambient temperature, 303K.

conducting species concentration, η and the charge carrier mobility, μ along with the type of charge carriers (cationic/anionic) and temperature (Hirankumar et al., 2005; Raphael et al., 2010). The evolution of the ambient temperature, 30°C (303K) ionic conductivity of polymer electrolyte films as a function of the AN concentration is shown in Figure 5. The ionic conductivity of polymer electrolyte films, σ were calculated using Equation 1:

$$\sigma = \frac{t}{R_b A} \qquad (1)$$

To attain the ionic conductivity, σ (Scm^{-1}), the thickness of the electrolyte, t (cm), the bulk resistance, R_b (Ω), and the area of electrolyte-electrodes contact, A (cm^2) should be known. Concentration dependence of ionic conductivity provides information on the specific interaction between the salt and polymer matrix.

As illustrated in Figure 5, the ionic conductivity increases gradually with the addition of AN concentration up to 45wt.%. The increment of ionic conductivity with increasing AN concentration can be related to the increasing number of mobile charge carriers and/or their mobility (Samsudin and Isa, 2012). This may also be attributed to an increase of the amorphous nature of polymer electrolytes as confirmed by the XRD analysis. The optimum AN concentration of 45 wt.% gives the highest ionic conductivity of (7.71 ± 0.04) $x10^{-3}$ Scm^{-1} compared to pure CMC film of (1.86 ± 0.03) $x10^{-8}$ Scm^{-1}. Beyond that, the ionic conductivity decreases. The subsequent decrease at AN concentration of 50 wt.% could be caused by the formation of ion multiples or ion aggregates which reduces the number of mobile ions and limits its mobility in the polymer electrolytes (Woo et al., 2011; Yahya and Arof, 2003). At such high concentration, polymer electrolyte film was observed mechanically unstable. This was evidenced by the reduction in film flexibility as a consequence of rapid increase in viscosity (Ferry et al., 1998). Table 2 summarizes the ionic conductivity of single polymer-ammonium salt system at room temperature as reported in the literatures. The list indicates that the ionic conductivity obtained from the present work is higher compared to the previous works reported for similar AN salt system.

Conclusion

The new solid polymer electrolytes based on CMC and AN have been successfully prepared by solution cast method. XRD studies confirmed that all films were predominantly amorphous. IR spectra indicated that the complexation of polymer electrolyte films occured in the frequency range of 1633 - 829 cm^{-1}. The study revealed that the concentration of AN greatly influenced the ionic conductivity of the polymer electrolyte films. The pure CMC film gives the ionic conductivity in the order of 10^{-8} Scm^{-1}. Upon addition of the salt, the ionic conductivity has increased to the optimum AN concentration (45 wt.%) in the order of 10^{-3} Scm^{-1} at ambient temperature. The present polymer electrolytes were identified as promising materials for electrochemical device applications.

Table 2. Ionic conductivity of solid polymer electrolytes based on single polymer and AN salt system measured at room temperature from literatures compared to the present work.

Polymer host	AN concentration (wt.%)	Solvent	Conductivity, σ (Scm^{-1})	Reference
Chitosan	45	Distilled water	2.53×10^{-5}	Majid and Arof (2005)
Chitosan	40	Acetic acid	8.38×10^{-5}	Ng and Mohamad (2006)
Methylcellulose	25	Distilled water	2.10×10^{-6}	Shuhaimi et al. (2010)
Tapioca starch	25	Oxalic acid	2.15×10^{-6}	Azlan and Isa (2011)
Tapioca starch	25	Distilled water	2.83×10^{-5}	Khiar and Arof (2010)
CMC	45	Distilled water	$(7.71 \pm 0.04) \times 10^{-3}$	Present work

ACKNOWLEDGEMENT

The authors would like to acknowledge laboratory assistants of Department of Physical Sciences, Universiti Malaysia Terengganu for the assistance during sample preparation and characterization. The authors also grateful to Universiti Malaysia Terengganu for financial support through *Geran Galakan Penyelidikan (GGP) 68007/2013/58.*

REFERENCES

Adinugraha MP, Marseno DW, Haryadi (2005). Synthesis and characterization of sodium carboxymethylcellulose from cavendish banana pseudo stem (*Musa cavendishii* LAMBERT). Carbohydr. Polym. 62:164-169.

Armand M (1994). The history of polymer electrolytes. Solid State Ionics. 69:309-319.

Azlan AL, Isa MIN (2011). Proton conducting biopolymer electrolytes based on tapioca starch- NH_4NO_3. Solid State Sci. Technol. Lett. 18:124-129.

Baril D, Michot C, Armand M (1997). Eletrochemistry ofliquid vs. Solid: Polymer electrolytes. Solid State Ionics 94:35-47.

Bhargav PB, Mohan VM, Sharma AK, Rao VVRN (2009). Investigations on electrical properties of (PVA:NaF) polymer electrolytes for electrochemical cell applications. Curr. Appl. Phys. 9:165-171.

Chai MN, Isa MIN (2013). The oleic acid composition effect on the carboxymethyl cellulose based biopolymer electrolyte. J. Crystallization Process Technol. 3:1-4.

Chellappa RS, Dattelbaum DM, Velisavljevic N, Sheffield S (2012). The phase diagram of ammonium nitrate. J. Chem. Phys. 137:064504.

Ferry A, Oradd G, Jacobsson P (1998). Ionic interactions and transport in a low-molecular-weight model polymer electrolyte. J. Chem. Phys. 108:7426-7433.

Hassan MA, Gouda ME, Sheha E (2010). Investigations on the electrical and structural properties of PVA doped with $(NH_4)_2SO_4$. J. Appl. Polym. Sci. 116:1213-1217.

Idris NK, Aziz NAN, Zambri MSM, Zakaria NA, Isa MIN (2009). Ionic conductivity studies of chitosan-based polymer electrolytes doped with adipic acid. Ionics. 15:643-646.

Jiang LY, Li YB, Zhang L, Wang XJ (2009). Preparation and characterization of a novel composite containing carboxymethyl cellulose used for bone repair. Mater. Sci. Eng. C. 29:193-198.

Kadir MFZ, Aspanut Z, Majid SR, Arof AK (2011). Ftir studies of plasticized poly(vinyl alcohol)–chitosan blend doped with NH_4NO_3polymer electrolyte membrane. Spectrochim. Acta, Part A. 78:1068-1074.

Khiar ASA, Arof AK (2010). Conductivity studies of starch-based polymer electrolytes. Ionics 16:123-129.

Kumar KK, Ravi M, Pavani Y, Bhavani S, Sharma AK, Rao VVRN (2011). Investigations on the effect of complexation of NaF salt with polymer blend (PEO/PVP) electrolytes on ionic conductivity and optical energy band gaps. Physica B. 406:1706-1712.

Lang AJ, Vyazovkin S (2008). Phase and thermal stabilization of ammonium nitrate in the form of PVP–AN glass. Mater. Lett. 62:1757-1760.

Lii C-Y, Tomasik P, Zaleska H, Liaw S-C, Lai VMF (2002). Carboxymethyl cellulose–gelatin complexes. Carbohydr. Polym. 50:19-26.

Ma J, Sahai Y (2013). Chitosan biopolymer for fuel cell applications. Carbohydr. Polym. 92:955-975.

Majid SR, Arof AK (2005). Proton-conducting polymer electrolyte films based on chitosan acetate complexed with NH_4NO_3 salt. Phys. B. 355:78-82.

Ng LS, Mohamad AA (2006). Protonic battery based on a plasticized chitosan-NH_4NO_3 solid polymer electrolyte. J. Power Sources 163:382-385.

Pushpamalar V, Langford SJ, Ahmad M, Lim YY (2006). Optimization of reaction conditions for preparing carboxymethyl cellulose from sago waste. Carbohydr. Polym. 64:12-18.

Quartarone E, Mustarelli P, Magistris A (1998). PEO-based composite polymer electrolytes. Solid State Ionics 110:1-14.

Ramesh S, Yahaya AH, Arof AK (2002). Dielectric behaviour of PVC-based polymer electrolytes. Solid State Ionics 152-153:291-294.

Samsudin AS, Isa MIN (2012). Structural and ionic transport study on CMC doped NH_4Br: A new types of biopolymer electrolytes. J. Appl. Sci. 12:174-179.

Selvasekarapandian S, Hema M, Kawamura J, Kamishima O, Baskaran R (2010). Characterization of PVA-NH_4NO_3 polymer electrolyte and its application in rechargeable proton battery. J. Phys. Soc. Jpn. 79SA:163-168.

Shuhaimi NEA, Teo LP, Majid SR, Arof AK (2010). Transport studies of NH_4NO_3 doped methyl cellulose electrolyte. Synth. Met. 160:1040-1044.

Sit YK, Samsudin AS, Isa MIN (2012). Ionic conductivity study on hydroxyethyl cellulose (HEC) doped with NH_4Br based biopolymer electrolytes. Res. J. Recent Sci. 1:16-21.

Subban RHY, Ahmad AH, Kamarulzaman N, Ali AMM (2005). Effects of plasticiser on the lithium ionic conductivity of polymer electrolyte PVC-$LiCF_3SO_3$. Ionics 11:442-445.

Tarascon JM, Armand M (2001). Issues and challenges facing rechargeable lithium batteries. Nature 414:359-367.

Woo HJ, Majid SR, Arof AK (2011). Conduction and thermal properties of a proton conducting polymer electrolyte based on poly (ε-caprolactone). Solid State Ionics 199-200:14-20.

Yahya MZA, Arof AK (2003). Effect of oleic acid plasticizer on chitosan–lithium acetate solid polymer electrolytes. Eur. Polym. J. 39:897-902.

Probing of dielectric properties of high density poly-ethylene/calcium carbonate (HDPE/$CaCO_3$) nano-micro composite

M. A. Al-Eshaikh and M. Iqbal Qureshi*

Research Center, College of Engineering, King Saud University, P. O. Box 800, Riyadh 11421, Saudi Arabia.

In this experimental investigation, high density poly-ethylene/calcium carbonate (HDPE-$CaCO_3$) micro/nano composite sheets were prepared by melt blending the master batch using twin screw extruder followed by injection molding. The dispersion characteristics were determined by field emission scanning electron microscope (FESEM) and were found to be reasonably uniform. These sheets were then subjected to electrical stresses in order to evaluate their dielectric performance. The dissipation factor (tanδ) and relative permittivity (ε_r) were evaluated as a function of temperature in a practical range of 23 to 80°C. The composite exhibited decreased ε_r and tanδ values compared with unfilled HDPE. These results support the multi-layer model proposed in literature for nano-composite. It is also clarified that this nano-micro composite is higher in both ac breakdown strength and partial discharge resistance than for the base polymer.

Key words: High density poly-ethylene/calcium carbonate (HDPE/$CaCO_3$) nano-micro composite, partial discharge, surface degradation, dielectric properties.

INTRODUCTION

The application of nanotechnologies in the field of electrical insulation for power and high voltage engineering requires extensive research and development of these "advanced" materials. The innovative materials created from the incorporation of nano-particles in a polymer matrix are commonly called nano-dielectrics. It is now accepted that nano-dielectrics insulation performs better in some applications in comparison with conventional insulating materials. These nano-dielectric materials require a detailed electrical, thermal, mechanical and chemical characterization prior to their application in the equipment that undergoes multiple stresses during its life time. Several reviews and original papers have been reported on the electrical properties of such composites (Castellon et al., 2011; Celebrese et al., 2011; Danikas and Tanaka, 2009; Tanaka et al., 2005; Tanaka, 2008; Iyer et al., 2011; Krivda et al., 2012; Nelson, 2010; Park et al., 2011; Wang et al., 2011; Working Group, 2011).

High density poly-ethylene (HDPE) is widely used as a commodity polymer due to its distinctive mechanical and physical properties. However, to improve its performance the HDPE has been reinforced with cheaply available fillers such as calcium carbonate ($CaCO_3$). Zebarjad and Sajjadi (2008) reported that nano-sized $CaCO_3$ has significant effect on crystallinity and melting point of HDPE. However, due to its polar nature it is difficult to evenly disperse and stabilize in a polymer matrix. In order to disperse inorganic nano-fillers into organic polymer matrices, the affinity between their surfaces in contact should be taken into consideration. The surface energy is a physical parameter that directly represents the affinity between the two substances. It is necessary to reduce the difference between the two in order to obtain a good composite. In this context, several techniques such as using compatibilizers and coupling agents, high shear force mixing, solution mixing etc. have been adopted to get better dispersion results (Celebrese et al., 2011; Nelson, 2010; Imai et al., 2006; Working Group, 2011). Nano-particles have much higher interfacial area per unit

*Corresponding author. E-mail: mqureshi1@ksu.edu.sa.

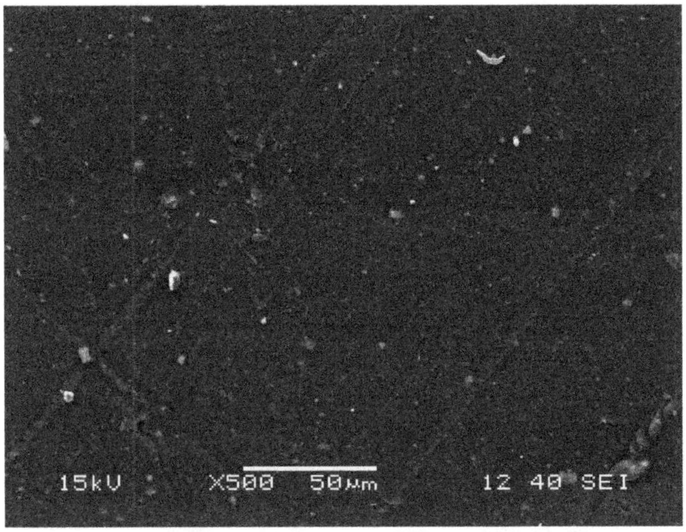

Figure 1. SEM surface micrograph of $CaCO_3$ filled specimen.

volume and in addition the interface between the particle and polymer is thought to have a finite thickness in which the properties of the polymer differ from the original matrix (Tanaka et al., 2008). In such systems the total interfacial volume can be very large and may control the dielectric response of the system (Lewis, 1994). However, there have been conflicting reports in the literature regarding the effect of nano-micro particles on the dielectric properties of polymer composites, since they have been shown to inflict increase and decrease in dielectric properties (Imai et al., 2006; Krivida et al., 2012; Singha and Thomas, 2008). While variations are to be expected in different systems, the effect of processing can also be a determining factor in the final properties. Therefore, it is important to carefully consider and control the processing conditions used when making such composites.

In the present experiments, HDPE composite with $CaCO_3$ was prepared using master batch approach. The master batch contained both the polymer carrier and nano-micro filler. The dispersion and distribution of $CaCO_3$ particles in the composite was characterized by scanning electron microscope (SEM). The impact of $CaCO_3$ filler on the salient electrical properties such as its relative permittivity (ε_r) dissipation factor (tanδ) was compared. Insulating materials used in electrical power apparatus are generally exposed to surface discharges in the course of their normal operation resulting in deterioration of the material's surface that can ultimately lead to their breakdown. In this context, the samples' ac breakdown strength (E_b) and partial discharge (PD) resistance were also investigated using the IEC(b) electrode system for several hours. Comparisons were also made for surface roughness observed by precision profilometer and PD resistance was evaluated based on the surface roughness of filled and unfilled specimens.

EXPERIMENTAL METHODS

Preparation of samples

HDPE-54 produced by Saudi Arabian Basic Industry Corporation (SABIC) was used as the base polymer and is of injection molding grade with a narrow molecular weight distribution and high flow ability. The master batch 'Micro Filler-0189' supplied by Wuxi Changhong Master-batches Co., China contained 80 ± 3% micro and ~20% nano $CaCO_3$ on LLDPE as a carrier with granule size of 20 nm to 2 µm. Its surface is not treated. HDPE was dry blended with 5% $CaCO_3$ filler and palletized using an intermeshing and co-rotating twin screw extruder which had both dispersive and distributive mixing elements. The melting was at 235°C under pressure of 7.0 bars. The extrudate was cooled in a water bath, dried and palletized for further use (Elleithy et al., 2010). Flat sheets 100 mm wide, 110 mm long and 1 mm thick were injection molded for the purpose of testing.

The morphological analysis was carried out using field emission scanning electron microscope (FESEM), JOEL model USM-7600F coupled with energy-dispersive X-ray spectroscopy (EDX) elemental analysis tool. To avoid electrostatic charging the surface of the samples was platinum sputter coated imparting thickness of ~10 µm before subjecting them to SEM observations.

The SEM surface profile of $CaCO_3$ filled specimen is shown in Figure 1 which shows a reasonable good distribution with minor tendency of larger particles that have agglomerated. Agglomeration is a well known phenomenon, its probability increases with the decrease in particle size (Celebrese et al., 2011; Park et al., 2011). Nevertheless, the particle size distribution in this molded sample is reasonably homogeneous. Figure 2 exhibits the EDX elemental distribution in the filled sample. It is clear that the presence of calcium is around ~0.63% in this sample, whereas both the filled and unfilled samples are clear of any spurious traces of any other impurities, thus, exhibiting cleanliness of the development process.

Measurements of relative permittivity and dissipation factor

The ε_r and tanδ of the samples was measured as a function of temperature using Tettex precision Schering bridge under electrical stress of 500 V. The samples were placed in test cell Tettex type 2604 in which the temperature was controlled with a precision of ±1°C.

Measurement of dielectric breakdown strength

Dielectric breakdown strength experiments were carried out in a Perspex test cell containing polished sphere to sphere electrode system as shown in Figure 3. In this case, 50 mm wide, 50 mm long and 1.0 mm thick samples were inserted one at a time between the pair of electrodes. Whole electrode system was immersed in transformer oil to enhance the surrounding insulation and suppress occurrence of any surface discharge. High voltage of 60 Hz was ramped at a rate of 1.0 kV/s till breakdown ensued. The power supply was a 90 kV/230 V transformer and the measuring system was microprocessor controlled. The oil was continuously stirred during the measurements to ensure that readings are not affected by the by-products of degradation. Electrodes were replaced after a set of ten measurements.

Measurement of surface erosion due to partial discharge

The IEC(b) electrode system as shown in Figure 4 was used to test these specimens. It consisted of a stainless steel rod and plane electrode. The rod electrode was connected to the high voltage and

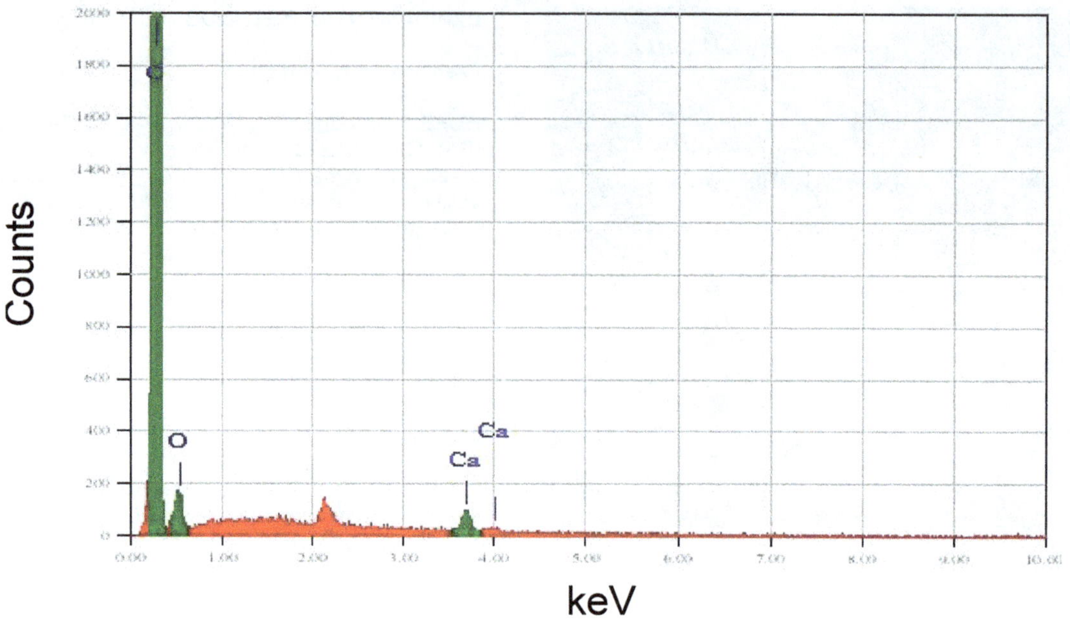

Element		(keV)	mass%	Error%	At%	Compound	mass%	Cation	K
C K		0.277	93.89	0.23	95.61			97.7391	
O K	*	0.525	5.49	4.57	4.20			1.4650	
Ca K	*	3.690	0.63	1.19	0.19			0.7959	
Total			100.00		100.00				

Figure 2. Elemental distribution in $CaCO_3$/HDPE composite.

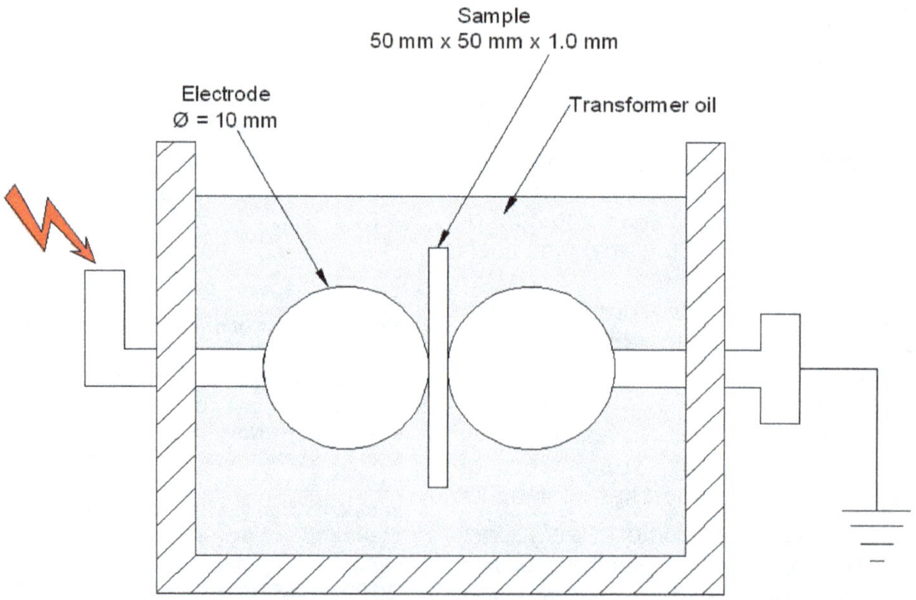

Figure 3. Test cell with sphere to sphere electrodes system for breakdown strength test.

was 6 mm in diameter with end curvature of 1.0 mm radius. The polished plane electrode was grounded. The rod-plane geometry was enclosed in a transparent Perspex cylindrical test cell maintained at room temperature and atmospheric pressure. The specimens, one at a time, were subjected to 60 Hz ac at 15 kV_{rms} for predetermined fixed times of 6, 19, and 48 h. The resistance towards PD was compared as the depth of the eroded surfaces from adjacent to the electrode edge to the end of the eroded surface using Ultra Surf Inductive Profiler type i-60 from Tylor Hobson, UK.

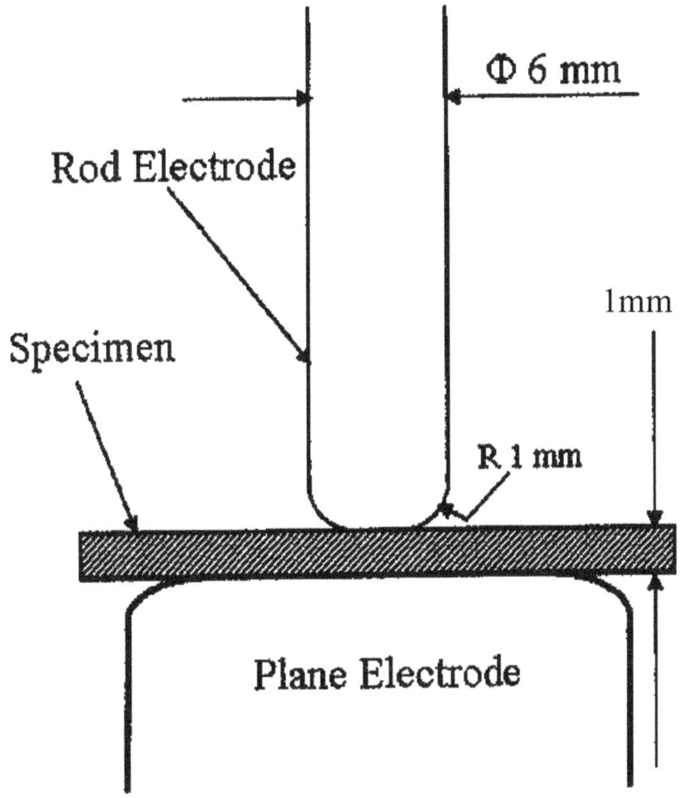

Figure 4. IEC(b) rod-plane electrode assembly.

RESULTS AND DISCUSSION

Higher Values of E_b and PD resistance of a polymer are its important features and pre-requisite for its application as an insulating material. The dielectric ε_r and tanδ have influence on the local field distribution and on resistance towards PD erosion of the specimen's surface when these are exposed to high electrical field and temperature. These salient features were, therefore, measured for both types of specimens investigated.

Figure 5 shows variation of ε_r and tanδ as a function of temperature. With the increase in temperature, the ε_r decreases slightly in $CaCO_3$ filled sample, while it increases in case of unfilled sample. Similarly, a decrease of almost 45% is observed in tanδ value with rise in temperature in filled sample in the measured temperature range. Opposite observations have been reported by Imai et al. (2006) in nano-micro silicate filled epoxy and Hoffmann et al. (2011) in nano-silica filled epoxy at power frequency. The increase in tanδ in unfilled samples can be attributed to higher polarization losses, whereas decrease in ε_r in filled HDPE could be due to blocking and trapping of free charge carriers at the filler interfaces. Singha and Thomas (2008) have reported decrease in ε_r depending on the interfacial polarization at particle interlayer; its relaxation mechanism and also due to blocking of dipole mobility. The presence of "free volume" has also been attributed to the reduction in ε_r. Introduction of nano-particles in the base matrix reduces the free volume (Nelson and Hu, 2004). Tanaka et al. (2005) have proposed a multi-layer model based on the presence of free volume and its interaction with the inter-layers of the nano-composite. Two kinds of players were proposed to be working in this scenario. The first one is the inner inter-layer which acts to block the motion of dipoles originating from some polar radicals, while outer loose layer acts to reduce the free volume. Both of these layers thus contribute in the reduction of ε_r, which is experimentally demonstrated here and thus, substantiates the hypothesis of multi-layer model (Tanaka et al., 2005). Both characters (that is, tanδ and ε_r) of $CaCO_3$ filled polymer exhibit here much beneficial dielectric response.

Figure 6 shows successive profiles, representing the progressive erosion of the surface of unfilled HDPE specimen in time lapse of 6, 19, 25 and 48 h, respectively. The surface was scanned on each sample along a radial line starting from the trace of the edge of the electrode up to a distance of 10 mm. Similarly, depth profiles were obtained for the filled specimens. Figure 7 compares the erosion depth profile for filled and unfilled composite samples after 25 h of aging. It is clear that PD erosion in unfilled sample is not only deeper but its pattern is also different. This needs further clarification as to describe the mechanisms that result into this difference. This

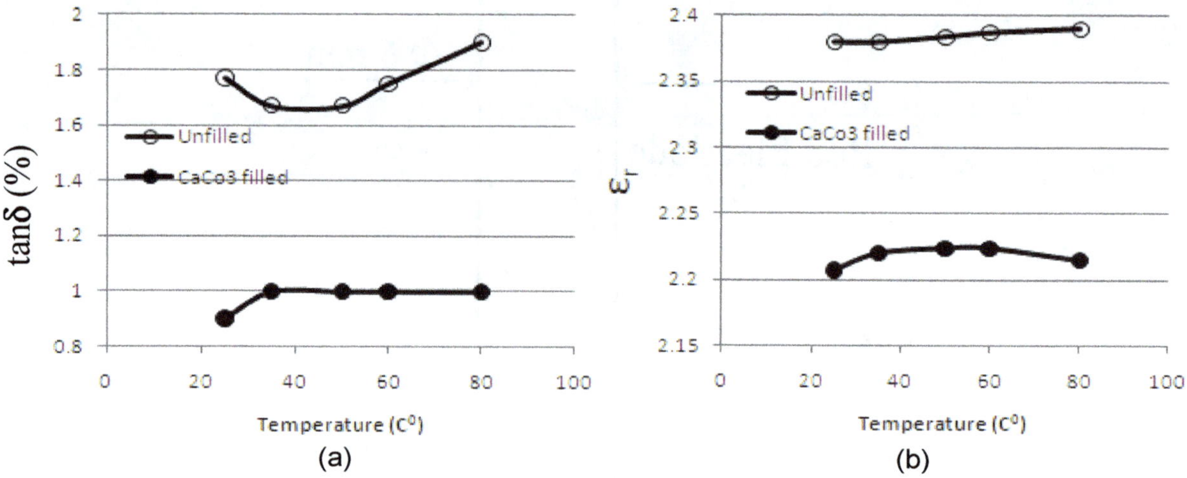

Figure 5. Variation of ε_r and tanδ as a function of temperature for HDPE samples with and without $CaCO_3$ particles.

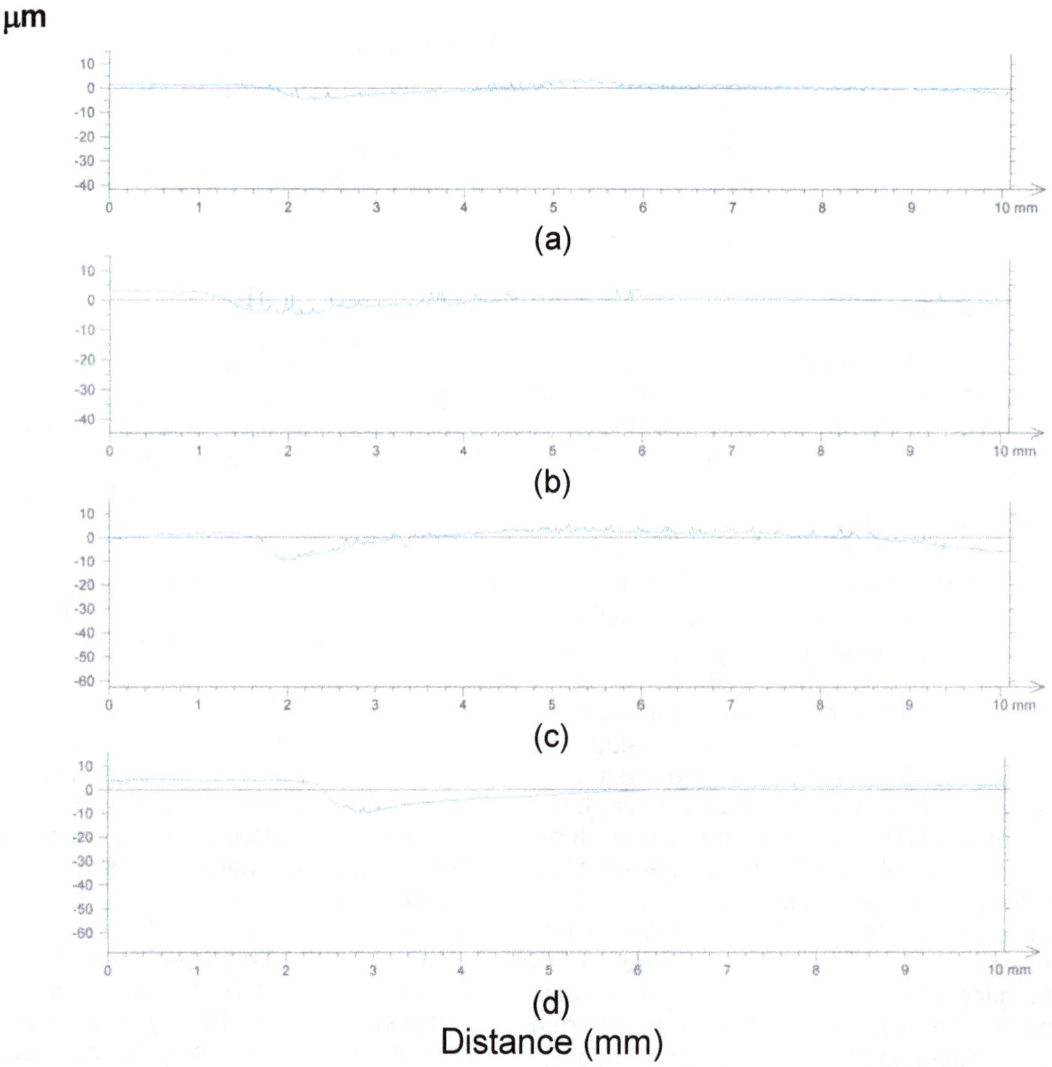

Figure 6. Surface erosion profiles of degraded unfilled specimen after aging of (a) 6 h, (b) 19 h, (c) 25 h, and (d) 48 h.

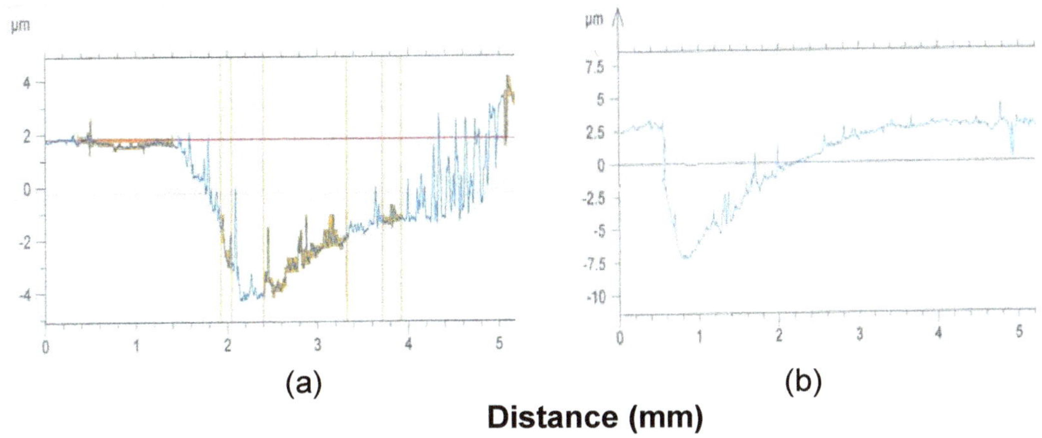

Figure 7. Comparison of erosion on 25 h aged, (a) unfilled, and (b) filled polymer composites.

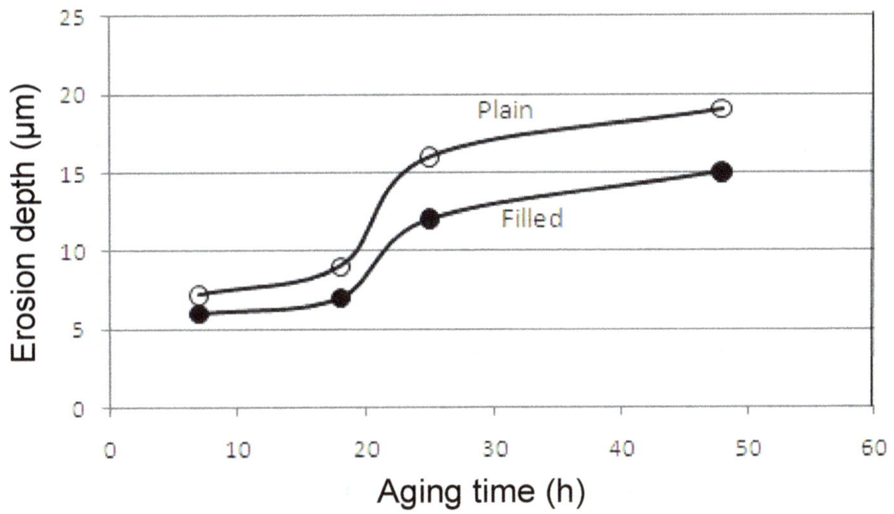

Figure 8. Erosion depth as a function of aging duration.

aspect will be further investigated to clarify erosion characteristics in near future. Figure 8 compares the erosion depth as a function of aging time. Here, the point data are the average of three repeated runs on samples of the same type. It is clear that the erosion in time is not linear but tends toward saturation after about 30 h of aging in both samples. However, it is obvious that $CaCO_3$ filled HDPE sample degrades much lesser. Saturation of erosion depth in the long run could be caused by factors such as change in the electric field at the electrode edge caused by the formation of eroded cavity and the physical change in the edge profile of the electrode.

Weibull statistical plots with 90% confidence bounds for E_b of the two compared systems are shown in Figure 9. The shape and scale parameters are summarized in Table 1. In here, the shape parameter could be obtained from the slope meaning the data distribution, and the scale parameter represented the E_b by which 63.2% (characteristics value) of the cumulative probability was expected to fail. The P_5 value refers to E_b value at which 5% would survive under a given electric field. This value is important for the design engineer. The analysis shows 31% increase in the characteristic value of the E_b of the filled composite, whereas it shows 24% increase in P_5 value, as well. Interestingly, there is not much effect on the shape parameter value which suggests that the filler was evenly dispersed in HDPE which resulted in improved dielectric characteristics.

Conclusion

In this study, HDPE composite was prepared with 5% volume of nano-micro sized $CaCO_3$ by the process of

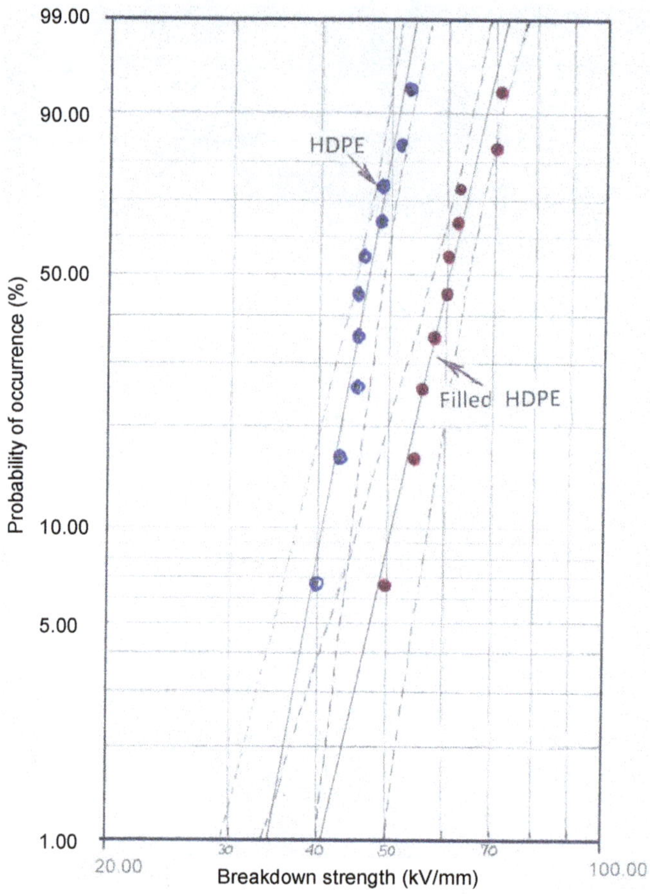

Figure 9. Weibull distribution plot (90% confidence bounds) for the unfilled and filled HDPE composites.

Table 1. Weibull parameters of ac breakdown strength for filled and unfilled HDPE.

Sample type	Scale parameters (kV/mm)	Shape parameter	P_5 (kV/mm)
HDPE	48.1	12.3	38.1
Filled HDPE	62.9	11.52	47.2

melt blending from master batch by extrusion and then followed by injection molding. The morphological analysis revealed a fairly homogeneous dispersion of particles. $CaCO_3$ filled samples exhibit much smaller value of ε_r and almost negligible increase in tanδ as compared to unfilled sample as a function of temperature. HDPE filled with nano-micro $CaCO_3$ is also superior in PD resistance as well as in its ac breakdown strength as compared to unfilled polymer.

ACKNOWLEDGEMENTS

This research was conducted under a financial grant #30/430 from the Deanship of Scientific Research, and Research Center, College of Engineering, King Saud University. The authors are indebted to SABIC polymer chair for preparation of samples used in this research.

REFERENCES

Castellon J, Nguyen HN, Agnel S, Tunreille A (2011). Electrical Properties Analysis of Micro and Nano Composite Epoxy Resin Materials. IEEE Trans. Dielectric Electr. Insul. 18:651-657.

Celebrese C, Hui L, Schadlere LS, Nelson JK (2011). A Review on the Importance of Nanocomposite Processing to Enhance Electrical Insulation. IEEE Trans. Dielectric Electr. Insul. 18:938-945.

Danikas MG, Tanaka T (2009). Nanocomposite: A Review on Electrical Treeing and Breakdown. IEEE Electr. Insul. Mag. 25:19-25.

Elleithy RH, Ilyas A, Ali MA, Al-Zahrani SM (2010). High Density Polyethylene / Micro-Calcium Carbonate Composite: A Study of the Morphological, Thermal and Viscoelastic Properties. J. Appl. Polym.

Sci. 117:2413-2421.

Hoffmann C, Weidner J, Muth T, Jenau F (2011). Treeing Growth and Lifetime of Epoxy Filled with 26% Silica Nano-Particles at 50 Hz AC Voltage. XVIIth ISHE, Germany. pp. 1207-1212.

Imai T, Sawa F, Ozaki T, Shimizn T, Kuge S, Kozako M, Tanaka T (2006). Effect of Epoxy/Filler Interface on Properties of Nano or Micro-Composites. IEEJ Trans. Fundam. Math. 26:84-91.

Iyer G, Gorur RS, Richert R (2011). Dielectric Properties of Epoxy Based Nanocomposite for High Voltage Insulation. IEEE Trans. Dielectric Electr. Insul. 18:659-666.

Krivida A, Tanaka T, Frechtte M, Castellon J, Fabiani D, Montanari GC, Gorur R, Morshuis P, Gubanski S, Kndersberger J, Vaughn A, Pelissou S, Tanaka Y, Schmidt LE, Iyer G, Andritsch T, Seiler J, Anglhuber M (2012). Characterization of Epoxy Microcomposite and Nanocomposite Materials for Power Engineering Applications. IEEE Electr. Insul. Mag. 28:38-51.

Lewis TJ (1994). Nanometeric Dielectrics. IEEE Trans. Dielectric Electr. Insul. 15:812-825.

Nelson JK, Hu Y (2004). The Impact of Nanocomposite Formulation on Electrical Endurance. Proc. IEEE Int. Conf. Solid Dielectric pp. 832-835.

Nelson JK Eds (2010). Dielectric Polymer Nanocomposites. Springer, New York, USA, pp. 1-30. ISBN 978-1-4419-1591-7.

Park J, Lee CH, Lee JY, Kim HD (2011). Preparation of Epoxy/Micro – Nano Composite by Electrical Field Dispersion Process and its Mechanical and Electrical Properties. IEEE Trans. Dielectric Electr. Insul. 18:667-674.

Singha S, Thomas MJ (2008). Dielectric Properties of Epoxy Nanocomposites. IEEE Trans. Dielectric Electr. Insul. 15:12-23.

Tanaka T (2008). Dielectric Nanocomposites with Insulating Properties. IEEE Trans. Dielectric Electr. Insul. 12:914-928.

Tanaka T, Kozako M, Fuse N, Ohki Y (2005). Proposal of a Multi-Core Model for Polymer Nano-Composite Dielectrics. IEEE Trans. Dielectric Electr. Insul. 12:669-681.

Tanaka T, Ohki Y, Ochi M, Harada M, Imai T (2008). Enhanced PD Resistance of Epoxy / Clay Nano-composite Prepared by Newly Developed Organic Modification and Solubalization Methods. IEEE Trans. Dielectric Electr. Insul. 11:81-89.

Wang Z, Izuka T, Kozako M, Ohki Y, Tanaka T (2011). Development of Epoxy / BN Composite with High Thermal Conductivity and Sufficient Dielectric Breakdown Strength. IEEE Trans. Dielectric Electr. Insul. 18:1973-1983.

Working Group D 1.24 (2011). Polymer Nanocomposites - Fundamentals and Possible Applications to Power Sector. CIGRE. Brochure P. 451.

Zebarjad SM, Sajjadi SA (2008). On the Strain Rate Sensitivity of HDPE/$CaCO_3$ Nanocomposite. Mater. Sci. Eng. A 475:365-367.

The ability to use light emitting diode (LED) as emergency, instead of gas lamp or tingistin lamp in home lighting

Jassim M. Najim

Department of Physics, KHAWLAN Faculty of Education Arts and Science, Sana'a University, Yemen.

An electric circuit was made and connected to an alternative current that was passing through lower transformer. This process would produce a current the value of which was 500 mA and a voltage of 6 volt was used to charge battery. When the voltage source (alternative current) was cut, the circuit started lighting. We measured the charge and discharge in addition to time lapses to compare them with the results of the experiment. The results were presented and discussed to compare them with the spent energy in tingistin lamp or gas lamp.

Key words: Electric circuit, light emtting diode (LED) as emergency, instead of gas lamp or tingistin lamp, low energy consumption.

INTRODUCTION

Light emitting diode (LED) lighting has been around since the 1960s, but is just now beginning to appear in the residential market for space lighting. A LED is a special type of semiconductor diode-like a normal diode, it consists of a chip of semicoducting material impregnated, or doped, with impurities to create a structure called a p-n junction. A LED is a semiconductor diode that emits light when an electrical current is applied in the forward direction of the device, as in a simple LED circuit. The effect is a form of electroluminescence where an incoherent and narrow-spectrum light is emitted from the p-n junction. When the voltage across the LED in this case is fixed for the emitted photons, this form releases, when charge-carriers (electron and hole) are created by an electric current passing through the junction, when electron meets a hole it falls into a lower energy level, and releases energy in the form of a photon. When the voltage across the p-n junction is in the correct direction, a significant current flows and the device is said to be forward-biased. But if the voltage is of the wrong polarity, the device is said to be revers biased, very little current flows, and no light is emitted (Jassim, 1999, 2008, 2009; Chih-Hsuan et al., 2008). The wavelength of the light emitted, and, therefore, its color, depends on the bandgap energy of materials forming the p-n junction. A normal diode, typically made of silicon or germanium, emits in a visible faro infrared light, but materials, used for a LED have bandgap energies corresponding to near-infrared, visible or near-ultraviolet light. White light can be produced by mixing differently colored light, the most common method is to use red, green, and blue (RGB). Hence the method is called multi-colored white LED, (sometimes referred to as RGB led) (Hilmi, 2007; Hirosaki et al., 2005). The rising cost of energy also makes the use of LEDs in commercial crop culture imminent. With their energy efficiency, LEDs have opened new perspectives for optimizing the energy conversion and the nutrient supply both on and off Earth.

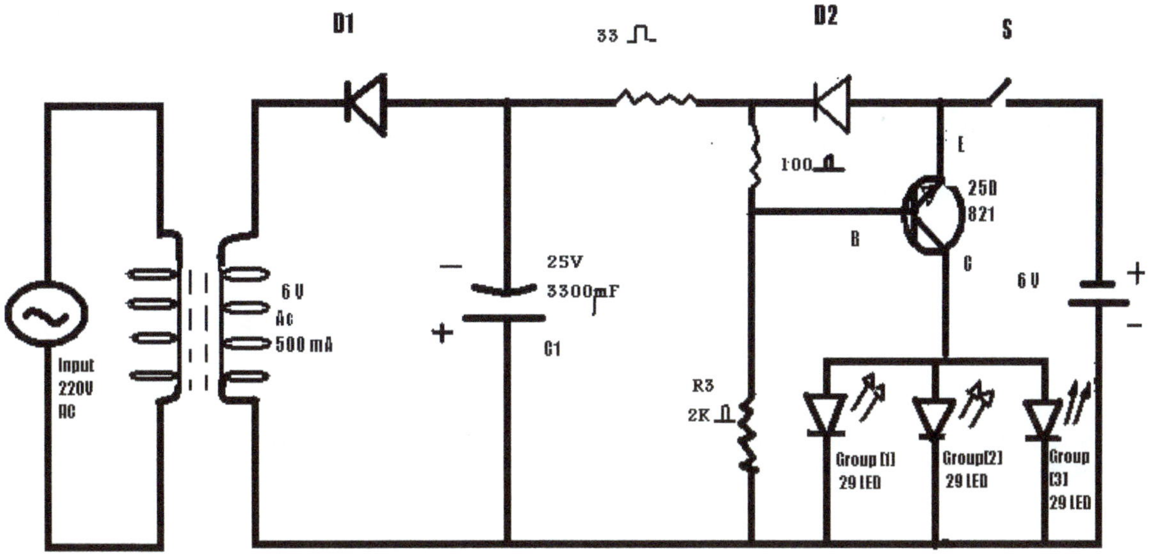

Figure 1. Electric circuit illustrate who we use lighting emitting Diode [LED] as emergency or instead of gas lamp or in home lighting.

EXPERIMENT

Step 1: Making an electric circuit

We made an electric circuit as illustrated in Figure 1. This circuit was connected with an alternative current of 220 voltage that passed through the lower transformer producing a current of 500 mA and 6 volt that passed through diode D1 and at the same time charged the capacity C1 and this capacity regulates the half –wave rectifier to get good direct current (D.C). After that it passed through diode D2. At this time diode D2 is in forward bias to charge the battery through resistance R1. This will result in inducing voltage on diode D2 and would be in the emitte-base of the transistor in off situation. In this case the light emitting diode (LED) would give light, when the source voltage was cut the transistor T1was changed to ON situation after providing the transistor T1 with current base through resistance R3, then the transistor would go to ON situation and light emitting diode (LED) would be in light situation (Figure 1). When the source voltage came again the transistor changed to Off situation and the group of LED changed through R1 and diode D2 and so on. This time we wanted to indicate that if we remove the battery and the electric circuit that was connected direct to alternative current through transformer, the current would be available all the time and emergency situation would be cancelled.

Step 2: Charging battery

As can be seen in Figure 2 the time needed to charge battery was 6 h, after that we see saturation. The time constant of the circuit was equal to the multiplication of C1 by R1, therefore, we must put a suitable value for C1 and R1 in accordance with entering time signal to charge the capacity.

Step3: Discharge statement

In this situation lighting extended to three groups of LED, each group contained 29 leds. Here, by lighting is meant the process of discharge of current coming from the battery charged of 6 volte.

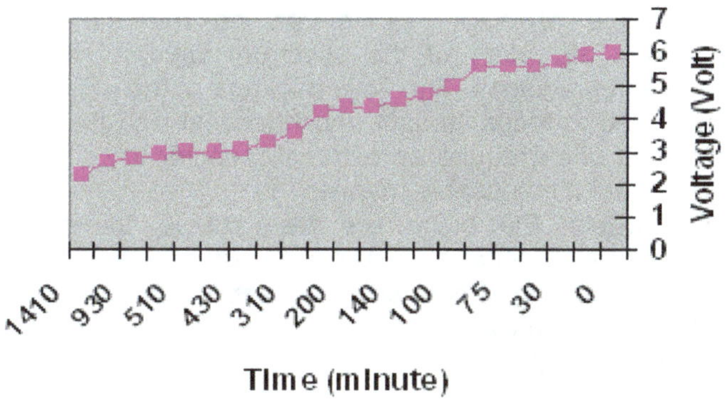

Figure 2. Charge battery.

The results (Figures 3 and 4) showed that we got a twenty three - hour discharge (lighting) which represented the amount of the battery energy spent by the three LED groups.

DISCUSSION

The experiment results, as shown in Figures 1, 2, 3 and 4, explain the possibility of using LED to produce light at home or in the street etc. The energy-saving capacity of LED lights means that they are better for the environment than traditional forms of lighting, and they keep energy bills lower. LEDs last for a long time, typically far longer than fluorescent bulbs. LED lamps have many advantages over traditional lighting methods, these include:

1. Low energy consumption (Figures 3 and 4) the energy

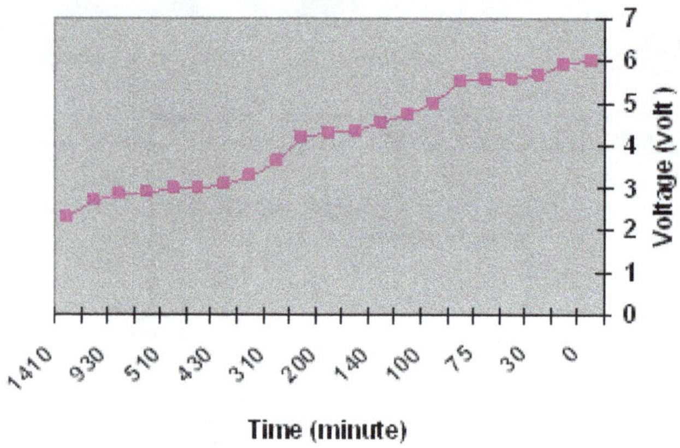

Figure 3. The relationship between the voltage and time (discharge), the decreasing it means LED lighting.

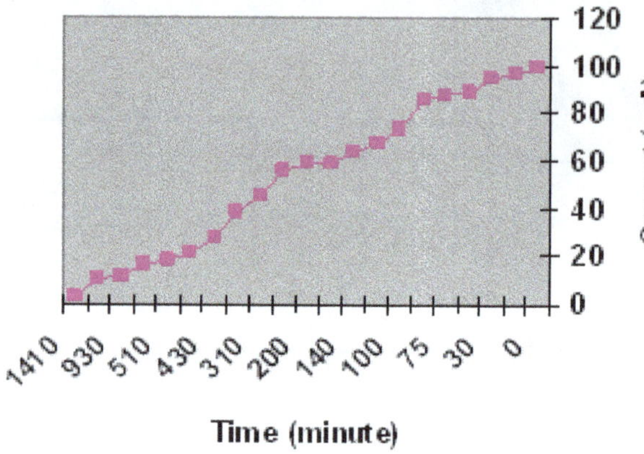

Figure 4. The relationship between the current and the time, the decreasing it means LED lighting.

consumption proximity was 3 watt.

2. LEDs can have a relatively long useful life than incandescent or fluorescent lighting; as we see in Figures 3 and 4 the discharge of battery proximity was 24 h. This electronic circuit is scientifically successful because it needs a long time to unload electric charge.

3. Range of color –LEDs can be manufactured to produce all colors of the spectrum without infared or ultraviolet radiation. The solid package of the LED can be designed to focus its light. Incandescent and fluorescent sources often require an external reflector to collect light and direct it in a usable manner.

4. Durable-LED bulbs are resistant to thermal and vibrational shocks and turn on instantly from -40 to 185°C, making them ideal for applications subject to frequent on-off cycling.

5. No mercury is used in the manufacturing of LEDs.

Conclusion

The results of this study proves that it is possible to use LED for generating light needed in any place. It is also found that in future other resources, other than electricity energy, can be used to produce light by resorting to LED, simply because it consumes very little amount of power to produce light. White LED is well known as a promising device for solid state lighting. It has the advantages of long life, good endurance of heavy impact, no mercury containing and potentially high efficiency. We measured the charge and discharge in addition to time lapses to compare them with the results of the experiment. The results were presented and discussed to compare them with the spent energy in tingistin lamp or gas lamp. However, light output from LED lamp is usually less than 3W, which is very small if compared with traditional light sources and the time of discharge is 23 h. In this article, we will present a new electronic circuit design LED module, with the normal size for lighting purpose. Because of the low power requirement for LEDs, using solar panels becomes more practical and less expensive than running an electric line or using a generator for lighting in remote or off-grid areas. LED light bulbs are also ideal for use with small portable generators which home owners use for backup power in emergencies. More than one – fifth of US electricity is used to power artificial lighting. Light-emitting diodes based on group III/nitride semiconductors are bringing about a revolution in energy-efficient lighting since the development of incandescent light bulbs in the late 1800s, various methods of producing white light more efficiently have been investigated. Of these, white-light sources based on light-emitting diodes (LEDs) look set to have a considerable impact on issues such as energy consumption, environment and even the health of individuals (Siddha et al., 2009).

REFERENCES

Chih-Hsuan T, Jui-Wen P, Wen-Shing S (2008). Simulating the illuminance and the efficiency of the LED and fluorescent lights used in indoor lighting design. Opt. Express, 16(23):18692-18701.

Hilmi VD (2007). White light generation tuned by dual hybridization of nanocrystals and conjugated polymer, New J. phys. 9:362.

Hirosaki N, Rong J, Kimoto K, Sekiguchi T (2005). Characterization and Properties of green- emitting beta-SiAlon:EU2+ Powder phosphor for white light emitting diode. Appl. phys. lett. (USA)86:211905-1-3.

Jassim MN (1999). Studying the effect Beta and Gamma radiation on the electrical properties of diode ,Zener diode and transistors, Iraqi. J. Sci. 4:40C.

Jassim MN (2008). Studying the effect of X-ray Radiation on the electric properties of diode 1N1405. Int. J. Nanoelectronics Mater. 1:35-39.

Jassim MN (2009). Studying the different effects of Gamma and X-ray Irradiation on Electrical Properties of Silicon diode type 1N1405, Int. J. Nanoelectronics Mater. 2:41-46.

Siddha P, James SS, Steven PDB, Shuji N (2009). Prospects for LED lighting. Nature photonics. 3:180-182.

9

Hydromagnetic flows of a mixture of two Newtonian fluids between two parallel plates

S. Barış and M. Ş. Demir

Department of Mechanical Engineering, Faculty of Engineering, Istanbul University, 34320 Avcilar-Istanbul, Turkey.

The paper aims to study the flow of a binary mixture of electrically conducting, incompressible and viscous fluids between two parallel plates in the presence of a transverse uniform magnetic field. The solution of such a flow model has many applications in magnetohydrodynamic (MHD) power generators, MHD pumps, MHD accelerators, and MHD flowmeters. Exact solutions have been obtained for the following four problems: (1) steady hydromagnetic Couette flow, (2) unsteady hydromagnetic Couette flow, (3) steady hydromagnetic Poiseuille flow, (4) unsteady hydromagnetic Poiseuille flow. The mean velocity of the mixture is drawn for different values of magnetic parameters and results are interpreted with the aid of graphs. The previous solutions involving single Newtonian fluid appear as the special cases of the present analysis.

Key words: Binary mixture, Newtonian fluid, magnetohydrodynamics (MHD), steady/unsteady flow, Couette/Poiseuille flow.

INTRODUCTION

The mixture theory finds important applications in various branches of engineering and technology. Familiar examples are suspensions, emulsions, multigrade oils, polycrystalline aggregates, granular media, bubbly liquids, liquid crystals, fluid filled porous elastic solids, composite elastic solids and alloys (Srivastava et al., 1982). The inadequacy of the basic theory for a single continuous media in predicting the behavior of such substances leads to developments in the continuum theory of mixtures. Historical discussion on the development of the subject is sufficiently available in the literature. Theoretical research on the modern formulation of the thermomechanics of interacting continua was initiated by Truesdell (1957). He presented the balance of mass, momentum, energy and the second law of thermodynamics in the context of the continuum theory. Review articles on the mixture theory by Bowen (1976), Atkin and Craine (1976) and Bedford and Drumheller (1983) are of particular interest. We also refer the reader to the books by Truesdell (1984), Samohyl (1987) and Rajagopal and Tao (1995) regarding the historical development of the theory and detailed analysis of various results on this subject.

Adkins (1963) formulated constitutive equations for the stresses in each constituent. He also examined some steady flows of compressible mixtures of non-Newtonian fluids. The continuum theory of compressible mixtures of Newtonian fluids was first considered by Green and Naghdi (1965). Müller (1968) also studied a thermomechanical theory for mixtures of fluids in which there are no chemical reactions. Eringen and Ingram (1965) and Ingram and Eringen (1967) studied mixtures of chemically reacting fluids. The constitutive equations for an incompressible mixture of Newtonian fluids were derived by Mills (1966) using the theory of Green and Naghdi (1965). Craine (1971) examined the flow induced by the steady oscillations of an infinite plate in a mixture of two incompressible Newtonian fluids. In his subsequent study (Craine, 1973), he considered the same problem for a binary mixture of incompressible Newtonian hemihedral fluids. Wilhelm and Van Der Werff (1977) investigated the flow of two miscible, viscous, incompressible fluids subject to oscillatory pressure gradient in a cylindrical tube. Beevers and Craine (1982) extended the list of known solutions for a mixture of two viscous fluids and discussed in a more detail methods for evaluating the response functions. Some exact solutions for flows of a binary mixture of viscous incompressible fluids in different geometries were obtained by Göğüş (1988, 1991, 1992a, b, 1994 and 1995). Many other authors including Al-Sharif et al. (1993), Chamniprasart et al. (1993), Wang et al. (1993), Barış (2005), and Massoudi (2008) worked on applications of the theory of two miscible fluids to practical problems within the context of the mixture theory. Recently, Barış and Demir (2012) have obtained the exact solutions in series form for the flow of a mixture of two incompressible Newtonian fluids in a semicircular duct.

The present paper aims to study the Hartmann problem for a binary mixture of Newtonian fluids and generate theoretical results. In Hartmann flow, the hydromagnetic analogue of Couette and Poiseuille flows, there is an imposed, uniform magnetic field normal to the surfaces. The flow may be induced by a pressure gradient or by relative motion of the two solid walls. Flows of this type are encountered in a variety of applications such as magnetohydrodynamics (MHD) power generators, MHD pumps, MHD accelerators, and MHD flowmeters, and they can also be expanded into various industrial uses. The study of the flow of immiscible fluids under the influence of a magnetic field was considered by various authors. Shail (1973) studied Hartmann flow of a conducting fluid and a non-conducting fluid layer in a channel. Mitra (1982) analyzed the unsteady flow of two electrically conducting fluids between two parallel plates. Lohrasbi and Sahai (1988) considered MHD two-phase flow and heat transfer in a horizontal channel and obtained analytical solutions for the case where one of two fluids was assumed to be electrically non-conducting. Malashetty and Leela (1992) analytically investigated the problem of two-phase MHD flow and heat transfer in a horizontal channel for which both phases are electrically conducting. Malashetty et al. (2001) examined the two-fluid MHD flow and heat transfer in an inclined channel. Umavathi et al. (2006) presented analytical solutions of an oscillatory Hartmann two-fluid flow and heat transfer in an horizontal channel. Recently, Umavathi et al. (2008), Nikodijevic et al. (2011), and Sivaraj et al. (2012) have studied the two-fluid MHD flow and heat transfer with various geometries. Most of the problems relating to the petroleum industry, geophysics, plasma physics, magneto-fluid dynamics, and so forth involve the two-fluid MHD flow situations.

The present investigations on the two-fluid MHD flow pertain to the mechanics of two immiscible fluids. Different from all studies mentioned above, the present paper deals with the flow of a binary mixture of viscous fluids between two parallel plates in the presence of a transversely magnetic field. The basic scientific method utilized in the present research is the mixture theory. We obtained the analytical solutions for the MHD Couette and Poiseuille flows of a mixture of two incompressible Newtonian fluids in a parallel plate channel.

BASIC THEORY

A brief review of the notation and basic equations of a mixture containing two incompressible Newtonian fluids is presented in this section. The reader should consult the articles by Atkin and Craine (1976) for more details.

The mixture of two viscous fluids is considered to be a purely mechanical system. That is, thermal effects and chemical reactions are ignored. The fluids in the mixture will be represented $s^{(1)}$ and $s^{(2)}$. If $\mathbf{v}^{(\beta)}$ denotes the velocity of $s^{(\beta)}$, the material time derivative $D^{(\beta)}/Dt$ is defined by

$$\frac{D^{(\beta)}}{Dt} = \frac{\partial}{\partial t} + v_k^{(\beta)} \frac{\partial}{\partial x_k} \tag{1}$$

where x_k 's are the spatial coordinates and the superscript β refers the β -th fluid. Here and henceforth β takes the values 1 and 2. The mean velocity $\mathbf{v}$ of the mixture is calculated from

$$\rho\mathbf{v} = \rho_1\mathbf{v}^{(1)} + \rho_2\mathbf{v}^{(2)} \tag{2}$$

where ρ_1 and ρ_2 are the current densities of $s^{(1)}$ and $s^{(2)}$ at time t after mixing. The reference densities ρ_{10} and ρ_{20} before the mixing are related to the current densities through $\rho_1 = \phi_1\rho_{10}$ and $\rho_2 = (1-\phi_1)\rho_{20}$, where ϕ_1 is the volume fraction of $s^{(1)}$. The mixture density ρ is given by the sum $\rho = \rho_1 + \rho_2$. In this work, we shall restrict our attention to a binary mixture of incompressible Newtonian fluids. For such a mixture, we can express the current densities in the form

$$\rho_1 = \frac{\rho_{10}(\rho_{20}-\rho)}{\rho_{20}-\rho_{10}}, \qquad \rho_2 = \frac{\rho_{20}(\rho-\rho_{10})}{\rho_{20}-\rho_{10}} \tag{3}$$

Assuming no interconversion of mass between the two fluids, conservation of mass for the two fluids are

$$\frac{\partial\rho_1}{\partial t} + (\rho_1 v_k^{(1)})_{,k} = 0, \qquad \frac{\partial\rho_2}{\partial t} + (\rho_2 v_k^{(2)})_{,k} = 0 \tag{4}$$

where a comma denotes partial differentiation with respect to x_k.

If $\boldsymbol{\sigma}^{(1)}$ and $\boldsymbol{\sigma}^{(2)}$ denote the partial stress tensors of the two-fluid, then the equations for the balance of linear momentum are given by

$$\rho_1 \frac{D^{(1)}v_k^{(1)}}{Dt} = \sigma_{ik,i}^{(1)} - f_k + \rho_1 F_k^{(1)}, \qquad \rho_2 \frac{D^{(2)}v_k^{(2)}}{Dt} = \sigma_{ik,i}^{(2)} + f_k + \rho_2 F_k^{(2)} \tag{5}$$

where f_k represents the mechanical interaction forces between the fluids and $F_k^{(\beta)}$ represents the body force per unit mass of the β th fluid. With these equations as the basis, in order to solve any problem related to binary mixture of fluids one needs to provide the constitutive relations for the interaction forces and stress tensors. The derivation of the constitutive equations appropriate to a binary mixture of incompressible Newtonian fluids has been outlined in Atkin and Craine (1976). If the mixture is considered to be purely mechanical system, the partial stress tensors in such a mixture are related to the motion in the following manner:

$$\sigma_{ik}^{(1)} = (-p_1 + \lambda_1 d_{jj}^{(1)} + \lambda_3 d_{jj}^{(2)})\delta_{ik} + 2\mu_1 d_{ik}^{(1)} + 2\mu_3 d_{ik}^{(2)} + \lambda_5 (w_{ik}^{(1)} - w_{ik}^{(2)}) \tag{6}$$

$$\sigma_{ik}^{(2)} = (-p_2 + \lambda_4 d_{jj}^{(1)} + \lambda_2 d_{jj}^{(2)})\delta_{ik} + 2\mu_4 d_{ik}^{(1)} + 2\mu_2 d_{ik}^{(2)} - \lambda_5 (w_{ik}^{(1)} - w_{ik}^{(2)}) \tag{7}$$

with the material coefficients satisfying the inequalities

$$\begin{aligned} &\lambda_5 \geq 0,\ \mu_1 \geq 0,\ \mu_2 \geq 0,\ \lambda_1 + \frac{2}{3}\mu_1 \geq 0,\ \lambda_2 + \frac{2}{3}\mu_2 \geq 0,\ (\mu_3 + \mu_4)^2 \leq 4\mu_1\mu_2, \\ &\left[\lambda_3 + \lambda_4 + \frac{2}{3}(\mu_3 + \mu_4)\right]^2 \leq 4\left(\lambda_1 + \frac{2}{3}\mu_1\right)\left(\lambda_2 + \frac{2}{3}\mu_2\right) \end{aligned} \tag{8}$$

where p_1, p_2, $d_{ik}^{(\beta)}$, and $w_{ik}^{(\beta)}$ are given by

$$p_1 = (\rho - \rho_{20})\left(\rho_1 \frac{dA_1}{d\rho} + \lambda\right), \qquad p_2 = (\rho - \rho_{10})\left(\rho_2 \frac{dA_2}{d\rho} - \lambda\right) \tag{9}$$

$$2d_{ik}^{(\beta)} = v_{i,k}^{(\beta)} + v_{k,i}^{(\beta)}, \qquad 2w_{ik}^{(\beta)} = v_{i,k}^{(\beta)} - v_{k,i}^{(\beta)} \tag{10}$$

In these equations, λ is a Lagrange multiplier associated with Equations (3) and (4), p_1 and p_2 the mechanical pressures, d_{ik} the deformation rate tensor, w_{ik} the spin tensor and A_1 and A_2 the Helmholtz free energy of the fluids. The mixture Helmholtz free energy A is defined by

$$\rho A = \rho_1 A_1 + \rho_2 A_2 \tag{11}$$

Note that, under isothermal conditions, the material coefficients in Equations (6) and (7) depend only on the mixture density.

Finally, we shall mention the interaction force f_k appearing in Equation (5). Deriving constitutive relations for the interaction forces is one of the most important issues of research in multifluid flows. Massoudi (2003) discussed a variety of possible forms of this term. For instance, for fluid-solid and fluid-fluid mixtures, the interaction force generally depends on densities, temperatures, velocity differences, velocity gradients and possibly other quantities. Such interactions play a very important role in the nature of solutions. To make the theory be of practical utility, we need to simplify the constitutive expression for the interaction force. A good starting point is the inclusion of the effects due to drag and density gradient, that is, (Atkin and Craine, 1976).

$$f_k = \alpha(v_k^{(1)} - v_k^{(2)}) - \lambda\rho_{,k} \tag{12}$$

where α is the interaction coefficient which is a function of the mixture density. Evaluation of α is indeed a difficult task. One simple approach for estimating the value of α is to make use of Hadamard-Rybczynski solution as a first approximation (Dai and Khonsari, 1994). Many authors like Craine (1971, 1973), Al-Sharif et al. (1993), Chamniprasart et al. (1993), Wang et al. (1993), Dai and Khonsari (1994), Göğüş (1988, 1991, 1992a, b, 1994, 1995) and Barış and Demir (2012) have benefited from Equation (12) to make calculation for various problems related to binary fluid mixtures.

In the next sections, we derive the dimensionless forms of the governing equations. To gain further insight into the influence of the parameters in these equations, we will analytically solve the simplified equations for some simple hydromagnetic flows of a binary fluid mixture between two long horizontal plates.

Steady hydromagnetic Couette flow

We want to examine the steady flow of binary mixture of electrically conducting incompressible and viscous fluids between two parallel insulated plates separated by a distance H in the presence of a transverse magnetic field. As Figure 1 shows, we select a rectangular Cartesian system with the x-axis in the direction of motion and the y-axis perpendicular to the plates. Two plates of the channel are of infinite extent in the x- and z- directions and the flow is fully developed, so the velocity depends only on y. The flow is caused by the motion of the plate at $y = 0$ with a constant velocity U_0 in the x- direction. An external uniform magnetic field B_0 is applied in the y- direction. The magnetic Reynolds number is assumed to be very small. In this case, the induced magnetic field produced by motion of fluid can be ignored in comparison to the applied one. In addition, the imposed and induced electric fields are assumed to be negligible, thus the electromagnetic body force per unit volume simplifies $\mathbf{F}_{em} = \sigma(\mathbf{v} \times \mathbf{B}) \times \mathbf{B}$, where $\mathbf{B}$ is the magnetic field vector and σ is the electrical conductivity. Due to the assumptions stated above, Maxwell's equations become redundant.

It should be mentioned here that a great deal of interest has been focused on the flow problems in the presence of a uniform applied magnetic field. But it can't be appropriate to make such an assumption in some engineering applications. So far a few researchers have worked on the flow of an electrically conducting fluid under a non-uniform space or time dependent magnetic field, due to the difficulty in obtaining the complete solution for the problem of this type. Recently, such an attempt was made by Asghar and Ahmad (2012). They found the analytical solution for unsteady Couette flow in the presence of an arbitrary non-uniform space dependent applied magnetic field. In this paper, for simplicity of analysis, we confined ourselves only to the uniform magnetic field case. It is felt that the uniform magnetic field findings will be a good starting point for shedding light on more complicated two-fluid MHD flows. We shall seek a solution of the form

$$u_{\beta S} = u_{\beta S}(y), \qquad \rho = \rho(y) \tag{13}$$

where the function $u_{\beta S}$ denotes the velocity component of the β th fluid in the x-direction for the case of steady flow. Under the above assumptions, substituting Equations (6) and (7) into the equations of motion (5) we obtain

$$M_1 u_{1S}'' + M_2 u_{2S}'' + M_1' u_{1S}' + M_2' u_{2S}' - \alpha(u_{1S} - u_{2S}) - \sigma_1 B_0^2 u_{1S} = 0 \tag{14}$$

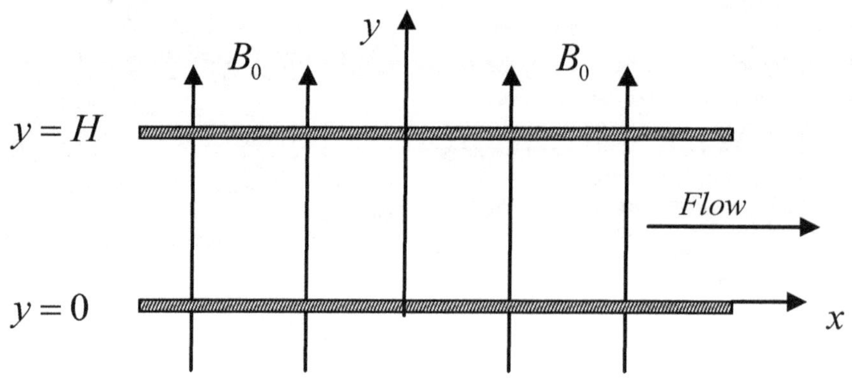

Figure 1. Sketch of flow geometry and coordinate system.

$$\lambda\rho' = p_1' \tag{15}$$

$$M_3 u_{1S}'' + M_4 u_{2S}'' + M_3' u_{1S}' + M_4' u_{2S}' + \alpha(u_{1S} - u_{2S}) - \sigma_2 B_0^2 u_{2S} = 0 \tag{16}$$

$$-\lambda\rho' = p_2' \tag{17}$$

where

$$M_1 = \mu_1 - \frac{\lambda_5}{2},\ M_2 = \mu_3 + \frac{\lambda_5}{2},\ M_3 = \mu_4 + \frac{\lambda_5}{2},\ M_4 = \mu_2 - \frac{\lambda_5}{2} \tag{18}$$

In the above equations, primes denote differentiation with respect to y and σ_β is the electrical conductivity of the β th fluid. The last terms on the left hand sides of Equations (14) and (16) result from electromagnetic body forces. Note that we neglect non-magnetic body forces. With the use of Equations (3), (9) and (11), elimination of λ' between Equations (15) and (17) yields

$$(\rho - \rho_{10})(\rho_{20} - \rho)\,\rho'\,\frac{d^2(\rho A)}{d\rho^2} = 0 \tag{19}$$

We deduce from the above equation that ρ is a constant. As a result, the coefficients M_1, M_2, M_3 and M_4 are constants. We shall now write the equations of motion in terms of a set of dimensionless variables. If $\bar{f}$ is used to denote the dimensionless form of a quantity f, it follows that

$$\bar{M}_i = \frac{M_i}{\mu},\quad \bar{\alpha} = \frac{\alpha H^2}{\mu},\quad \bar{u}_{\beta S} = \frac{u_{\beta S}}{U_0},\quad \bar{y} = \frac{y}{H},\quad Ha_\beta = B_0 H\left(\frac{\sigma_\beta}{\mu}\right)^{1/2} \tag{20}$$

where μ is the viscosity of the mixture. Thus the equations of motion in non-dimensional form become,

$$\bar{M}_1 \bar{u}_{1S}'' + \bar{M}_2 \bar{u}_{2S}'' - \bar{\alpha}(\bar{u}_{1S} - \bar{u}_{2S}) - Ha_1^2 \bar{u}_{1S} = 0 \tag{21}$$

$$\bar{M}_3 \bar{u}_{1S}'' + \bar{M}_4 \bar{u}_{2S}'' + \bar{\alpha}(\bar{u}_{1S} - \bar{u}_{2S}) - Ha_2^2 \bar{u}_{2S} = 0 \tag{22}$$

The magnetic parameter Ha_β in the above equations is often referred to as the Hartmann number. The boundary conditions for the velocity components are

$$\bar{u}_{\beta S}(0) = 1, \qquad \bar{u}_{\beta S}(1) = 0 \tag{23}$$

The solutions of Equations (21) and (22) can be given as

$$\bar{u}_{1S} = C_1 e^{s_1 \bar{y}} + C_2 e^{-s_1 \bar{y}} + C_3 e^{s_2 \bar{y}} + C_4 e^{-s_2 \bar{y}} \tag{24}$$

$$\bar{u}_{2S} = C_5 e^{s_1 \bar{y}} + C_6 e^{-s_1 \bar{y}} + C_7 e^{s_2 \bar{y}} + C_8 e^{-s_2 \bar{y}} \tag{25}$$

where

$$s_1 = \sqrt{\frac{b - \sqrt{b^2 - 4ac}}{2a}},\qquad s_2 = \sqrt{\frac{b + \sqrt{b^2 - 4ac}}{2a}} \tag{26}$$

with

$$a = \bar{M}_1\bar{M}_4 - \bar{M}_2\bar{M}_3,\quad b = \bar{\alpha}(\bar{M}_1 + \bar{M}_2 + \bar{M}_3 + \bar{M}_4) + \bar{M}_1 Ha_1^2 + \bar{M}_4 Ha_2^2,$$
$$c = \bar{\alpha}(Ha_1^2 + Ha_2^2) + Ha_1^2 Ha_2^2 \tag{27}$$

Applying the boundary conditions (23) to Equations (24) and (25), we find the constants $C_1, \ldots, C_8$ as follows:

$$C_1 = \frac{k_2 - 1}{k_2 - k_1}\frac{e^{-s_1}}{e^{-s_1} - e^{s_1}},\quad C_2 = \frac{k_2 - 1}{k_2 - k_1}\frac{e^{s_1}}{e^{s_1} - e^{-s_1}},\quad C_3 = \frac{k_1 - 1}{k_1 - k_2}\frac{e^{-s_2}}{e^{-s_2} - e^{s_2}},$$
$$C_4 = \frac{k_1 - 1}{k_1 - k_2}\frac{e^{s_2}}{e^{s_2} - e^{-s_2}},\quad C_5 = k_1 C_1,\quad C_6 = k_1 C_2,\quad C_7 = k_2 C_3,\quad C_8 = k_2 C_4 \tag{28}$$

where

$$k_1 = \frac{\bar{\alpha} + Ha_1^2 - \bar{M}_1 s_1^2}{\bar{\alpha} + \bar{M}_2 s_1^2},\quad k_2 = \frac{\bar{\alpha} + Ha_1^2 - \bar{M}_1 s_2^2}{\bar{\alpha} + \bar{M}_2 s_2^2} \tag{29}$$

Unsteady hydromagnetic Couette flow

Now is the time to investigate unsteady Couette flow of a mixture of two incompressible Newtonian fluids between two parallel plates. A uniform magnetic field is applied in a direction perpendicular to the flow of the binary fluid mixture. The mixture and two plates are initially at rest.

The lower plate is suddenly accelerated from rest and moves in its own plane with constant velocity U_0, while the upper plate is held stationary. It is assumed that the flow is entirely driven by the motion of the lower plate. It seems reasonable to assume that the velocity components and densities of fluids are of the form

$$u_\beta = u_\beta(y,t), \qquad \rho_\beta = \rho_\beta(y,t) \tag{30}$$

Substitution of Equation (30) into Equation (4) yields $\partial\rho_\beta/\partial t = 0$. Thus $\partial\rho/\partial t = 0$ and $\rho = \rho(y)$. As made in the preceding section, elimination of λ' between y - components of the equations of motion give Equation (19), which implies that ρ is a constant. Since ρ has been proved to be a constant, all of the material coefficients are constants. As a result, the dimensionless equations of motions are as follows:

$$\bar{\rho}_1\frac{\partial \bar{u}_1}{\partial \bar{t}} = \bar{M}_1\frac{\partial^2 \bar{u}_1}{\partial \bar{y}^2} + \bar{M}_2\frac{\partial^2 \bar{u}_2}{\partial \bar{y}^2} - \bar{\alpha}(\bar{u}_1 - \bar{u}_2) - Ha_1^2\bar{u}_1 \tag{31}$$

$$\bar{\rho}_2\frac{\partial \bar{u}_2}{\partial \bar{t}} = \bar{M}_3\frac{\partial^2 \bar{u}_1}{\partial \bar{y}^2} + \bar{M}_4\frac{\partial^2 \bar{u}_2}{\partial \bar{y}^2} + \bar{\alpha}(\bar{u}_1 - \bar{u}_2) - Ha_2^2\bar{u}_2 \tag{32}$$

where

$$\bar{M}_i = \frac{M_i}{\mu}, \quad \bar{\alpha} = \frac{\alpha H^2}{\mu}, \quad \bar{u}_\beta = \frac{u_\beta}{U_0}, \quad \bar{y} = \frac{y}{H}, \quad \bar{\rho}_\beta = \frac{\rho_\beta}{\rho}, \quad \bar{t} = \frac{\mu t}{\rho H^2}, \quad Ha_\beta = B_0 H\left(\frac{\sigma_\beta}{\mu}\right)^{1/2} \tag{33}$$

The boundary and initial conditions are

$$\bar{u}_\beta(0,\bar{t}) = 1; \quad \bar{t} > 0, \qquad \bar{u}_\beta(1,\bar{t}) = 0; \quad \bar{t} \ge 0, \tag{34}$$

$$\bar{u}_\beta(\bar{y},0) = 0; \quad 0 < \bar{y} \le 1 . \tag{35}$$

We first have to transform the problem so that the boundary conditions $(34)_1$ are homogeneous. This can be achieved by decomposing $\bar{u}_\beta(\bar{y},\bar{t})$ into the steady Couette velocity field $\bar{u}_{\beta S}(\bar{y})$ and the transient component $f_\beta(\bar{y},\bar{t})$:

$$\bar{u}_\beta(\bar{y},\bar{t}) = \bar{u}_{\beta S}(\bar{y}) - f_\beta(\bar{y},\bar{t}) \tag{36}$$

The transient components satisfy the following differential equations

$$\bar{\rho}_1\frac{\partial f_1}{\partial \bar{t}} = \bar{M}_1\frac{\partial^2 f_1}{\partial \bar{y}^2} + \bar{M}_2\frac{\partial^2 f_2}{\partial \bar{y}^2} - \bar{\alpha}(f_1 - f_2) - Ha_1^2 f_1 \tag{37}$$

$$\bar{\rho}_2\frac{\partial f_2}{\partial \bar{t}} = \bar{M}_3\frac{\partial^2 f_1}{\partial \bar{y}^2} + \bar{M}_4\frac{\partial^2 f_2}{\partial \bar{y}^2} + \bar{\alpha}(f_1 - f_2) - Ha_2^2 f_2 \tag{38}$$

that are consistent with the boundary and initial conditions

$$f_1(0,\bar{t}) = 0, \quad f_2(0,\bar{t}) = 0, \quad f_1(1,\bar{t}) = 0, \quad f_2(1,\bar{t}) = 0 \tag{39}$$

$$f_1(\bar{y},0) = \bar{u}_{1S}(\bar{y}), \quad f_2(\bar{y},0) = \bar{u}_{2S}(\bar{y}) \tag{40}$$

Finite Fourier sine transform will be used to solve the simultaneous partial differential equations (37) and (38) with the conditions (39) and (40). Finite Fourier sine transform of a function $f(y)$ defined for $0 < y < a$ is

$$F_s\{f(y)\} = \tilde{f}(n) = \int_0^a f(y)\sin(ny)dy, \qquad n = 1,2,3,... \tag{41}$$

with inverse transform

$$F_s^{-1}\{\tilde{f}(n)\} = f(y) = \frac{2}{a}\sum_{n=1}^{\infty}\tilde{f}(n)\sin(ny) \tag{42}$$

With the aid of Equation (39), application of the Fourier sine transform to Equations (37) and (38) gives

$$\frac{d\tilde{f}_1}{dt} = -l_1\tilde{f}_1 - l_2\tilde{f}_2 \tag{43}$$

$$\frac{d\tilde{f}_2}{dt} = -l_3\tilde{f}_1 - l_4\tilde{f}_2 \tag{44}$$

where

$$l_1 = \frac{\bar{M}_1 n^2\pi^2 + \bar{\alpha} + Ha_1^2}{\bar{\rho}_1}, \quad l_2 = \frac{\bar{M}_2 n^2\pi^2 - \bar{\alpha}}{\bar{\rho}_1}, \quad l_3 = \frac{\bar{M}_3 n^2\pi^2 - \bar{\alpha}}{\bar{\rho}_2}, \quad l_4 = \frac{\bar{M}_4 n^2\pi^2 + \bar{\alpha} + Ha_2^2}{\bar{\rho}_2} \tag{45}$$

Taking the Fourier sine transform of Equation (40) results in:

$$\tilde{f}_1(n,0) = \frac{n\pi}{s_1^2 + n^2\pi^2}\left\{(1-(-1)^n e^{s_1})C_1 + (1-(-1)^n e^{-s_1})C_2\right\} + \frac{n\pi}{s_2^2 + n^2\pi^2}\left\{(1-(-1)^n e^{s_2})C_3 + (1-(-1)^n e^{-s_2})C_4\right\} \tag{46}$$

$$\tilde{f}_2(n,0) = \frac{n\pi}{s_1^2 + n^2\pi^2}\left\{(1-(-1)^n e^{s_1})C_5 + (1-(-1)^n e^{-s_1})C_6\right\} + \frac{n\pi}{s_2^2 + n^2\pi^2}\left\{(1-(-1)^n e^{s_2})C_7 + (1-(-1)^n e^{-s_2})C_8\right\} \tag{47}$$

Solving Equations (43) and (44) simultaneously and using the conditions (46) and (47), we obtain

$$\tilde{f}_1(n,\bar{t}) = \frac{(l_1 + r_2)\tilde{f}_1(n,0) + l_2\tilde{f}_2(n,0)}{r_2 - r_1}e^{r_1\bar{t}} - \frac{(l_1 + r_1)\tilde{f}_1(n,0) + l_2\tilde{f}_2(n,0)}{r_2 - r_1}e^{r_2\bar{t}} \tag{48}$$

$$\tilde{f}_2(n,\bar{t}) = \frac{l_1 + r_1}{l_2}\frac{(l_1 + r_2)\tilde{f}_1(n,0) + l_2\tilde{f}_2(n,0)}{r_1 - r_2}e^{r_1\bar{t}} - \frac{l_1 + r_2}{l_2}\frac{(l_1 + r_1)\tilde{f}_1(n,0) + l_2\tilde{f}_2(n,0)}{r_1 - r_2}e^{r_2\bar{t}} \tag{49}$$

where

$$r_{1,2} = -\frac{l_1 + l_4}{2} \mp \frac{\sqrt{(l_1 - l_4)^2 + 4l_2 l_3}}{2} \tag{50}$$

With the help of Equation (42), inverting Equations (48) and (49) and then substituting of the results into Equation(36), we find the following solution for $\bar{u}_\beta(\bar{y},\bar{t})$

$$\bar{u}_\beta(\bar{y},\bar{t}) = \bar{u}_{\beta S}(\bar{y}) - 2\sum_{n=1}^{\infty} \tilde{f}_\beta(n,\bar{t})\sin(n\pi\bar{y}) \tag{51}$$

Steady hydromagnetic Poiseuille flow

In this section we consider the steady flow of the binary mixture under consideration between two parallel plates in the presence of a transverse magnetic field. The flow is driven by an externally imposed constant pressure gradient in the x-direction, namely $-\partial p/\partial x = p_x > 0$. We seek solutions in which the velocity of the β th fluid and the mixture density are assumed to have the form:

$$w_{\beta S} = w_{\beta S}(y), \quad \rho = \rho(y) \tag{52}$$

As previously stated, it is proved that the total density and material coefficients become constants. Consequently, the equations of motion in the x-direction reduce to

$$\bar{M}_1 \bar{w}_{1S}'' + \bar{M}_2 \bar{w}_{2S}'' - \bar{\alpha}(\bar{w}_{1S} - \bar{w}_{2S}) - Ha_1^2 \bar{w}_{1S} = -\phi_1 \tag{53}$$

$$\bar{M}_3 \bar{w}_{1S}'' + \bar{M}_4 \bar{w}_{2S}'' + \bar{\alpha}(\bar{w}_{1S} - \bar{w}_{2S}) - Ha_2^2 \bar{w}_{2S} = \phi_1 - 1 \tag{54}$$

where

$$\bar{M}_i = \frac{M_i}{\mu}, \quad \bar{\alpha} = \frac{\alpha H^2}{\mu}, \quad \bar{w}_{\beta S} = \frac{w_{\beta S}\mu}{p_x H^2}, \quad \bar{y} = \frac{y}{H}, \quad Ha_\beta = B_0 H\left(\frac{\sigma_\beta}{\mu}\right)^{1/2} \tag{55}$$

The adherence boundary conditions of the problem are

$$\bar{w}_{\beta S}(0) = 0, \qquad \bar{w}_{\beta S}(1) = 0 \tag{56}$$

The velocity fields can be obtained by solving Equations (53) and (54) under the relevant boundary conditions as follows:

$$\bar{w}_{1S} = D_1 e^{s_1\bar{y}} + D_2 e^{-s_1\bar{y}} + D_3 e^{s_2\bar{y}} + D_4 e^{-s_2\bar{y}} + D_5 \tag{57}$$

$$\bar{w}_{2S} = k_1 D_1 e^{s_1\bar{y}} + k_1 D_2 e^{-s_1\bar{y}} + k_2 D_3 e^{s_2\bar{y}} + k_2 D_4 e^{-s_2\bar{y}} + D_6 \tag{58}$$

where

$$D_1 = \frac{(D_6 - D_5 k_2)(1 - e^{-s_1})}{(k_2 - k_1)(e^{s_1} - e^{-s_1})}, \qquad D_2 = \frac{(D_6 - D_5 k_2)(e^{s_1} - 1)}{(k_2 - k_1)(e^{s_1} - e^{-s_1})},$$
$$D_3 = \frac{(D_6 - D_5 k_1)(1 - e^{-s_2})}{(k_1 - k_2)(e^{s_2} - e^{-s_2})}, \qquad D_4 = \frac{(D_6 - D_5 k_1)(e^{s_2} - 1)}{(k_1 - k_2)(e^{s_2} - e^{-s_2})} \tag{59}$$

with

$$D_5 = \frac{\bar{\alpha} + \phi_1 Ha_2^2}{\bar{\alpha}(Ha_1^2 + Ha_2^2) + Ha_1^2 Ha_2^2}, \qquad D_6 = \frac{\bar{\alpha} + (1-\phi_1)Ha_1^2}{\bar{\alpha}(Ha_1^2 + Ha_2^2) + Ha_1^2 Ha_2^2} \tag{60}$$

Unsteady hydromagnetic Poiseuille flow

Finally, we study the problem of unsteady flow of a mixture of two incompressible Newtonian fluids between two parallel plates. There is an external magnetic field of constant strength in the y-direction. The mixture is initially at rest. The mixture begins to flow due to the sudden imposition of a constant pressure gradient in the x-direction. We look for a solution of the form

$$w_\beta = w_\beta(y,t), \quad \rho_\beta = \rho_\beta(y,t) \tag{61}$$

As made in the case of unsteady hydromagnetic Couette flow, it is proved ρ is a constant. This is why all the coefficients in the constitutive equations are constants. Thus, the dimensionless governing equations are as follows:

$$\bar{\rho}_1 \frac{\partial \bar{w}_1}{\partial \bar{t}} = \bar{M}_1 \frac{\partial^2 \bar{w}_1}{\partial \bar{y}^2} + \bar{M}_2 \frac{\partial^2 \bar{w}_2}{\partial \bar{y}^2} - \bar{\alpha}(\bar{w}_1 - \bar{w}_2) - Ha_1^2 \bar{w}_1 + \phi_1 \tag{62}$$

$$\bar{\rho}_2 \frac{\partial \bar{w}_2}{\partial \bar{t}} = \bar{M}_3 \frac{\partial^2 \bar{w}_1}{\partial \bar{y}^2} + \bar{M}_4 \frac{\partial^2 \bar{w}_2}{\partial \bar{y}^2} + \bar{\alpha}(\bar{w}_1 - \bar{w}_2) - Ha_1^2 \bar{w}_2 + 1 - \phi_1 \tag{63}$$

where

$$\bar{M}_i = \frac{M_i}{\mu}, \quad \bar{\alpha} = \frac{\alpha H^2}{\mu}, \quad \bar{w}_\beta = \frac{w_\beta \mu}{p_x H^2}, \quad \bar{y} = \frac{y}{H}, \quad \bar{\rho}_\beta = \frac{\rho_\beta}{\rho}, \quad \bar{t} = \frac{\mu t}{\rho H^2}, \quad Ha_\beta = B_0 H\left(\frac{\sigma_\beta}{\mu}\right)^{1/2} \tag{64}$$

The boundary and initial conditions are

$$\bar{w}_\beta(0,\bar{t}) = 0; \quad \bar{t} \ge 0, \qquad \bar{w}_\beta(1,\bar{t}) = 0; \quad \bar{t} \ge 0 \tag{65}$$

$$\bar{w}_\beta(\bar{y},0) = 0; \quad 0 \le \bar{y} \le 1 \tag{66}$$

Note that all of the above conditions are homogeneous, yet there exists a non-trivial solution, since the governing equations are non-homogeneous. We attempt to find a solution of the form

$$\bar{w}_\beta(\bar{y},\bar{t}) = \bar{w}_{\beta S}(\bar{y}) - g_\beta(\bar{y},\bar{t}) \tag{67}$$

The components $g_\beta(\bar{y},\bar{t})$ must satisfy Equations (37) and (38) and the boundary conditions (39) by writing $g_\beta(\bar{y},\bar{t})$ in place of $f_\beta(\bar{y},\bar{t})$, but with modified initial conditions which now are:

$$g_\beta(\bar{y},0)=\bar{w}_{\beta S}(\bar{y}) \quad (68)$$

The procedure for determining $g_\beta(\bar{y},\bar{t})$ is the same as that used in the case of unsteady hydromagnetic Couette flow, so it is not repeated here. As expected, the second part of the solution given in Equation (51) is also valid for $g_\beta(\bar{y},\bar{t})$ provided $\tilde{f}_\beta(n,0)$ is replaced with $\tilde{g}_\beta(n,0)$ which are given by the following analytical expressions

$$\tilde{g}_1(n,0)=\frac{D_1 n\pi}{s_1^2+n^2\pi^2}(1-(-1)^n e^{s_1})+\frac{D_2 n\pi}{s_1^2+n^2\pi^2}(1-(-1)^n e^{-s_1})$$
$$+\frac{D_3 n\pi}{s_2^2+n^2\pi^2}(1-(-1)^n e^{s_2})+\frac{D_4 n\pi}{s_2^2+n^2\pi^2}(1-(-1)^n e^{-s_2})+\frac{D_5(1-(-1)^n)}{n\pi} \quad (69)$$

$$\tilde{g}_2(n,0)=\frac{k_1 D_1 n\pi}{s_1^2+n^2\pi^2}(1-(-1)^n e^{s_1})+\frac{k_1 D_2 n\pi}{s_1^2+n^2\pi^2}(1-(-1)^n e^{-s_1})$$
$$+\frac{k_2 D_3 n\pi}{s_2^2+n^2\pi^2}(1-(-1)^n e^{s_2})+\frac{k_2 D_4 n\pi}{s_2^2+n^2\pi^2}(1-(-1)^n e^{-s_2})+\frac{D_6(1-(-1)^n)}{n\pi} \quad (70)$$

We now obtain the solution for the velocity field of the β th fluid by going back through the various substitutions:

$$\bar{w}_\beta(\bar{y},\bar{t})=\bar{w}_{\beta S}(\bar{y})-2\sum_{n=1}^{\infty}\tilde{g}_\beta(n,\bar{t})\sin(n\pi\bar{y}) \quad (71)$$

NUMERICAL RESULTS AND DISCUSSION

Some simple unidirectional hydromagnetic flows of a binary mixture of Newtonian fluids between two parallel plates are studied theoretically. The two miscible fluids are assumed to be incompressible and electrically conducting, having different viscosities and electrical conductivities. The resulting differential equations are solved analytically. The analytical solutions are made possible under very special conditions when all material properties are assumed to be constants, and the only interaction force is drag resulting from relative velocity in a linear fashion. Removal of these assumptions will make the governing equations highly nonlinear and necessitate a complex numerical solution.

To make predictions based on the foregoing analysis, it is necessary to know material coefficients in the constitutive equations. For a mixture composed of water and oil with water volume fraction ϕ_1, we benefit from the formulae suggested by Sampaio and Williams (1977) to assign the reasonable values to the material coefficients. In all the computations presented here, the following values of the dimensionless parameters are used (Barış and Demir, 2012):

$$\bar{M}_1=0.4868,\quad \bar{M}_2=\bar{M}_3=0.2497,\quad \bar{M}_4=0.5132,$$
$$\bar{\alpha}=10^6,\quad \bar{\rho}_1=0.8108,\quad \bar{\rho}_2=0.1892,\quad \phi_1=0.75 \quad (72)$$

Now we want to discuss the reliability of the series solutions given in Equations (51) and (71). As expected, these series are rapidly convergent for large values of time but slowly convergent for small values of time. However, it is important to note that these series solutions can also be used for small values of time provided number of terms in the series expansions is enough to yield satisfactory accuracy. For example, in the case of unsteady Couette flow with $Ha_1/Ha_2=1000$, $Ha_1=2$ and $\bar{t}=0.1$, the fourth term is the first one in the series expansion, absolute value of which is less than 10^{-12}. Therefore, the sum of the first four term will give the velocity values of the fluids with an error of less than 10^{-12}. On the other hand, it is necessary to take the first nine term into account for the same order of accuracy in the case of $\bar{t}=0.02$.

The analytical solutions in the present work include those corresponding to pure Newtonian fluid as a special case. If one sets $\bar{M}_1=\bar{M}_2=\bar{M}_3=\bar{M}_4=1/4$, $\bar{\rho}_1=\bar{\rho}_2=\phi_1=1/2$, and $Ha_1=Ha_2=Ha/\sqrt{2}$ in Equations (51) and (71), these are obtained as follows:

Unsteady hydromagnetic Couette flow between two parallel plates:

$$\bar{u}_N(\bar{y},\bar{t})=\frac{\sinh[Ha(1-\bar{y})]}{\sinh[Ha]}-2\sum_{n=1}^{\infty}\frac{n\pi}{n^2\pi^2+Ha^2}\sin[n\pi\bar{y}]e^{-(n^2\pi^2+Ha^2)\bar{t}} \quad (73)$$

Unsteady hydromagnetic Poiseuille flow between two parallel plates:

$$\bar{w}_N(\bar{y},\bar{t})=\frac{1-\cosh[Ha\bar{y}]}{Ha^2}+\sinh[Ha\bar{y}]\frac{\cosh[Ha]-1}{Ha^2\sinh[Ha]}-2\sum_{n=1}^{\infty}\frac{1-(-1)^n}{n\pi(n^2\pi^2+Ha^2)}\sin[n\pi\bar{y}]e^{-(n^2\pi^2+Ha^2)\bar{t}} \quad (74)$$

The limiting solutions mentioned above give us confidence regarding our analytical calculations. To demonstrate the influence of the applied magnetic field on the velocity profiles, numerical evaluations of the analytical solutions are performed and results are plotted in Figures 2 to 5. In these figures the material parameters $\bar{\rho}_1,\bar{\rho}_2,\phi_1,\bar{M}_1,\bar{M}_2,\bar{M}_3,\bar{M}_4,\bar{\alpha}$ and the ratio Ha_1/Ha_2 are kept the constant values. It is clear from these figures that the main effect of the magnetic field on the flow is to decrease the velocity. This is expected since the application of a transverse magnetic field normal to the flow direction has a tendency to create a drag-like

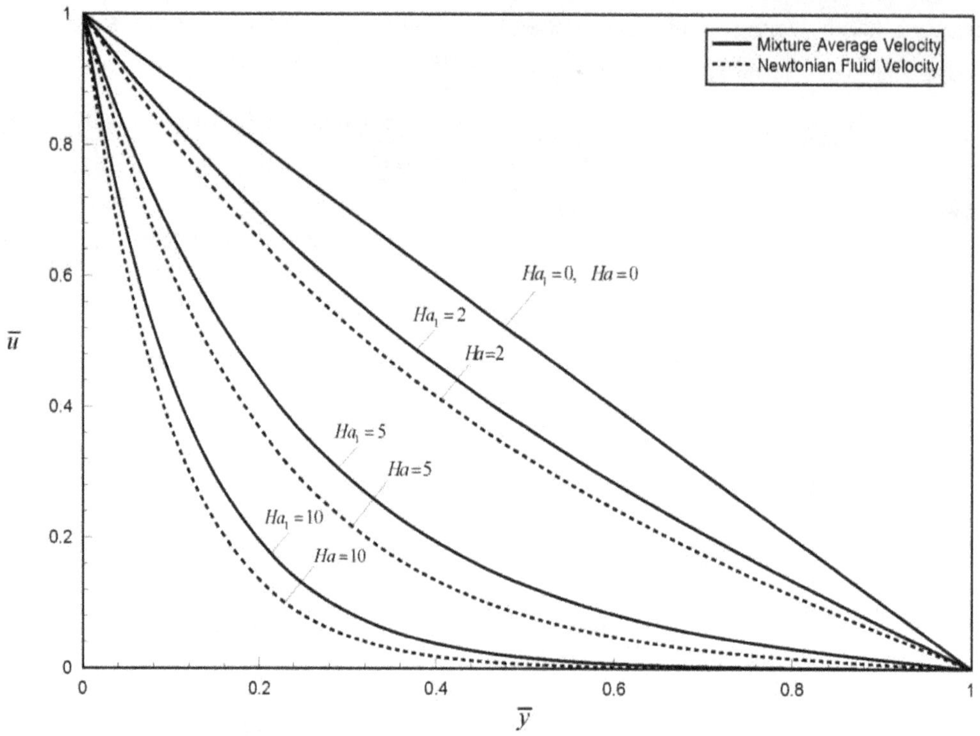

Figure 2. Velocity profiles of steady hydromagnetic Couette flow between parallel plates for $Ha_1/Ha_2 = 1000$.

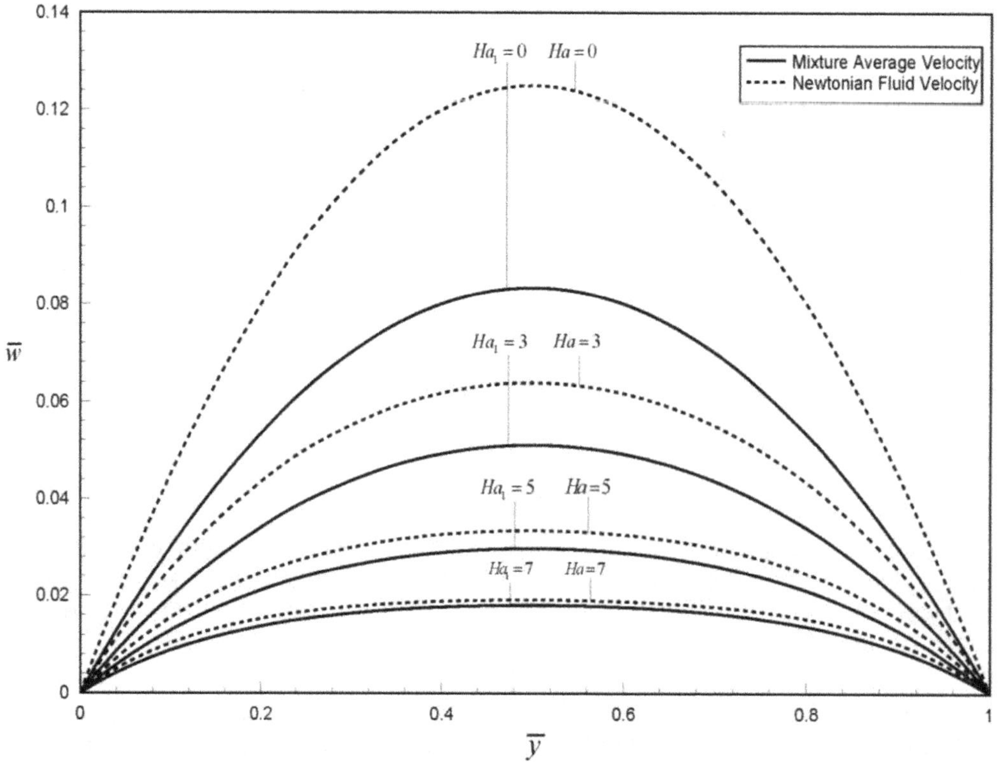

Figure 3. Velocity profiles of steady hydromagnetic Poiseuille flow between parallel plates for $Ha_1/Ha_2 = 1000$.

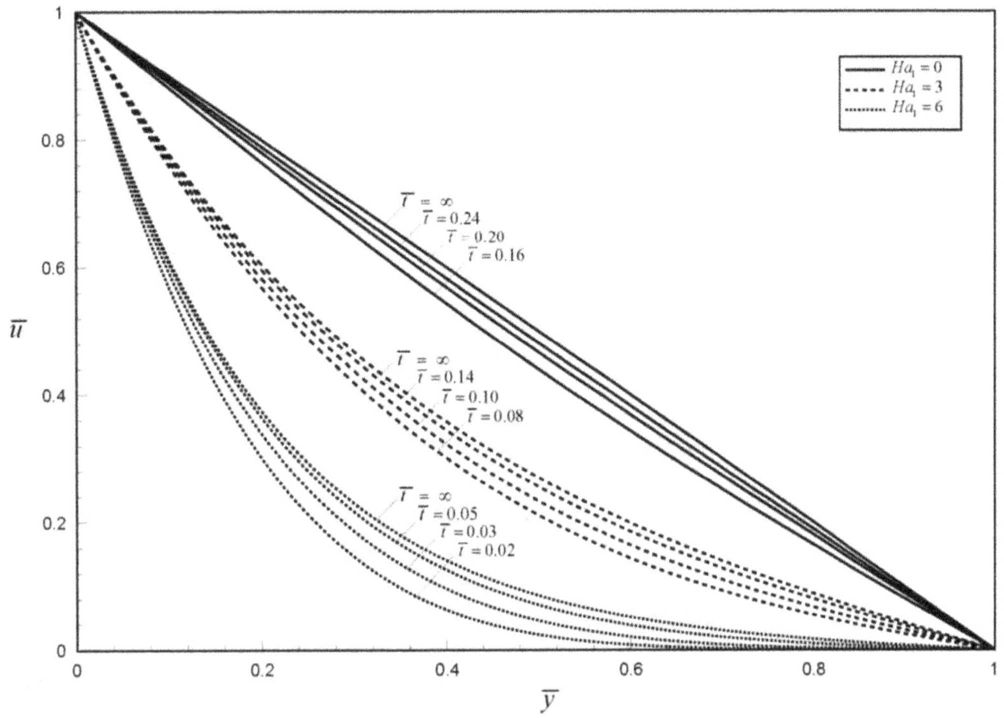

Figure 4. Velocity profiles of unsteady hydromagnetic Couette flow between parallel plates for $Ha_1/Ha_2 = 1000$.

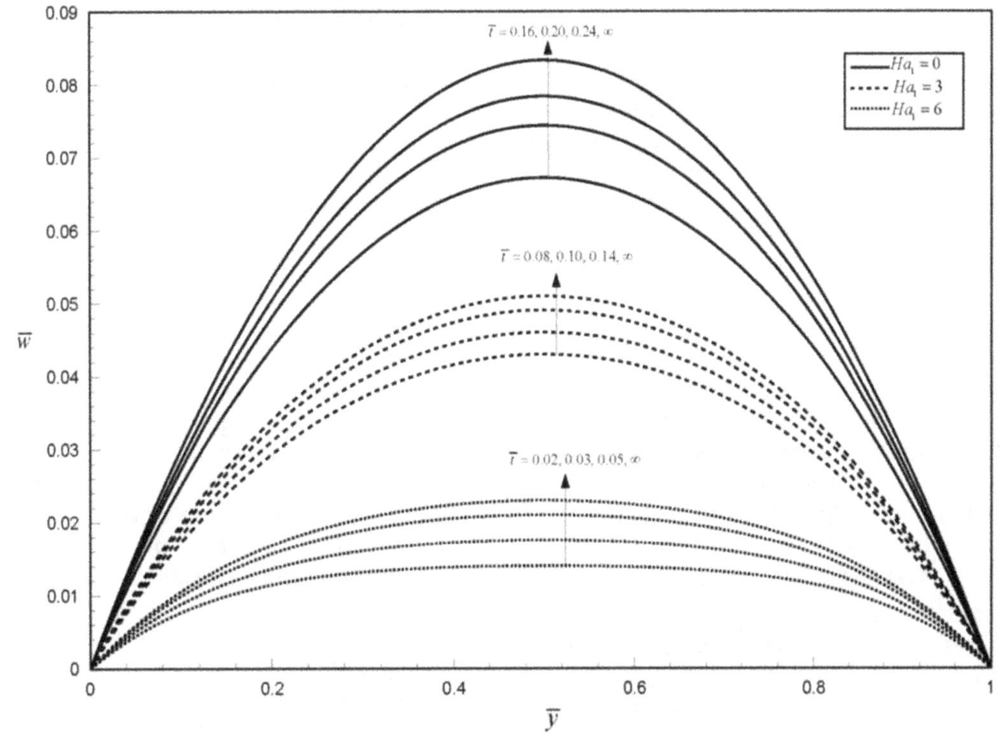

Figure 5. Velocity profiles of unsteady hydromagnetic Poiseuille flow between parallel plates for $Ha_1/Ha_2 = 1000$.

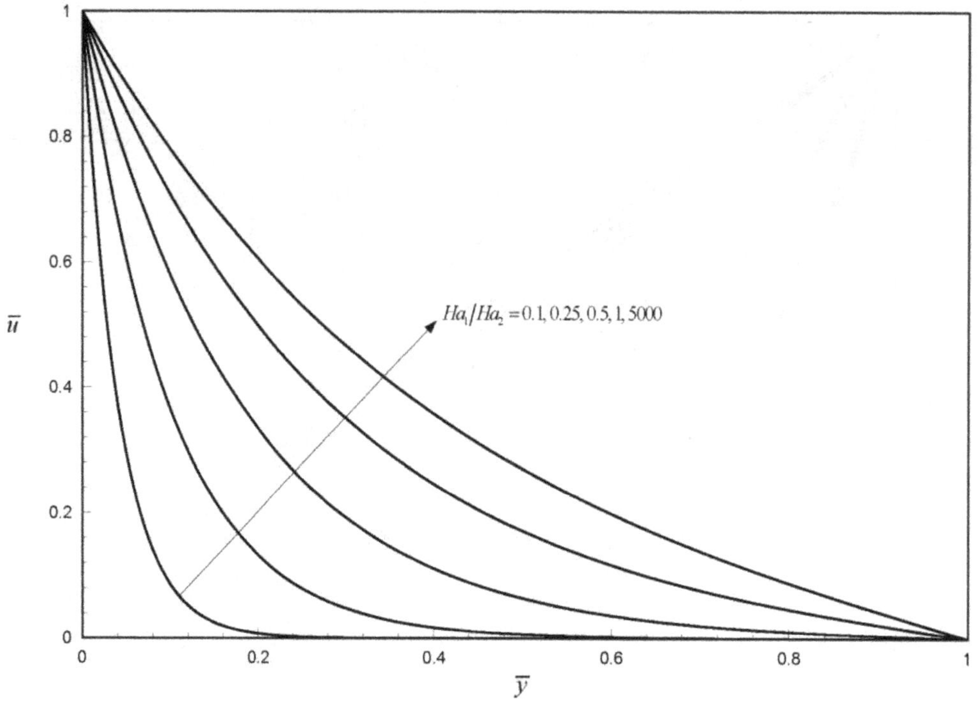

Figure 6. Velocity profiles of steady hydromagnetic Couette flow between parallel plates for different values of Ha_1/Ha_2 ($Ha_1 = 3$).

Lorentz force. This force has a decreasing effect on the velocity.

We observe from Figure 2 that when the Hartmann number increases, the velocity gradient at the moving plate increases and hence the force necessary to move this plate is greater. Figure 3 shows that as the strength of the applied magnetic field increases, the velocity profiles are flattened over the greater part of the cross-section. In other words, the magnetic field causes the shear stresses in the fluid in the vicinity of the plates to become larger. Figures 4 and 5 illustrate the time histories of hydromagnetic Couette and Poiseuille flows for various values of the Hartmann number, respectively. These figures exhibit the same transient behavior, namely the velocity gradually increases with in time and it reaches the steady-state values. Again from these figures, we arrive at the conclusion that the application of the magnetic field speeds up the transition from the unsteady-state to the steady-state. For example, in the case of Poiseuille flow with $Ha_1 = 6$, the transient behavior lasts about one-sixth as long as it does in the non-mhd case. The effect of the ratio of Hartmann numbers (Ha_1/Ha_2) on the velocity field is shown in Figures 6 to 9. It is found that the effect of decreasing Ha_1/Ha_2 is to decrease the velocity field.

Validations of the results presented in this paper can be judged by comparing them with the experimental results. Unfortunately, no comparisons with experimental data were performed due to a lack of existence of such data. For this reason, it is not possible to comment with any certainty on the relative merits of the constitutive equations used here. It is hoped that the exact results presented in this paper can be useful as a benchmark for validating the numerical solutions to more complicated two-fluid MHD flows.

Conclusions

Couette and Poiseuille flows of a binary fluid mixture between two infinitely long parallel plates in the presence of a transverse magnetic field were investigated. Under very special conditions stated previously, steady-state and transition solutions were obtained analytically by the usual methods of solving these kinds of equations. Numerical evaluations of the analytical solutions were performed and graphical results for the velocity distributions were presented to illustrate the influence of the magnetic field on the solutions. It was found that, owing the presence of a transverse magnetic field, increases in the values of the Hartmann number have the tendency to slow the motion of the fluid mixture. In addition, as the Hartmann number increased, the transition to the steady-state became faster. It should be noted that the velocity distributions for the fluid mixture

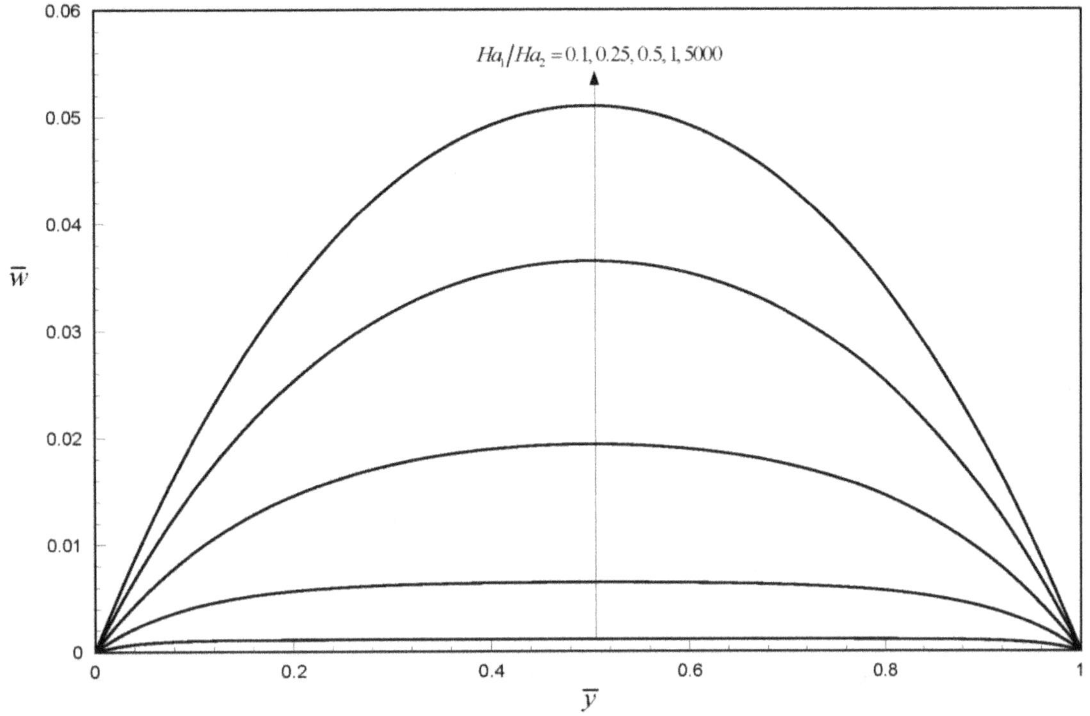

Figure 7. Velocity profiles of steady hydromagnetic Poiseuille flow between parallel plates for different values of Ha_1/Ha_2 ($Ha_1 = 3$).

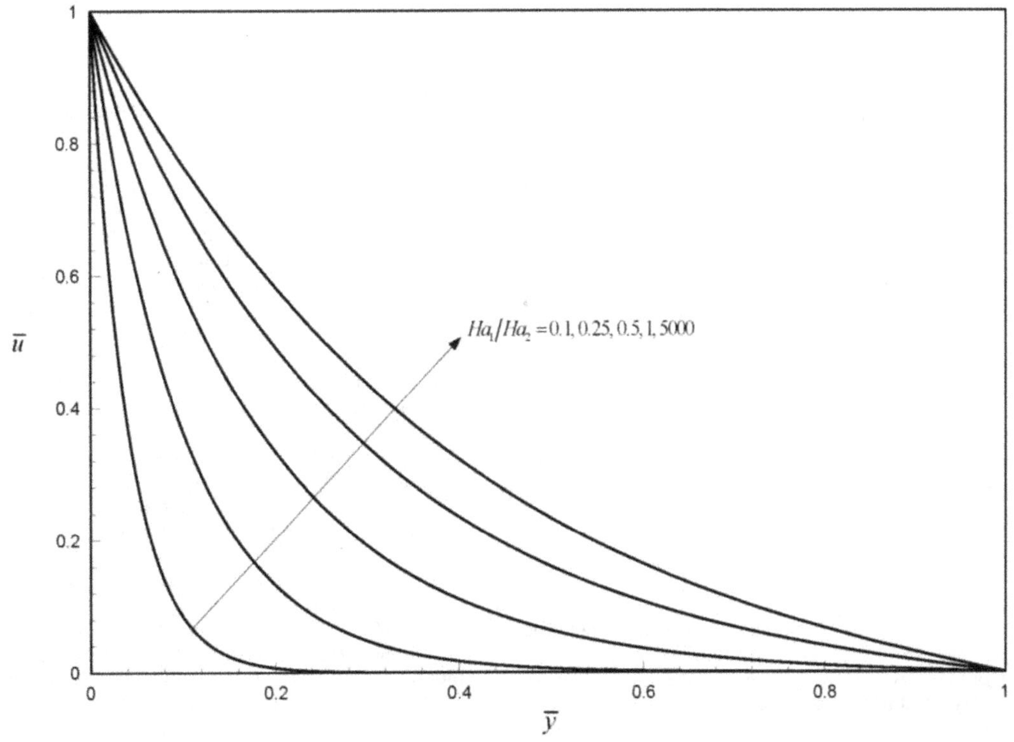

Figure 8. Velocity profiles of unsteady hydromagnetic Couette flow between parallel plates for different values of Ha_1/Ha_2 ($Ha_1 = 3$, $\overline{t} = 0.1$).

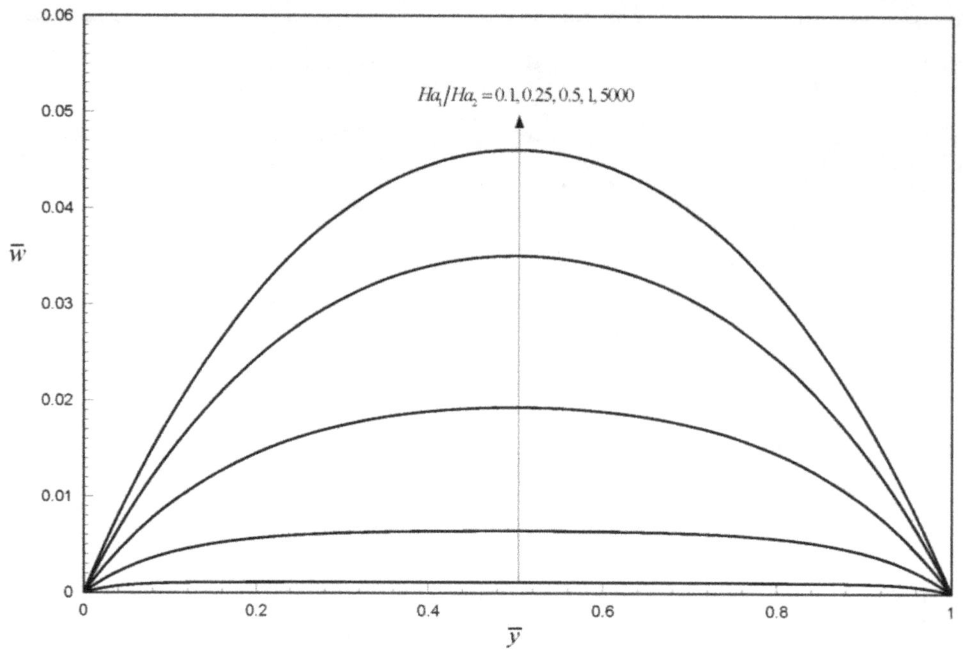

Figure 9. Velocity profiles of unsteady hydromagnetic Poiseuille flow between parallel plates for different values of Ha_1/Ha_2 ($Ha_1 = 3$, $\bar{t} = 0.1$).

have qualitatively the same characteristics as those exhibited by a single Newtonian fluid. Also, all the results corresponding to $\bar{M}_1 = \bar{M}_2 = \bar{M}_3 = \bar{M}_4 = 1/4$, $\bar{\rho}_1 = \bar{\rho}_2 = \phi_1 = 1/2$, and $Ha_1 = Ha_2 = Ha/\sqrt{2}$ reduce the classical solutions of a single Newtonian fluid. This provides a useful check. A further check on the validity of the theoretical results presented here can be accomplished by comparisons with experimental data. As far as the authors are aware, such a quantitative comparison has been hampered by a lack of reliable experimental data. For this reason, the researcher of necessity has to rely on the mixture theory to produce the correct results qualitatively at least.

ACKNOWLEDGEMENTS

We would like to thank the editor and referees for their useful comments and suggestions regarding an earlier version of this paper.

REFERENCES

Adkins JE (1963). Non-linear diffusion, 1. Diffusion and flow of mixtures of fluids.Phil. Trans. Roy. Soc. London, A. 255:607-633.

Al-Sharif A, Chamniprasart K, Rajagopal KR, Szeri AZ (1993). Lubrication with binary mixtures: Liquid-liquid emulsion. J. Tribol. 115:46-55.

Asghar S, Ahmad A (2012). Unsteady Couette flow of viscous fluid under a non-uniform magnetic field. Appl. Mat. Lett. 25:1953-1958.

Atkin RJ, Craine RE (1976). Continuum theories of mixtures: Applications. J. Inst. Maths. Appl. 17:153-207.

Barış S (2005). Unsteady flows of a binary mixture of incompressible Newtonian fluids in an annulus. Int. J. Eng. Sci. 43:1471-1485.

Barış S, Demir MŞ (2012). Flow of a binary mixture of fluids in a semicircular duct. Int. J. Phys. Sci. 7:1333-1345.

Bedford A, Drumheller DS (1983). Theory of immiscible and structured mixtures. Int. J. Eng. Sci. 21:863-960.

Beevers CE, Craine RE (1982). On the determination of response functions for a binary mixture of incompressible Newtonian fluids. Int. J. Eng. Sci. 20:737-745.

Bowen RM (1976). Theory of mixtures.In: A.C. Eringen (Ed.), Continuum Physics, Vol. III, Academic Press, New York.

Chamniprasart K, Al-Sharif A, Rajagopal KR, Szeri AZ (1993). Lubrication with binary mixtures: Bubbly oil. J. Tribol. 115:253-260.

Craine RE (1971). Oscillations of a plate in a binary mixture of incompressible Newtonian fluids. Int. J. Eng. Sci. 9:1177-1192.

Craine RE (1973). An oscillating plate in a binary mixture of hemihedral fluids. Z. Angew. Math. Phys. 24:365-374.

Dai F, Khonsari MM (1994). A theory of hydrodynamic lubrication involving the mixture of two fluids. J. Appl. Mech. 61:634-641.

Eringen AC, Ingram JD (1965). A continuum theory of chemically reacting media-I. Int. J. Eng. Sci. 3:197-212.

Göğüş MŞ (1988). The influence of vibrating plates on Poiseuille flow of a binary mixture.Int. J. Eng. Sci. 26:313-323.

Göğüş MŞ (1991). The influence of a vibrating plate on the flow between parallel plates of a binary mixture. Int. J. Eng. Sci. 29:1651-1659.

Göğüş MŞ (1992a). The influence of a longitudinal vibrating pipe on Poiseuille flow of a binary mixture. Int. J. Eng. Sci. 30:141-151.

Göğüş MŞ (1992b). The steady flow of a binary mixture between two rotating parallel non-coaxial disks. Int. J. Eng. Sci. 30:665-677.

Göğüş MŞ (1994). The velocity profile in pulsatile flow of a binary mixture. Int. J. Eng. Sci. 32: 705-714

Göğüş MŞ (1995). The steady flow of a binary mixture between two rotating parallel coaxial disks. Int. J. Eng. Sci. 33:611-624.

Green AE, Naghdi PM (1965). A dynamical theory of interacting continua. Int. J. Eng. Sci. 3: 231-241.

Ingram JD, Eringen AC (1967). A continuum theory of chemically reacting media-II. Constitutive equations of reacting fluid mixtures. Int. J. Eng. Sci. 4:289-322.

Lohrasbi J, Shai V (1988). Magnetohydrodynamic heat transfer in two-phase flow between parallel plates. App. Sci. Res. 45:53-66.

Malashetty MS, Leela V (1992). Magnetohydrodynamic heat transfer in two-phase flow. Int. J. Eng. Sci. 30:371-377.

Malashetty MS, Umavathi JC, Kumar JP (2001). Convective MHD two-fluid flow and heat transfer in an inclined channel. Heat Mass Trans. 37:259-264.

Massoudi M (2003). Constitutive relations for the interaction force in multicomponent particulate flows. Int. J. Non-Linear Mech. 38:313-336.

Massoudi M (2008). Flow of a binary mixture of linearly incompressible viscous fluids between two horizontal parallel plates. Mech. Res. Comm. 35:603-608.

Mills N (1966). Incompressible mixtures of Newtonian fluids. Int. J. Eng. Sci. 4:97-112.

Mitra P (1982). Unsteady flow of two electrically conducting fluids between two rigid prallel plates. Bull. Calc. Math. Soc. 74:87-95.

Müller I (1968). A thermodynamic theory of mixtures of fluids. Arc. Rat. Mech. Anal. 28:1-39.

Nikodijevic D, Stamenkovic Z, Milenkovic D, Blagojevic B, Nikodijevic J (2011). Flow and heat transfer of two immiscible fluids in the presence of uniform inclined magnetic field. Math. Prob. Eng. Article ID:132302, 18 P.

Rajagopal KR, Tao L (1995). Mechanics of Mixtures, World Scientific Publishing, New Jersey.

Samohyl I (1987). Thermomechanics of Irreversible Processes in Fluid Mixtures, Tuebner, Liepzig.

Sampaio R, Williams WO (1977). On the viscosities of liquid mixtures. Z. Angew. Math. Phys. 28:607-613.

Shail R (1973). On laminar two-phase flows in magnetohydrodynamics. Int. J. Eng. Sci. 11:1103-1108.

Sivaraj R, Kumar BR, Prakash J (2012). MHD mixed convective flow of viscoelastic and viscous fluids in a vertical porous channel. Applic. Appl. Math. (AAM) 7:99-116.

Srivastava BM, Agarval RP, Srivastava VP (1982). Unsteady flow of suspension through a rectangular pipe. Z. Angew Math. Mech. (ZAMM). 62:261-264.

Truesdell C (1957). Sulle basi della thermomeccanica. Rend. Accad. Naz. Lincei. 8:158-166.

Truesdell C (1984). Rational Thermodynamics (second edition), Springer-Verlag, New York.

Umavathi JC, Mateen A, Chamkha AJ, Mudhaf AA (2006). Oscillatory Hartman two-fluid flow and heat transfer in a horizontal channel. Int. J. Appl. Mech. Eng. 11:155-178.

Umavathi JC, Chamkha AJ, Mateen A, Kumar JP (2008). Unsteady magnetohydrodynamic two-fluid flow and heat transfer in a horizontal channel. Int. J. Heat Tech. 26:121-133.

Wang SH, Al-Sharif A, Rajagopal KR, Szeri AZ (1993). Lubrication with binary mixtures: Liquid-liquid emulsion in an EHL conjunction. J. Tribol. 115:515-522.

Wilhelm HE, Van Der Werff TJ (1977). Oscilating flows of miscible fluids. Appl. Sci. Res. 33:339-352.

10

Transient mass transfer flow past an impulsively started infinite vertical plate with ramped plate velocity and ramped temperature

N. Ahmed and M. Dutta

Department of Mathematics, Gauhati University, Guwahati – 781014, Assam, India.

An exact solution to the problem of a transient free convective mass transfer flow of a Newtonian non-Grey optically thin fluid past a suddenly started infinite vertical plate embedded in a porous medium with ramped wall temperature as well as ramped plate velocity in presence of appreciable thermal radiation and first-order chemical reaction is presented. The resulting system of equations governing the flow is solved by employing Laplace Transform technique in closed form. Detailed computations of the influence of ramped velocity parameter A, radiation parameter Q, Reynolds number Re, Schmidt number Sc, porosity parameter S and chemical reaction parameter K on the variations in the fluid velocity, fluid temperature, fluid concentration, and skin friction, Nusselt number and Sherwood number at the plate are demonstrated graphically. The results show that the effect of the ramped parameter A accelerates the fluid flow substantially. Further, our investigation reveals the fact that the viscous drag at the plate gets increased due to chemical reaction in case of ramped plate temperature. Comparison of some of the results of the present work is made with previously published results under special cases, and shows a good agreement.

Key words: Ramped plate velocity, ramped plate temperature, thermal radiation, chemical reaction, optically thin, porosity.

INTRODUCTION

Natural or free convection is a physical process of heat and mass transfer involving fluids which originates when the temperature as well as species concentration change causes density variations inducing buoyancy forces to act on the fluid. Such flows exist abundantly in nature, and due to its applications in engineering and geophysical environments, these have been studied extensively in practice. The heating of rooms inside buildings using radiators is an example of application of heat transfer by free convection. Detailed areas of applications of free convection flow are found in Ghoshdastidar (2004) and Nield and Bejan (2006).

Welty et al. (2007) defines mass transfer as the transport of one constituent from a region of higher concentration to that of a lower concentration. Mass transfer is the basis for many biological and chemical processes. Biological processes include the oxygenation of blood and the transport of ions across membranes within the kidney. Chemical processes include the chemical vapour deposition of Silane (SiH_4) onto a silicon wafer, the doping of silicon wafer to form a semiconducting thin film, the aeration of wastewater, and the purification of ores and isotopes. Mass transfer also occurs in many other processes such as absorption, adsorption, drying, precipitation, membrane filtration and distillation.

Another process of heat transfer is radiation through electromagnetic waves. Radiative convective flows are encountered in several industrial and environmental processes. Some of these include heating and cooling chambers, evaporation from large open water reservoirs, solar power technology and astrophysical flows. The

*Corresponding author. E-mail: saheel_nazib@yahoo.com.

study of radiation interaction with convection for heat and mass transfer in fluids is quite significant. Various problems on steady and unsteady fluid flow past a moving plate in the presence of free convection and radiation has been studied by Cess (1966), England and Emery (1969), Raptis and Perdikis (1999), Muthucumaraswamy et al. (2001b) and Chamkha et al. (2001). These model studies have been performed on flows in a non-porous medium. Among the studies on radiative and free-convective flow past a vertical plate in porous medium, works of Raptis (1998), Sattar et al. (2000) and Hossain and Pop (2001) are significant.

In several occasions it is observed that a foreign mass reacts with the fluid and in such situations chemical reaction plays an important role in heat and mass transfer problems. In particular, the presence of foreign mass in air or water causes some kind of chemical reaction. Bird et al. (2001) states that during a chemical reaction between two species, heat is also generated. The reaction rate in chemical reactions generally depends on the concentration of the species itself. In Cussler (2009), a reaction is defined to be of first order if the rate of reaction is directly proportional to the concentration of the species. The effect of the presence of foreign mass on the free convection flow past a semi-infinite vertical plate was studied by Gebhart and Pera (1971). Several investigators have studied the effect of chemical reaction on different convective heat and mass transfer flows, among whom Anjalidevi and Kandasamy (1999), Muthucumaraswamy and Ganesan (2001a), Muthucumaraswamy and Shankar (2011) are worth mentioning.

Recently, Das et al. (2011) have studied the radiation effect on natural flow of an optically thin viscous incompressible fluid near a suddenly moving vertical plate with ramped wall temperature by adopting Cogley-Vincenti-Gilles equilibrium model introduced by Cogley et al. (1968). An optically thin fluid has no self absorption property, but it can absorb radiation emitted by the boundaries. In the investigation done by Das et al. (2011), a particular characteristic time $t_0 = \frac{\nu}{U_0^2}$ depending on the kinematic viscosity and the plate velocity has been considered.

In the present work an attempt has been made to study the effects of ramped plate velocity and chemical reaction on a natural flow of an optically thin viscous incompressible radiating non-Grey fluid past an impulsively started infinite vertical plate embedded in a porous media, with ramped wall temperature. The acceleration of the plate is sustained for a finite time interval and thereafter it moves with uniform velocity. In this proposed work, not a specific characteristic time t_0 is considered. Due to an arbitrary choice of t_0, the Reynolds number Re gets introduced into the problem.

Subsequently, the influence of the Reynolds number on the flow field in addition to other similar parameters is investigated in the preview of current study.

BASIC EQUATIONS

The equations governing the motion of an incompressible, viscous, radiating and chemically reacting fluid past a solid surface in porous medium are:

Equation of continuity:

$$\vec{\nabla}\cdot\vec{q}=0 \tag{1}$$

Momentum equation:

$$\rho\left[\frac{\partial\vec{q}}{\partial t'}+\left(\vec{q}\cdot\vec{\nabla}\right)\vec{q}\right]=-\vec{\nabla}p+\rho\vec{g}+\mu\nabla^2\vec{q}-\frac{\mu}{k^*}\vec{q} \tag{2}$$

Energy equation:

$$\rho C_p\left[\frac{\partial T'}{\partial t'}+\left(\vec{q}\cdot\vec{\nabla}\right)T'\right]=K_T\nabla^2T'+\varphi-\frac{\partial q_r}{\partial n'} \tag{3}$$

Species continuity equation:

$$\frac{\partial C'}{\partial t'}+\left(\vec{q}\cdot\vec{\nabla}\right)C'=D_M\nabla^2C'+\bar{K}\left(C'_\infty-C'\right) \tag{4}$$

All the physical quantities are defined in the list of symbols. Consider a transient, radiative and chemically reactive natural flow of an incompressible, Newtonian, non-Grey and optically thin fluid past an impulsively started vertical plate embedded in a porous medium with ramped plate velocity and temperature. For the sake of idealization of the model, the following assumptions are made:

(i) All the fluid properties are considered constants except the influence of the variation in density in the buoyancy force term.
(ii) The viscous dissipation of energy is negligible.
(iii) Permeability of the medium (k^*) is considered to be constant.
(iv)The flow is assumed to be one dimensional (parallel to the vertical plate).
(v) The radiation heat flux (q_r) in the direction of the plate velocity is considered negligible in comparison to that in the normal direction.
(vi) The chemical reaction is considered to be homogeneous and of first order.

A coordinate system (x', y', z') is now introduced with X-axis along the plate in the upward vertical direction, Y-axis normal to the plate directed into the fluid region and Z-axis along the width of the plate (Figure 1).

Initially the plate and the surrounding fluid were at rest and at the same temperature T'_∞ and concentration C'_∞. At time $t'>0$, the plate is suddenly moved in its own plane with initial velocity U_0 and uniform acceleration $A\frac{U_0}{t_0}$, and the wall temperature raises to $T'_\infty+(T'_W-T'_\infty)\frac{t'}{t_0}$ and the concentration changes to $C'_\infty+(C'_W-C'_\infty)\frac{t'}{t_0}$ for $0<t'\le t_0$, and a constant temperature

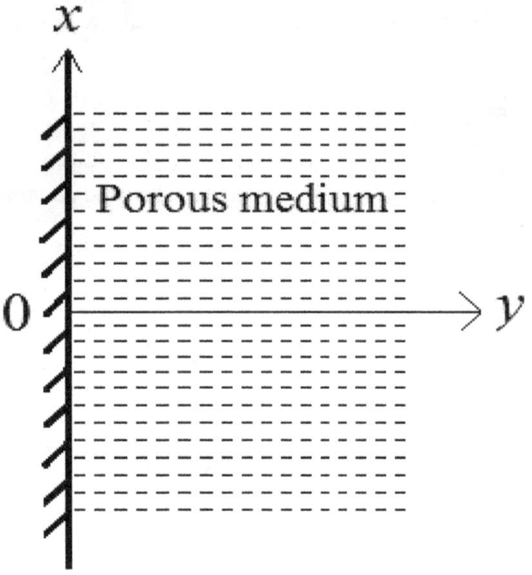

Figure 1. Geometry of the problem.

T'_W $(T'_W > T'_\infty)$, a constant concentration C'_W $(C'_W > C'_\infty)$, and a uniform plate velocity $(A+1)U_0$ is maintained at $t' > t_0$. Let $\vec{q} = (u',0,0)$ denote the fluid velocity at the point (x',y',z',t') in the fluid. Then Equation (1) reduces to $\frac{\partial u'}{\partial x'} = 0$ which yields

$$u' = u'(y',t') \tag{5}$$

In light of the above coordinate system, Equation (2) takes the form

$$\rho\frac{\partial u'}{\partial t'} = -\frac{\partial p}{\partial x'} - \rho g + \mu\frac{\partial^2 u'}{\partial y'^2} - \frac{\mu}{k^*}u' \tag{6}$$

and $\frac{\partial p}{\partial y'} = 0$ (7)

From Equation (7) it is inferred that p is independent of y' indicating the fact that the pressure near the plate is same as that far away from the plate in normal direction. This observation establishes the result that as one moves far away from the plate,

$$\frac{\partial p}{\partial x'} = -\rho_\infty g \tag{8}$$

Elimination of $\frac{\partial p}{\partial x'}$ from Equations (6) and (8), then yields

$$\rho\frac{\partial u'}{\partial t'} = \mu\frac{\partial^2 u'}{\partial y'^2} + (\rho_\infty - \rho)g - \frac{\mu}{k^*}u' \tag{9}$$

The equation of state according to classical Boussinesq approximation is:

$$\rho = \rho_\infty\left[1 - \beta(T' - T'_\infty) - \bar{\beta}(C' - C'_\infty)\right] \tag{10}$$

Coupling Equations (9) and (10) together and accomplishing the fact that $\frac{\rho_\infty}{\rho} \cong 1$, the following linear partial differential equation is obtained:

$$\frac{\partial u'}{\partial t'} = \nu\frac{\partial^2 u'}{\partial y'^2} + g\beta(T' - T'_\infty) + g\bar{\beta}(C' - C'_\infty) - \frac{\nu}{k^*}u' \tag{11}$$

Further, since the plate is of infinite length and in view of assumption (II), the following reduced form of the Equations (3) and (4) are obtained:

$$\rho C_p\frac{\partial T'}{\partial t'} = K_T\frac{\partial^2 T'}{\partial y'^2} - \frac{\partial q_r}{\partial y'} \tag{12}$$

$$\frac{\partial C'}{\partial t'} = D_M\frac{\partial^2 C'}{\partial y'^2} + \bar{K}(C'_\infty - C') \tag{13}$$

The initial and boundary conditions to be satisfied by Equations (11), (12) and (13) are:

$$u' = 0,\ T' = T'_\infty,\ C' = C'_\infty \ \text{ for } y' \ge 0 \text{ and } t' \le 0 \tag{14}$$

$$\begin{aligned} &u' = U_0 + A\frac{U_0}{t_0}t',\ T' = T'_\infty + (T'_W - T'_\infty)\frac{t'}{t_0}, \\ &C' = C'_\infty + (C'_W - C'_\infty)\frac{t'}{t_0} \ \text{ at } y' = 0 \text{ and } 0 < t' \le t_0 \end{aligned} \tag{15}$$

$$\begin{aligned} &u' = (A+1)U_0,\ T' = T'_W,\ C' = C'_W \ \text{ at } y' = 0 \\ &\qquad\qquad \text{and } t' > t_0 \end{aligned} \tag{16}$$

$$u' \to 0,\ T' \to T'_\infty,\ C' \to C'_\infty \ \text{ as } y' \to \infty \text{ for } t' > 0 \tag{17}$$

Cogley et al. (1968) emphasized that the rate of radiative heat flux in optically thin limit for a non-Grey gas near equilibrium is given by:

$$\frac{\partial q_r}{\partial y'} = 4I(T' - T'_\infty) \tag{18}$$

Where,

$$I = \int_0^\infty (K_\lambda)_W \left(\frac{\partial e_{\lambda h}}{\partial T'}\right)_W d\lambda \tag{19}$$

In view of Equation (18), Equation (12) becomes

$$\rho C_p\frac{\partial T'}{\partial t'} = K_T\frac{\partial^2 T'}{\partial y'^2} - 4I(T' - T'_\infty) \tag{20}$$

In order to normalize the flow model, the following non-dimensional variables and parameters are introduced:

$$\left.\begin{aligned}
&u=\frac{u'}{U_0},\ y=\frac{y'U_0}{\nu},\ t=\frac{t'}{t_0},\ \theta=\frac{T'-T'_\infty}{T'_W-T'_\infty},\\
&\phi=\frac{C'-C'_\infty}{C'_W-C'_\infty},\ \Pr=\frac{\mu C_p}{K_T},\ \mathrm{Q}=\frac{4I\nu}{\rho C_p U_0^2},\\
&\mathrm{Re}=\frac{U_0^2 t_0}{\nu},\ \mathrm{Gr}=\frac{\nu g\beta(T'_W-T'_\infty)}{U_0^3},\\
&\mathrm{Gm}=\frac{\nu g\overline{\beta}(C'_W-C'_\infty)}{U_0^3},\ \mathrm{S}=\frac{k^*U_0^2}{\nu^2},\\
&\mathrm{Sc}=\frac{\nu}{D_M},\ \mathrm{K}=\frac{\overline{K}\nu}{U_0^2}
\end{aligned}\right\}\tag{21}$$

All physical quantities are defined in the list of symbols as mentioned earlier. Utilizing the transformations and definitions Equation (21), the Equations (11), (13) and (20) become

$$\frac{\partial u}{\partial t}=\mathrm{Re}\left[\frac{\partial^2 u}{\partial y^2}+\mathrm{Gr}\theta+\mathrm{Gm}\phi-\frac{1}{\mathrm{S}}u\right]\tag{22}$$

$$\frac{\partial \theta}{\partial t}=\mathrm{Re}\left[\frac{1}{\Pr}\frac{\partial^2 u}{\partial y^2}-\mathrm{Q}\theta\right]\tag{23}$$

$$\frac{\partial \phi}{\partial t}=\mathrm{Re}\left[\frac{1}{\mathrm{Sc}}\frac{\partial^2 \phi}{\partial y^2}-\mathrm{K}\phi\right]\tag{24}$$

subject to the relevant initial and boundary conditions:

$$u=0,\ \theta=0,\ \phi=0 \text{ for } y\ge 0,\ t\le 0\tag{25}$$

$$u=\mathrm{A}t+1,\ \theta=t,\ \phi=t \text{ at } y=0,\ 0<t\le 1\tag{26}$$

$$u=\mathrm{A}+1,\ \theta=1,\ \phi=1 \text{ at } y=0,\ t>1\tag{27}$$

$$u\to 0,\ \theta\to 0,\ \phi\to 0 \text{ as } y\to\infty \text{ for } t>0\tag{28}$$

METHODS OF SOLUTION

Taking Laplace transform of the Equations (22), (23) and (24), the following equations are obtained

$$\frac{d^2\bar{u}}{dy^2}-\left(\frac{1}{\mathrm{Re}}s+\frac{1}{\mathrm{S}}\right)\bar{u}=-\mathrm{Gr}\bar{\theta}-\mathrm{Gm}\bar{\phi}\tag{29}$$

$$\frac{d^2\bar{\theta}}{dy^2}-\Pr\left(\frac{1}{\mathrm{Re}}s+\mathrm{Q}\right)\bar{\theta}=0\tag{30}$$

$$\frac{d^2\bar{\phi}}{dy^2}-\mathrm{Sc}\left(\frac{1}{\mathrm{Re}}s+\mathrm{K}\right)\bar{\phi}=0\tag{31}$$

Where

$$\left.\begin{aligned}
&\bar{u}(y,s)=\int_0^\infty u(y,t)e^{-st}dt\\
&\bar{\theta}(y,s)=\int_0^\infty \theta(y,t)e^{-st}dt\\
&\bar{\phi}(y,s)=\int_0^\infty \phi(y,t)e^{-st}dt
\end{aligned}\right\}\tag{32}$$

The corresponding boundary conditions for $\bar{u}$, $\bar{\theta}$ and $\bar{\phi}$ are:

$$\begin{aligned}
&\bar{u}=\mathrm{A}\frac{1}{s^2}\left(1-e^{-s}\right)+\frac{1}{s},\ \bar{\theta}=\frac{1}{s^2}\left(1-e^{-s}\right),\\
&\bar{\phi}=\frac{1}{s^2}\left(1-e^{-s}\right) \text{ at } y=0
\end{aligned}\tag{33}$$

$$\bar{u}\to 0,\ \bar{\theta}\to 0,\ \bar{\phi}\to 0 \text{ as } y\to\infty\tag{34}$$

The Equations (29), (30) and (31) are ordinary second order differential equations, the solutions of which subject to the conditions (33) and (34) are as follows:

$$\begin{aligned}
\bar{u}(y,s)=&\mathrm{A}\frac{1}{s^2}\left(1-e^{-s}\right)e^{-y\sqrt{P_1}}+\frac{1}{s}e^{-y\sqrt{P_1}}+\\
&\frac{P_2}{P_1-P_3^2}\left(e^{-y\sqrt{P_1}}-e^{-yP_3}\right)+\\
&\frac{P_4}{P_1-P_5^2}\left(e^{-y\sqrt{P_1}}-e^{-yP_5}\right)
\end{aligned}\tag{35}$$

$$\bar{\theta}(y,s)=\frac{1}{s^2}\left(1-e^{-s}\right)e^{-yP_3}\tag{36}$$

$$\bar{\phi}(y,s)=\frac{1}{s^2}\left(1-e^{-s}\right)e^{-yP_5}\tag{37}$$

Where

$$\left.\begin{aligned}
&P_1=\frac{1}{\mathrm{Re}}s+\frac{1}{\mathrm{S}},\ P_2=-\mathrm{Gr}\frac{1}{s^2}\left(1-e^{-s}\right),\\
&P_3=\sqrt{\Pr\left(\frac{1}{\mathrm{Re}}s+\mathrm{Q}\right)},\ P_4=-\mathrm{Gm}\frac{1}{s^2}\left(1-e^{-s}\right),\\
&P_5=\sqrt{\mathrm{Sc}\left(\frac{1}{\mathrm{Re}}s+\mathrm{K}\right)}
\end{aligned}\right\}\tag{38}$$

Taking inverse Laplace transform of Equations (35), (36) and (37), the representative velocity, temperature and concentration fields are obtained and are as below:

$$u=\psi_1-\overline{\psi_1}+\omega_1-\frac{\mathrm{ReGr}}{1-\Pr}E_4-\frac{\mathrm{ReGm}}{1-\mathrm{Sc}}E_8\tag{39}$$

$$\theta=\psi_2-\overline{\psi_2}\tag{40}$$

$$\phi=\psi_3-\overline{\psi_3}\tag{41}$$

Where

$$\psi_1=\mathrm{A}\times\psi\left(\mathrm{Re},\mathrm{S},1,y,t\right),\ \overline{\psi_1}=\mathrm{A}\times\psi\left(\mathrm{Re},\mathrm{S},1,y,t-1\right)\times\mathrm{H}(t-1),$$

$$\psi_2=\psi\left(\mathrm{Re},\frac{1}{\mathrm{Q}},\Pr,y,t\right),\ \overline{\psi_2}=\psi\left(\mathrm{Re},\frac{1}{\mathrm{Q}},\Pr,y,t-1\right)\times\mathrm{H}(t-1),$$

$$\psi_3=\psi\left(\mathrm{Re},\frac{1}{\mathrm{K}},\mathrm{Sc},y,t\right),\ \overline{\psi_3}=\psi\left(\mathrm{Re},\frac{1}{\mathrm{K}},\mathrm{Sc},y,t-1\right)\times\mathrm{H}(t-1),$$

$$\omega_1=\omega\left(\mathrm{Re},\mathrm{S},y,t\right),$$

$$F_1 = F(\text{Re},\text{S},a,1,y,t),\ F_2 = F\left(\text{Re},\frac{1}{\text{Q}},a,\text{Pr},y,t\right),$$

$$E_1 = \frac{1}{2a^2}e^{at}(F_1 - F_2),$$

$$G_1 = G(\text{Re},\text{S},a,1,y,t),\ G_2 = G\left(\text{Re},\frac{1}{\text{Q}},a,\text{Pr},y,t\right),$$

$$E_2 = \frac{1}{2a}(G_2 - G_1),$$

$$E_3 = E_1 + E_2,$$

$$\overline{F_1} = F(\text{Re},\text{S},a,1,y,t-1),\ \overline{F_2} = F\left(\text{Re},\frac{1}{\text{Q}},a,\text{Pr},y,t-1\right),$$

$$\overline{E_1} = \frac{1}{2a^2}e^{a(t-1)}\left(\overline{F_1} - \overline{F_2}\right)\times \text{H}(t-1),$$

$$\overline{G_1} = G(\text{Re},\text{S},a,1,y,t-1),\ \overline{G_2} = G\left(\text{Re},\frac{1}{\text{Q}},a,\text{Pr},y,t-1\right),$$

$$\overline{E_2} = \frac{1}{2a}\left(\overline{G_2} - \overline{G_1}\right)\times \text{H}(t-1),\ \overline{E_3} = \overline{E_1} + \overline{E_2},\ E_4 = E_3 - \overline{E_3}$$

$$F_3 = F(\text{Re},\text{S},b,1,y,t),\ F_4 = F\left(\text{Re},\frac{1}{\text{K}},b,\text{Sc},y,t\right),$$

$$E_5 = \frac{1}{2b^2}e^{bt}(F_3 - F_4),$$

$$G_3 = G(\text{Re},\text{S},b,1,y,t),\ G_4 = G\left(\text{Re},\frac{1}{\text{K}},b,\text{Sc},y,t\right),$$

$$E_6 = \frac{1}{2b}(G_4 - G_3),$$

$$E_7 = E_5 + E_6,$$

$$\overline{F_3} = F(\text{Re},\text{S},b,1,y,t-1),\ \overline{F_4} = F\left(\text{Re},\frac{1}{\text{K}},b,\text{Sc},y,t-1\right),$$

$$\overline{E_5} = \frac{1}{2b^2}e^{b(t-1)}\left(\overline{F_3} - \overline{F_4}\right)\times \text{H}(t-1),$$

$$\overline{G_3} = G(\text{Re},\text{S},b,1,y,t-1),\ \overline{G_4} = G\left(\text{Re},\frac{1}{\text{K}},b,\text{Sc},y,t-1\right),$$

$$\overline{E_6} = \frac{1}{2b}\left(\overline{G_4} - \overline{G_3}\right)\times \text{H}(t-1),\ \overline{E_7} = \overline{E_5} + \overline{E_6},\ E_8 = E_7 - \overline{E_7},$$

$$a = \frac{\text{Re}}{1-\text{Pr}}\left[\text{Pr}\,\text{Q} - \frac{1}{\text{S}}\right],$$

$$b = \frac{\text{Re}}{1-\text{Sc}}\left[\text{KSc} - \frac{1}{\text{S}}\right]$$

The real valued functions ψ, ω, F and G are defined as follows:

$$\psi(\xi,\eta,\alpha,y,t) = \frac{1}{2}\left[\left(t + \frac{y}{2\xi}\sqrt{\alpha\eta}\right)e^{y\sqrt{\frac{\alpha}{\eta}}}\,erfc\left(\frac{y}{2}\sqrt{\frac{\alpha}{\xi t}} + \sqrt{\frac{\xi t}{\eta}}\right) + \left(t - \frac{y}{2\xi}\sqrt{\alpha\eta}\right)e^{-y\sqrt{\frac{\alpha}{\eta}}}\,erfc\left(\frac{y}{2}\sqrt{\frac{\alpha}{\xi t}} - \sqrt{\frac{\xi t}{\eta}}\right)\right]$$

$$\omega(\xi,\eta,y,t) = \frac{1}{2}\left[e^{\frac{y}{\sqrt{\eta}}}\,erfc\left(\frac{y}{2\sqrt{\xi t}} + \sqrt{\frac{\xi t}{\eta}}\right) + e^{\frac{-y}{\sqrt{\eta}}}\,erfc\left(\frac{y}{2\sqrt{\xi t}} - \sqrt{\frac{\xi t}{\eta}}\right)\right]$$

$$F(\xi,\eta,\alpha,\delta,y,t) = e^{y\sqrt{\frac{\alpha\delta}{\xi}+\frac{\delta}{\eta}}}\,erfc\left(\frac{y}{2}\sqrt{\frac{\delta}{\xi t}} + \sqrt{\left(\alpha + \frac{\xi}{\eta}\right)t}\right) + e^{-y\sqrt{\frac{\alpha\delta}{\xi}+\frac{\delta}{\eta}}}\,erfc\left(\frac{y}{2}\sqrt{\frac{\delta}{\xi t}} - \sqrt{\left(\alpha + \frac{\xi}{\eta}\right)t}\right)$$

$$G(\xi,\eta,\alpha,\delta,y,t) = \left(t + \frac{y}{2\xi}\sqrt{\delta\eta} + \frac{1}{\alpha}\right)e^{y\sqrt{\frac{\delta}{\eta}}}\,erfc\left(\frac{y}{2}\sqrt{\frac{\delta}{\xi t}} + \sqrt{\frac{\xi t}{\eta}}\right) + \left(t - \frac{y}{2\xi}\sqrt{\delta\eta} + \frac{1}{\alpha}\right)e^{-y\sqrt{\frac{\delta}{\eta}}}\,erfc\left(\frac{y}{2}\sqrt{\frac{\delta}{\xi t}} - \sqrt{\frac{\xi t}{\eta}}\right)$$

Where ξ, η, α, δ, y, t are real variables.

LOCAL CO-EFFICIENT OF SKIN FRICTION

The viscous drag at the plate per unit area in the direction of the plate velocity is given by the Newton's law of viscosity in the form:

$$\tau' = \mu\frac{\partial u'}{\partial y'}\bigg]_{y'=0} = \mu\frac{U_0^2}{\nu}\frac{\partial u}{\partial y}\bigg]_{y=0} \tag{42}$$

The co-efficient of skin friction at the plate is given by

$$\tau = \frac{\tau'}{\frac{\mu U_0^2}{\nu}} = \frac{\partial u}{\partial y}\bigg]_{y=0} \tag{43}$$

LOCAL CO-EFFICIENT OF RATE OF HEAT TRANSFER

The heat flux q^* from the plate to the fluid is given by Fourier's law of conduction in the form

$$q^* = -K_T\frac{\partial T'}{\partial y'}\bigg]_{y'=0} = -K_T(T'_W - T'_\infty)\frac{U_0}{\nu}\frac{\partial \theta}{\partial y}\bigg]_{y=0} \tag{44}$$

The co-efficient of the rate of heat transfer from the plate to the fluid in terms of Nusselt number is given by

$$\text{Nu} = \frac{q^*\nu}{K_T(T'_W - T'_\infty)U_0} = -\frac{\partial \theta}{\partial y}\bigg]_{y=0} \tag{45}$$

LOCAL CO-EFFICIENT OF MASS TRANSFER

The mass flux from the plate to the fluid is governed by Fick's law in the following form:

$$\mathrm{M_W} = -D_M \frac{\partial C'}{\partial y'}\bigg]_{y'=0} = -D_M\left(C'_W - C'_\infty\right)\frac{U_0}{\nu}\frac{\partial \phi}{\partial y}\bigg]_{y=0} \quad (46)$$

The rate of mass transfer from the plate to the fluid in terms of the Sherwood number Sh is given by

$$\mathrm{Sh} = \frac{\mathrm{M_W}\nu}{D_M\left(C'_W - C'_\infty\right)U_0} = -\frac{\partial \phi}{\partial y}\bigg]_{y=0} \quad (47)$$

Detailed computations of skin friction, Nusselt number and Sherwood number are obtained but not presented here for the sake of brevity.

RESULTS AND DISCUSSION

In order to get clear insight of the physical problem, numerical calculations from the analytical solutions for the representative velocity field, temperature field, concentration field, the co-efficient of skin friction, the rate of heat transfer at the plate in terms of Nusselt number and the rate of mass transfer near the plate in terms of Sherwood number have been carried out by assigning some arbitrarily chosen specific values to the physical parameters like ramped velocity parameter A, Reynolds number Re, Schmidt number Sc, the radiation parameter Q, the porosity parameter S, and the chemical reaction parameter K. The numerical values computed from the analytical solutions of the problem have been visualized in Figures 2 to 13. Comparison has been made in Figure 14 with Figure 15 (viz. Figure 4 of Das et al., 2011). Figure 14 clearly shows how the problem taken by Das et al. (2011) is a special case of the current problem for the parameters A = 0, Gm = 0, Re = 1, Pr = 0.71, Sc = 0.6, Q = 2, S = 0.04, K = 0.001 and t = 0.1. Both Figures 14 and 15 establish the conformity of the two problems (viz. the current problem and the work of Das et al. (2011)) in the special case when the effect of A, Gm and K (that is, ramped velocity parameter, solutal Grash of number and chemical reaction) are absent. These figures uniquely indicate that an increase in thermal Grash of number causes an overall increase in fluid velocity. Further the velocity profiles for both the problems are almost identical, thereby showing an excellent agreement between the results obtained by Das et al. (2011) and the present authors.

The velocity profiles under the influence of A and K are exhibited in Figures 2 and 3. Figure 2 shows that an increase in ramped velocity parameter A causes the fluid velocity to increase steadily indicating the fact that the fluid motion is accelerated under the influence of parameter A. This figure further establishes the fact that the fluid velocity first increases in a thin layer adjacent to the plate and thereafter it decreases asymptotically as we move away from the plate. It is found that an increase in Reynolds number leads the velocity to increase gradually. We recall that an increase in Reynolds number means a decrease in internal friction. That is, the fluid velocity increases for small viscosity and this phenomenon is in excellent agreement with the physical fact that fluid become freer to move when the internal friction gets decreased. It is observed that our chemical reaction decelerates the flow as is evident from Figure 3.

The temperature profiles have been depicted in Figures 4 and 5. From Figure 4, we notice that an increase in Reynolds number raises the fluid temperature considerably. The thermal radiation has an inhibiting effect on the temperature distribution (Figure 5) which is supported by the phenomena that radiation causes a faster dissipation of heat and consequently lowers the temperature.

Figures 6 and 7, corresponds to concentration profiles versus normal coordinate y under the effect of the parameters Sc and K. It is inferred from Figures 6 and 7 that an increase in Sc or K leads the concentration level of the fluid to drop steadily. We recall that an increase in Schmidt number Sc means a fall in mass diffusivity. It interprets that an increase in mass diffusivity causes the species concentration of the fluid to rise upwards. This observation is in agreement with the outcome of Figure 6. Moreover, chemical reaction results in gradual decrease in the concentration level of the fluid as validated by Figure 7.

The variation in skin friction versus Reynolds number under the influence of Q, S and K are presented in Figures 8 to 10. All these three figures uniquely indicate that an increase in Q or S or K causes a corresponding increase in skin friction. It indicates that the viscous drag at the plate gets increased under the influence of radiation, porosity of the medium and chemical reaction. Further, we observe that the effect of the above parameters on the skin friction are more pronounced for small Reynolds number and for higher Reynolds number these parameters ceases to affect τ. This phenomenon is well supported by the fact that for large Reynolds number (that is, for small viscosity) the internal friction decreases significantly. The trend of behaviour of skin friction under Q, S and K is identical in case of ramped plate temperature and isothermal plate temperature.

Figure 11 shows how the rate of heat transfer from the plate to fluid in terms of Nusselt number Nu is influenced by Q and Re. An interesting observation is made in this figure that for ramped wall temperature, the Nusselt number decreases at a faster rate for small Reynolds number and it decreases slowly and steadily for higher Reynolds number that is, for small viscosity; on the other hand in case of uniform temperature (isothermal plate) Nu becomes steady after an initial increase for large viscosity.

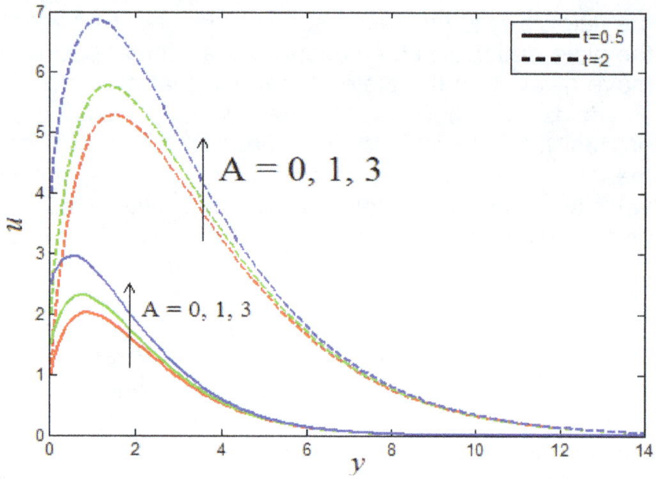

Figure 2. Velocity versus y for Gr = 10, Gm = 5, Re = 10, Pr = 0.71, Sc = 0.60, Q = 5, S = 4, K = 0.4.

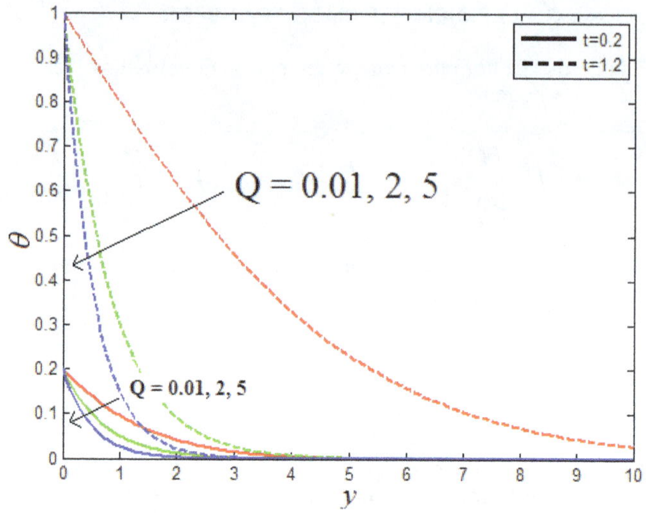

Figure 5. Temperature versus y for Re = 10, Pr = 0.71.

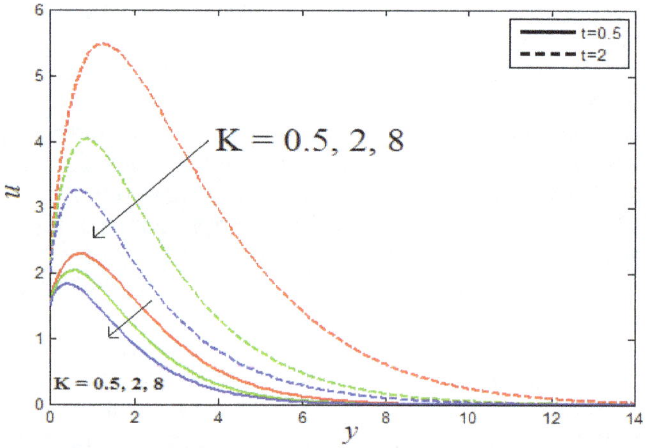

Figure 3. Velocity versus y for A = 1, Gr = 10, Gm = 5, Re = 10, Pr = 0.71, Sc = 0.60, Q = 5, S = 4.

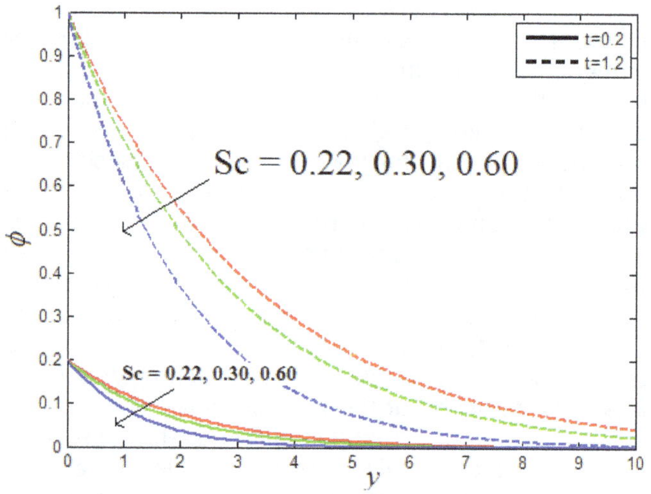

Figure 6. Concentration versus y for Re =10, K = 0.4.

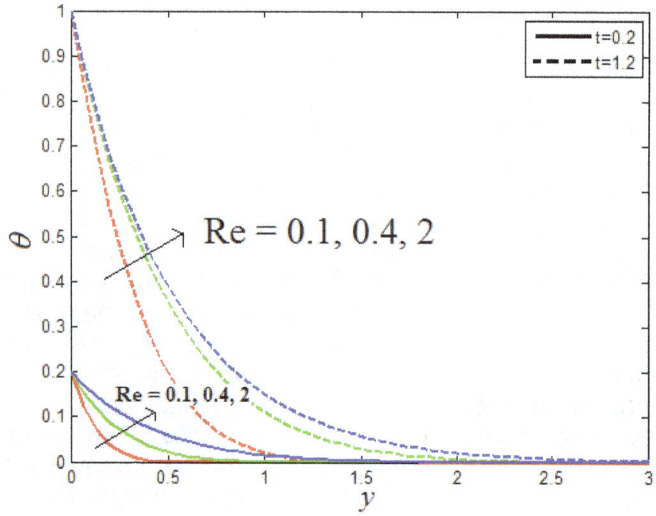

Figure 4. Temperature versus y for Pr = 0.71, Q = 5.

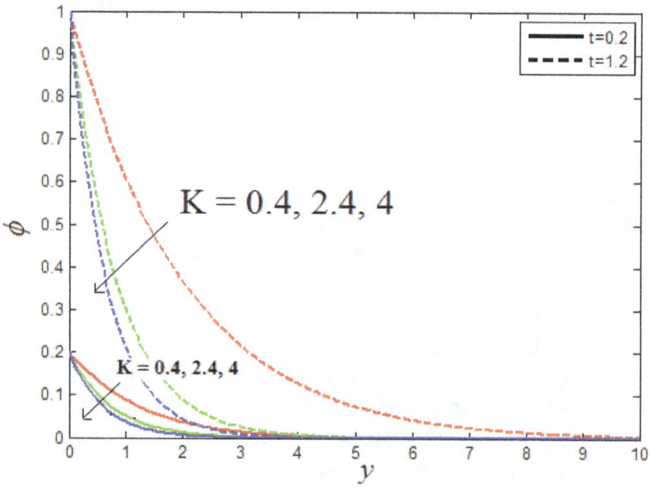

Figure 7. Concentration versus y for Re = 10, Sc = 0.60.

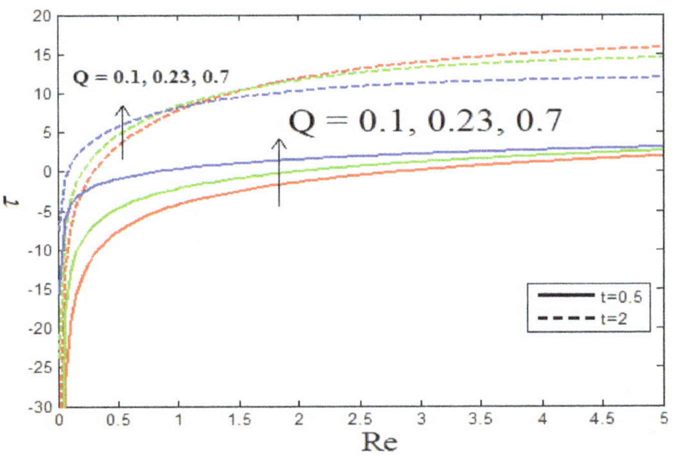

Figure 8. Skin friction versus Re for A = 1, Gr = 10, Gm = 5, Pr = 0.71, Sc = 0.60, S = 4, K = 0.4.

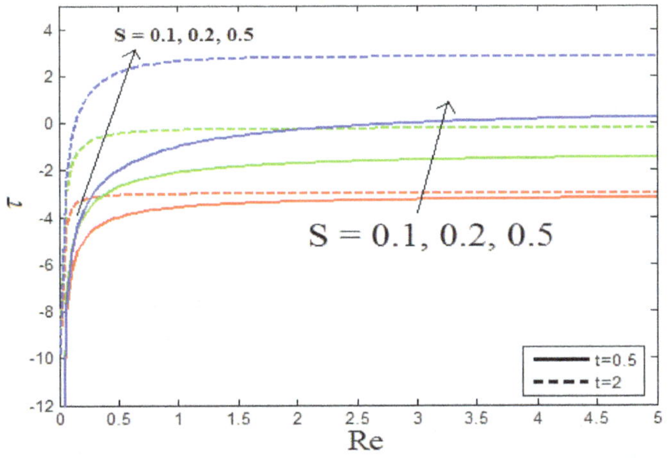

Figure 9. Skin friction versus Re for A = 1, Gr = 10, Gm = 5, Pr = 0.71, Sc = 0.60, Q = 5, K = 0.4.

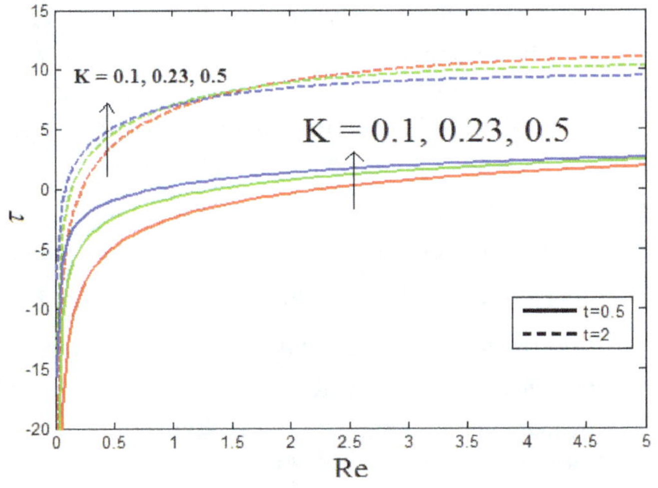

Figure 10. Skin friction versus Re for A = 1, Gr = 10, Gm = 5, Pr = 0.71, Sc = 0.60, Q = 2, S = 4.

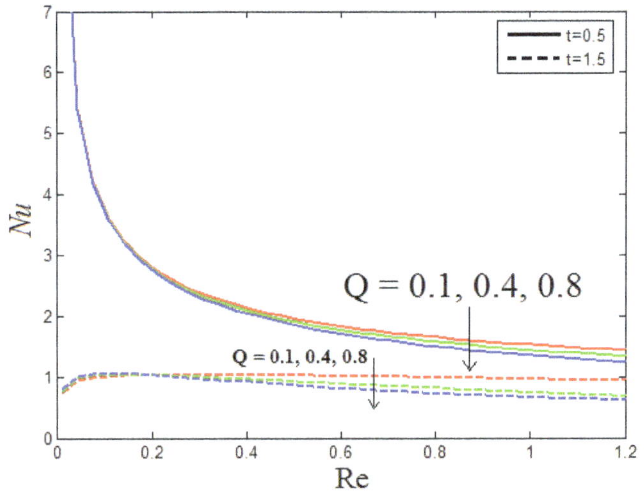

Figure 11. Nusselt number versus Re for Pr = 0.71.

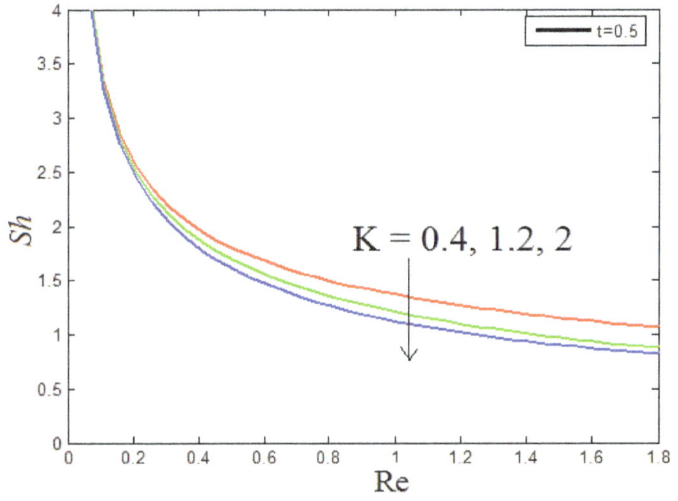

Figure 12. Sherwood number versus Re for Sc = 0.60, t = 0.5.

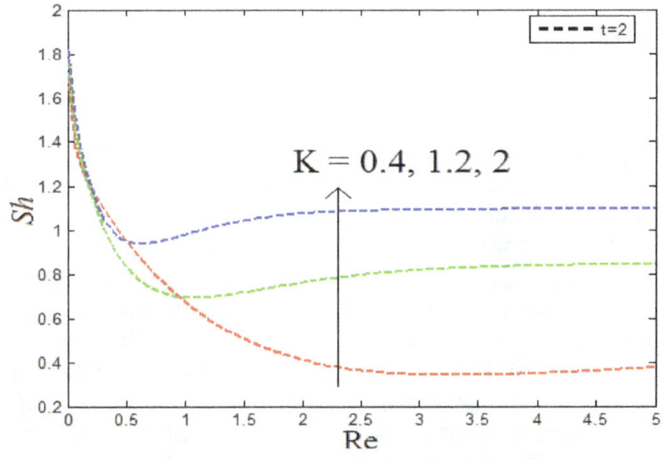

Figure 13. Sherwood number versus Re for Sc = 0.60, t = 2.

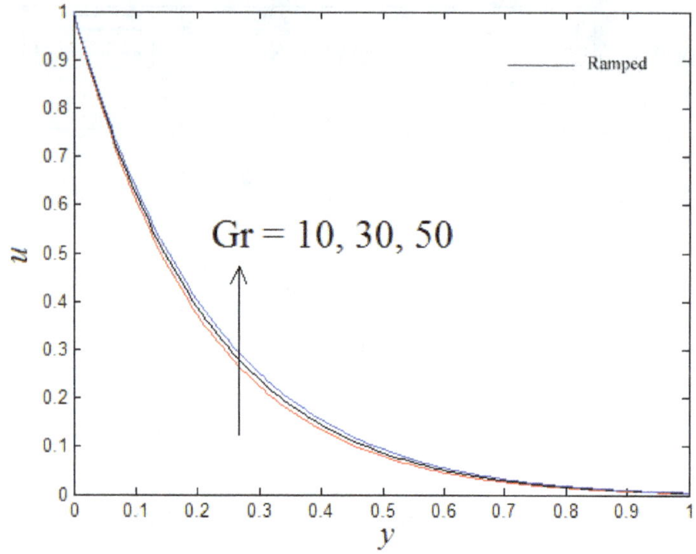

Figure 14. Velocity versus y for A = 0, Gm = 5, Re = 1, Pr = 0.71, Sc = 0.6, Q = 2, S = 0.04, K = 0.001, t = 0.1.

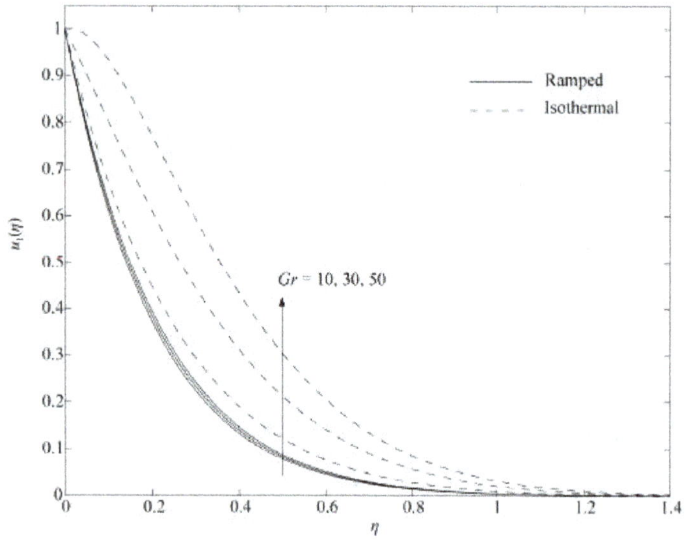

Figure 15. Figure 4 of the work of Das et al. (2011).

Figures 12 to 13 exhibit the change of behavior of the rate of mass transfer from the plate to the fluid in terms of Sherwood number Sh for $t<1$ (ramped temperature) and $t>1$ (isothermal temperature) respectively. Under chemical reaction, the mass transfer rate falls down in case of ramped plate temperature; but a reverse trend of behaviour is observed in case of isothermal plate temperature. For small Reynolds number, this rate is very sharp in either case. However, in case of uniform temperature of the plate, for small value of Reynolds number, Sh sharply decreases and as Reynolds number increases an opposing behavior is marked.

Conclusions

The fluid flow is accelerated with a corresponding increase in ramped velocity parameter A. Chemical reaction has a retarding effect on the fluid flow. The fluid temperature decreases with increasing thermal radiation and it increases as viscosity drops. The species concentration in the fluid decreases with increasing Schmidt number. The viscous drag at the plate gets increased under the influence of radiation, porosity of the medium and chemical reaction. In case of velocity, temperature and concentration profiles, as well as the skin friction, the trend of behavior is similar for both isothermal $(t>1)$ and ramped $(t<1)$ plate temperature. For ramped wall temperature, the Nusselt number decreases at a faster rate for small Re and it decreases slowly and steadily for higher Re; however, in case of uniform temperature Nu becomes steady after an initial increase for large viscosity. Under chemical reaction, the mass transfer behavior is grossly different for ramped and isothermal plate temperature.

NOMENCLATURE

A, Ramped velocity parameter (dimensionless); C_p, specific heat at constant pressure; C'_W, Reference concentration; C'_∞, concentration far away from plate; C', fluid concentration; D_M, mass diffusivity; $e_{\lambda h}$, Plank function; $\vec{g}$, gravitational acceleration vector; g, acceleration due to gravity; Gr, thermal Grashof number; Gm, solutal Grashof number; k^*, permeability of porous medium; K_T, thermal conductivity; $\overline{K}$, rate of first order homogeneous chemical reaction; $(K_\lambda)_W$, absorption co-efficient; K, chemical reaction parameter; p, pressure; Pr, Prandtl number; $\vec{q}$, fluid velocity vector; q_r, radiative heat flux; Q, radiation parameter; Re, Reynolds number; S, porosity parameter; Sc, Schmidt number; t', time; T', fluid temperature; T'_W, reference temperature; T'_∞, temperature far away from the plate; t, non-dimensional time; t_0, characteristic time; U_0, plate velocity; u', x-component of $\vec{q}$; u, non-dimensional fluid velocity; (x',y',z'), cartesian coordinates; y, non-dimensional y-coordinate; $\bar{f}(y,s)$, Laplace transform of function $f(y,t)$; ρ, fluid density; ρ_∞, fluid density far away from plate; μ, co-efficient of viscosity; φ, viscous dissipation of energy per unit volume; $\delta n'$, an element of the outward normal; β, co-efficient of volume expansion for heat transfer; $\overline{\beta}$, co-efficient of volume expansion for mass transfer; ν, kinematic viscosity; θ, non-dimensional temperature; ϕ, non-dimensional concentration; λ, wavelength; $(\text{Subscript})_W$ refers to the values of the physical quantities at the plate; $(\text{Subscript})_\infty$, refers to the values of the physical quantities far away from the plate.

REFERENCES

Anjalidevi SP, Kandasamy R (1999). Effects of chemical reaction, heat and mass transfer on laminar flow along a semi-infinite horizontal plate. Heat Mass Transf. 35(6):465-467.

Bird RB, Stewart WE, Lightfoot EN (2001). Transport phenomena. John Wiley and Sons, USA. P. 551.

Cess RD (1966). The interaction of thermal radiation with free convection heat transfer. Int. J. Heat Mass Transf. 9(11):1269-1277.

Chamkha AJ, Takhar HS, Soundalgekar VM (2001). Radiation effects on free convection flow past a semi-infinite vertical plate with mass transfer. Chem. Eng. J. 84(3):335-342.

Cogley AC, Gilles SE, Vincenti WG, Ishimoto S (1968). Differential approximation for radiative heat transfer in a non grey gas near equilibrium. Am. Inst. Aeron. Astron. 6(3):551-553.

Cussler EL (2009). Diffusion: Mass transfer in fluid systems. Cambridge University Press, UK. pp. 460-461.

Das S, Jana M, Jana RN (2011). Radiation effect on natural convection near a vertical plate embedded in a porous medium with ramped wall temperature. Open J. Fl. Dyn. 1(1):1-11.

England WG, Emery AF (1969). Thermal radiation effects on the laminar free convection boundary layer of an absorbing gas. J. Heat Transf. 91(1):37-44.

Gebhart B, Pera L (1971). The nature of vertical natural convection flows resulting from the combined buoyancy effects of thermal and mass diffusion. Int. J. Heat Mass Transf. 14(12):2025-2050.

Ghoshdastidar PS (2004). Heat transfer. Oxford University Press, Oxford. P. 225.

Hossain MA, Pop I (2001). Radiation effects on free convection over a vertical flat plate embedded in a porous medium with high porosity. Int. J. Therm. Sci. 40:289-295.

Muthucumaraswamy R, Ganesan P (2001a). First order chemical reaction on flow past an impulsively started vertical plate with uniform heat and mass flux. Acta Mech. 147:45-57.

Muthucumaraswamy R, Ganesan P, Soundalgekar VM (2001b). Heat and mass transfer effects on flow past an impulsively started vertical plate. Acta. Mech. 146:1-8.

Muthucumaraswamy R, Shankar MR (2011). First order chemical reaction and thermal radiation effects on unsteady flow past an accelerated isothermal infinite vertical plate. Ind. J. Sci. Technol. 4(5):573-577.

Nield DA, Bejan A (2006). Convection in porous media. Springer, USA. pp. 94-97.

Raptis A (1998). Radiation and free convection flow through a porous medium. Int. Comm. Heat Mass Transf. 25(2):289-295.

Raptis A, Perdikis C (1999). Radiation and free convection flow past a moving plate. Int. J. Appl. Mech. Eng. 4(4):817-821.

Sattar MA, Rahman MM, Alam MM (2000). Free convection flow and heat transfer through a porous vertical flat plate immersed in a porous medium with variable suction. J. Energy Heat Mass Transf. 21(2):17-21.

Welty JR, Wicks CE, Wilson RE, Rorrer GL (2007). Fundamentals of momentum, heat and mass transfer. John Wiley and Sons, USA. P. 398.

Photovoltaic solar cell simulation of shockley diode parameters in matlab

Awodugba, A. O., Sanusi, Y. K., and Ajayi, J. O.

Department of Pure and Applied Physics, Ladoke Akintola University of Technology, Ogbomoso, Nigeria.

In this work, the Shockley diode parameters were simulated using the Matlab software package with the solar cell I - V and P - V characteristics in focus. Our model has been based on previous studies while the effects of varying cell temperature, series resistance, R_s ambient irradiation and diode quality factor were put into consideration and the output current and power characteristics of the PV model simulated, all for 10 iterations using the Newton - Raphson algorithm. The effect of shunt is neglected throughout the simulation while the results showed expected trends as reported by previous authors. This work therefore will be very useful for users, especially researchers in this field.

Key words: Photovoltaic, solar cell, Matlab, simulation, photocurrent.

INTRODUCTION

The quest for alternative energy sources and requirements in the next decades has attracted a lot of attention because there is a strong worldwide demand for energy, not to mention the fact that there is dwindling fossil-fuel reserves and global warming stemming from climate change that has resulted in various degrees of natural disaster like flooding, extreme heat /drought and rise in ocean levels.

One of the candidates to replace pollutant fossil-fuel energy sources and which is also inexhaustible is the solar energy which can be harvested using the solar cells, modules and arrays.

When exposed to light, photons with energy greater than the bandgap energy of the semiconductor are absorbed and create an electron-hole pair. These carriers are swept apart under the influence of the internal electric fields of the p-n junction and thus creating a photogenerated current proportional to the incident radiation.

In the dark, the I-V output characteristic of a solar cell has an exponential characteristic of a solar cell has an exponential characteristic similar to that of a diode. When the cell is short circuited, this current flows in the external circuit, and when open circuited, this current is shunted internally by the intrinsic p-n junction diode. The characteristics of this diode therefore set the open circuit voltage characteristics of the cell Geoff (2000).

In all solar power systems, efficient simulations including PV panel are required before any experimental verification is carried out. To this end, several authors (Akihiro, 2005; Ashish and Rakesh, 2011; Abdulkadir et al., 2012; Gonzalez-Longatt, 2010; 2005; Salmi et al 2012; Mohammed 2009; Gow and Manning; 1999) have carried out simulations aimed at harvesting greater amount of solar energy or simply to improve on the efficiency of the solar cell.

Also, Bourdoucen and Gastli (2007) developed an analytical model for PV panels and arrays based on extracted physical parameters of solar cells with an advantage of simplifying mathematical modelling of different cells and panel configurations without losing necessary accuracy of system operation.

This paper therefore focuses on the simulation of an improved model of the solar cell based on the Shockley

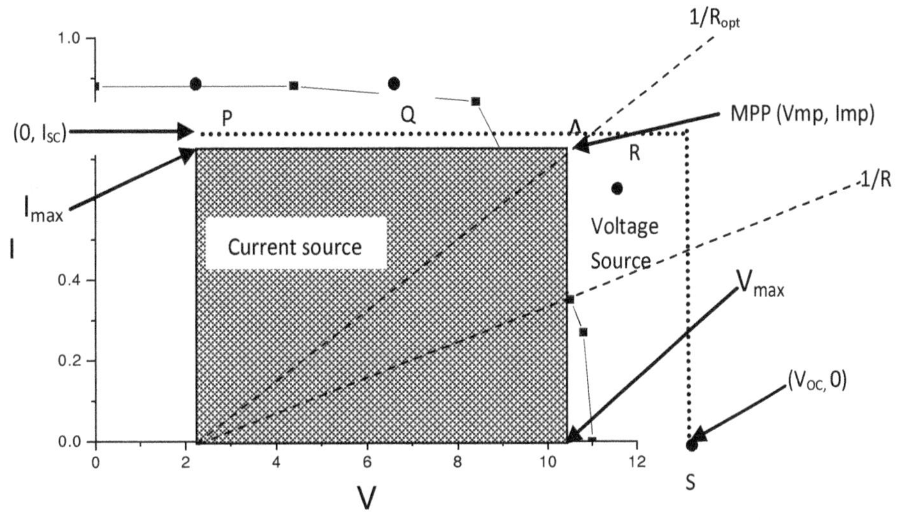

Figure 1. I - V Curve from the constructed and mounted module.

diode parameters using the Newton-Raphson algorithm in conjunction with Matlab software package.

METHODOLOGY

Mathematical formulation of the photovoltaic cell

The solar cell which is the basic unit of a photovoltaic module consists of a p-n junction fabricated in a thin wafer or layer of semiconductor. Though the electrical characteristics differ very little from a diode, it is represented by the Shockley Equation (1). In an ideal solar cell, $R_s = R_{sh} = 0$, a very relatively common assumption (Ramos et al., 2010). The Shockley equation is given as:

$$I_D = I_0\left(e^{\frac{V_C q}{nkT_{CK}}} - 1\right) \quad (1)$$

Where:
"I_D" is the dark current (A)
"I_0" is the saturation current of the diode (A)
"V_C" is a cell voltage (V)
"q" is the electronic charge ($1.6.10^{-19}$(Coul.)
"n" is the diode ideality constant
"k" is the Boltmann constant ($1.38.10^{-23}$J/K
"T_{CK}" is the cell temperature.
The net current "I" which is the difference between the photogenerated current (I_L) and the normal diode current

$$I = I_L - I_D \quad (2)$$

A simplified model [10] is given as:

$$I = I_L - I_0\left(e^{\frac{(V_C + IR_S)q}{nkT_{CK}}} - 1\right) \quad (3)$$

Where R_S is the series resistance in Ohms.

Equation (3) does not represent the I - V characteristics of a practical PV module, thus, inclusion of additional parameter R_{SH} (Shunt). Thus, Equation (3) becomes:

$$I = I_L - I_0\left(e^{\frac{(V_C + IR_S)q}{nkT_{CK}}} - 1\right) - \left(\frac{V_C + IR_S}{R_{SH}}\right) \quad (4)$$

The Series resistance R_S and the shunt R_{SH} are included in the real operation of the solar module to cater for the losses that exist in the cells and between the cells in the module. Thus, the simplest PV model of a solar cell is as shown in Figure 1.

$$I_L = I_L(T_1) + K_0(T - T_1) \quad (5)$$

$$I_L(T_1) = I_{SC}(T_{1,nom})\frac{G}{G_{nom}} \quad (6)$$

$$K_0 = \frac{I_{SC}(T_2) - I_{SC}(T_1)}{T_2 - T_1} \quad (7)$$

$$I_0 = I_0(T_1) \times \left[\frac{T}{T_1}\right]^{\frac{3}{n}} \times e^{\frac{qV_q(T_1)}{nk\left(\frac{1}{T} - \frac{1}{T_1}\right)}} \quad (8)$$

$$I_0(T_1) = \frac{I_{SC}(T_1)}{\left(e^{\frac{qV_{OC}(T_1)}{nkT_1}} - 1\right)} \quad (9)$$

$$R_S = -\frac{dV}{dI_{Voc}} - \frac{1}{X_V} \quad (10)$$

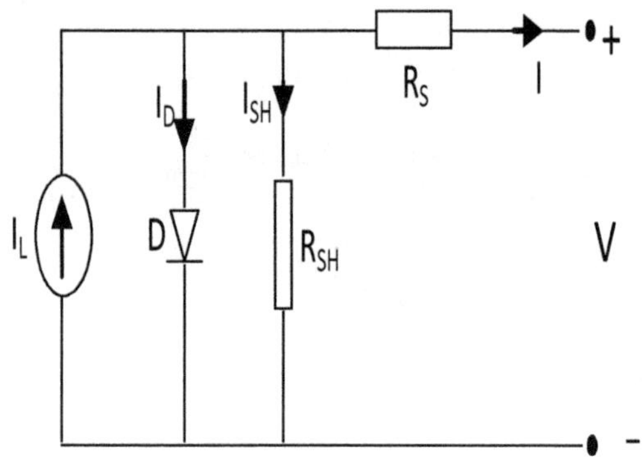

Figure 2. Single diode model of a solar cell.

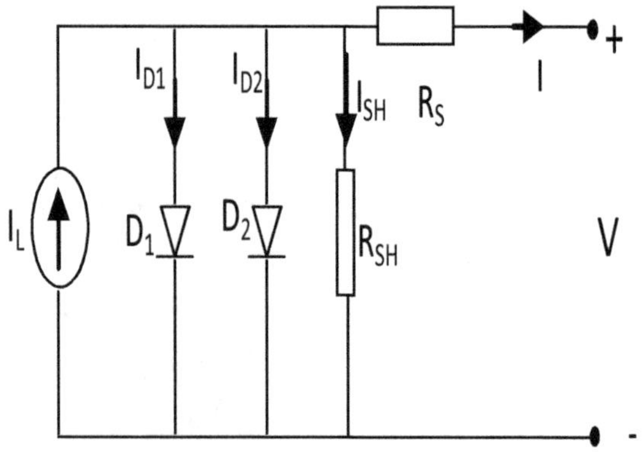

Figure 3. Double diode model of a solar cell.

$$X_V = I_o(T_1)\frac{q}{nkT_1}e^{\frac{qV_{OC}(T_1)}{nkT_1}} - \frac{1}{X_V} \qquad (11)$$

In all these equations that is, (5) to (11), the shunt resistance has been neglected while all the constants can be determined by examining the manufacturers' ratings of the PV cell, and then the measured I - V curves of the module.

Important features of a Current – Voltage (I - V) Curve for a solar cell

The plot of current against voltage for a typical solar cell is shown in Figure 1.

1). Short-circuit current, I_{SC}. This is the greatest value of the current generated by a cell and is produced when V = 0, the short circuit conditions.

2). Open circuit voltage, V_{OC}. This is the voltage drop across the diode (p-n junction) when photogenerated current, I_{ph} = 0, that is, at night when there is no illumination at all. This is represented mathematically as:

$$V_{OC} = \frac{nkT}{q}\ln\left[\frac{I_L}{I_O}\right] = V_t \ln\left[\frac{I_L}{I_O}\right] \qquad (12)$$

Where $V_t = \left[\frac{mkT_C}{e}\right]$ is known as thermal voltage and T is the absolute cell temperature.

3). Maximum power point, A. This is the point (V_{max}, I_{max}) at which the power dissipated in theresistive load is maximum: $P_{max} = V_{max} I_{max}$.

4). Maximum efficiency, η. The ratio between the maximum power and the incident light power.

$$\eta = \frac{P_{max}}{P_{in}} = \frac{I_{max} V_{max}}{AG_a} \qquad (13)$$

Where G_a is the ambient irradiation and A, is the cell area.

5). Fill factor, FF. The ratio of the maximum power that can be delivered to the load and the product of I_{SC} and V_{OC}.

$$FF = \frac{P_{max}}{V_{OC} I_{SC}} = \frac{I_{max} V_{max}}{V_{OC} I_{SC}} \qquad (14)$$

The fill factor is a measure of the real I - V characteristic with a value higher than 0.7 connoting a good cell. This value diminishes as the cell temperature is increased.

It has also been established, Gonzalez-Longatt (2010) that for a resistive load, the load characteristic is a straight line with slope 1/V = 1/R. The power delivered to the load depends on the value of the resistance only. If this load R is small, the cell operates in the region P - Q of the curve where the cell behaves as a constant current source, almost equal to the short circuit current. On the other hand, if the load R is large, the cell operates on the region R – S of the curve where the cell behaves more as constant voltage source, almost equal to the open-circuit voltage, Hansen et al. (2000).

Modeling of the solar cell

Various models have been developed and utilized in order to study solar cell behavior. The simplest and the most widely used equivalent circuit of a solar cell is a current source in parallel with a diode as shown in Figure 2. The double diode equivalent circuit Nishioka et al. (2007) shown in Figure 3 has also been reported.

In this work the single diode model is adapted because of its simplicity to program using the Matlab package.

This model is based on the previous works (Geoff, 2000; Gonzalez – Longatt, 2005; Ramos et al., 2010; Akihiro, 2005; Savita et al., 2010).

All the authors used Matlab Software which requires 3 values to calculate the net operating current of the module. These are:

1). Va = Module operating voltage
2). G = [Suns] = Irradiance with 1 suns = 1000 W/m^2
3). TaC = Module temperature, in °C

The body of the program is divided in 4 main parts:

1). Definition of constants (Boltzmann's constant, k, electronic charge, q, diode –ideality quality factor, A, irradiance, G, etc
2). Definition of variables
3). Calculation of I_L, I_o and R_s (Equations 5 – 10 are used)
4). Calculation of I

Newton – Raphson iterative method is exploited in this work to solve the double –exponential Equation (4) because of its quick convergence. In this work, ten iterations were performed to reduce error. The algorithm of the adopted MPPT logic is shown in figure 4 while The Matlab script file used to generate the simulation results in this work are listed below:

```
function Itt = calcur (Vd,Suns,Temp)
%k = 2 + Temp;
% Given voltage, illumination and temperature
% Itt,Vd = array current, voltage
 % G = num of Suns (1 Sun = 1000 W/m^2)
% T = Temp in Deg C
k = 1.38e-23; % Boltzman's const
q = 1.60e-19; % charge on an electron
Voo = Va;
tr = 3.016;
cap = 10;
Iph = Voo / tr * exp(- 1/(tr * cap));
A = 1.2; % "diode quality" factor, =2 for crystaline, <2 for
amorphous
Vg = 1.12; % band gap voltage, 1.12eV for xtal Si, ˜1.75 for
amorphous Si.
Ns = 36; % number of series connected cells (diodes)
 T1 = 273 + 25;
Voc_T1 = 21.06 /Ns; % open cct voltage per cell at temperature T1
Isc_T1 = 3.80; % short cct current per cell at temp T1
T2 = 273 + 75;
disp(T2);
Voc_T2 = 17.05 /Ns; % open cct voltage per cell at temperature T2
Isc_T2 = 3.92; % short cct current per cell at temp T2
TaK = 273 + Temp; % array working temp
TrK = 273 + 25; % reference temp
% constant "a" can be determined from Isc vs T
Iph_T1 = Isc_T1 * Suns;
aa =Isc_T2 - Isc_T1;
 ab = Isc_T1 * 1/(T2 - T1);
a = aa/ab;
Vt_T1 = k * T1 / q; % = A * kT/q
 Ir_T1 = Isc_T1 / (exp(Voc_T1/(A*Vt_T1))-1);
Ir_T2 = Isc_T2 / (exp(Voc_T2/(A*Vt_T1))-1);
b = Vg * q/(A*k);
 rvl = exp(-b *(1/TaK - 1/T1);
 Ir = Ir_T1 * (TaK/T1).^(3/A) * rvl);
X2v = Ir_T1/(A*Vt_T1) * exp(Voc_T1/(A*Vt_T1));
dVdI_Voc = - 1.15/Ns / 2; % dV/dI at Voc per cell --
% from manufacturers graph
Rs = - dVdI_Voc - 1/X2v; % series resistance per cell
% Itt = 0:0.01:Iph;
Vt_Ta = A * 1.38e-23 * TaK / 1.60e-19; % = A * kT/q
% solve for Itt: f(Itt) = Iph - Itt - Ir.*( exp((Vc+Itt.*Rs)./Vt_Ta) -1) = 0;
% Newton's method: Itt2 = Itt1 - f(Itt1)/f'(Itt1)
Vc = Vd/Ns;
 Itt = zeros(size(Vc));
% Ittv = Itt;
for j=1:10;
Itt = Itt - (Iph - Itt - Ir.*( exp((Vc+Itt.*Rs)/Vt_Ta) -1))/ (-1 -
(Ir.*(exp((Vc+Itt.*Rs)./Vt_Ta) -1)).*Rs./Vt_Ta);
end.
```

RESULTS

1). Effect of varying cell temperature.
The effect of varying cell temperature at solar irradiance G = 1Suns on the open-circuit voltage V_{oc} and the short circuit current I_{sc} is shown in Figures 5 and 6.

2). Effect of varying series resistance, R_s
Figures 7 and 8 showed the effect of the variation of R_s,

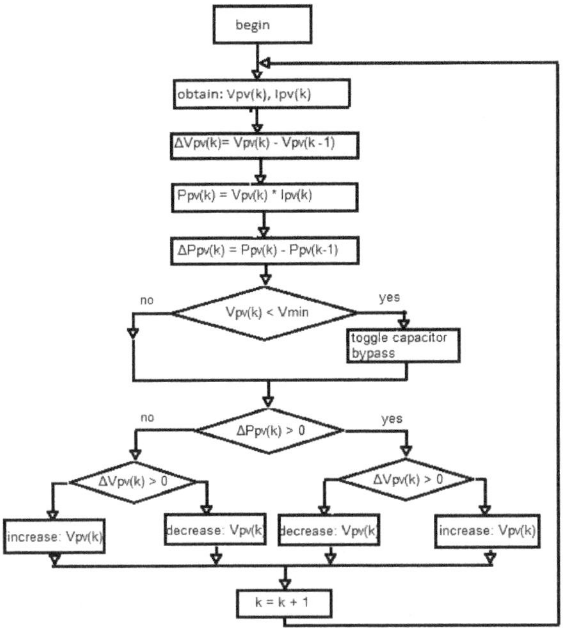

Figure 4. Algorithm of the adopted MPPT logic.

the series resistance on the slope angle of the I - V curves.

3). Effect of ambient irradiation
Figure 9 showed the effect of irradiance variation 0.2 – 1.5 Suns at constant temperature 25°C (Pandiarajan et al., 2011; Atlas and Sharaf, 2007; Kumari et al., 2012; Yushaizad, 2004).

4). Effect of diode quality factor
The response of I_{sc} and V_{oc} with the diode quality factor is shown in Figure 10.

DISCUSSION

With G = 1 Suns, the open circuit voltage V_{oc} dropped slightly with increasing cell temperature while the short circuit current, I_{SC} increased, the cell being thus less efficient. For instance, when the cell temperature is 24°C at an irradiance of G = 1 suns, the open circuit voltage is about 15.05 V. However, if the temperature is reduced to 0°C at the same irradiance, the open circuit voltage increased to 20.1 V. This behaviour is presented in Figures 5 and 6.

Also, the variation of R_s, the series resistance of the PV affected the slope angle of theI - V curves with the attendant deviation from the maximum power point as shown in Figures 7 and 8. This series resistance which is always very low, and some cases be neglected (Tsai et al., 2008; Savita et al., 2010), was included in this work so as to enable the model to be used for any given PV cell.

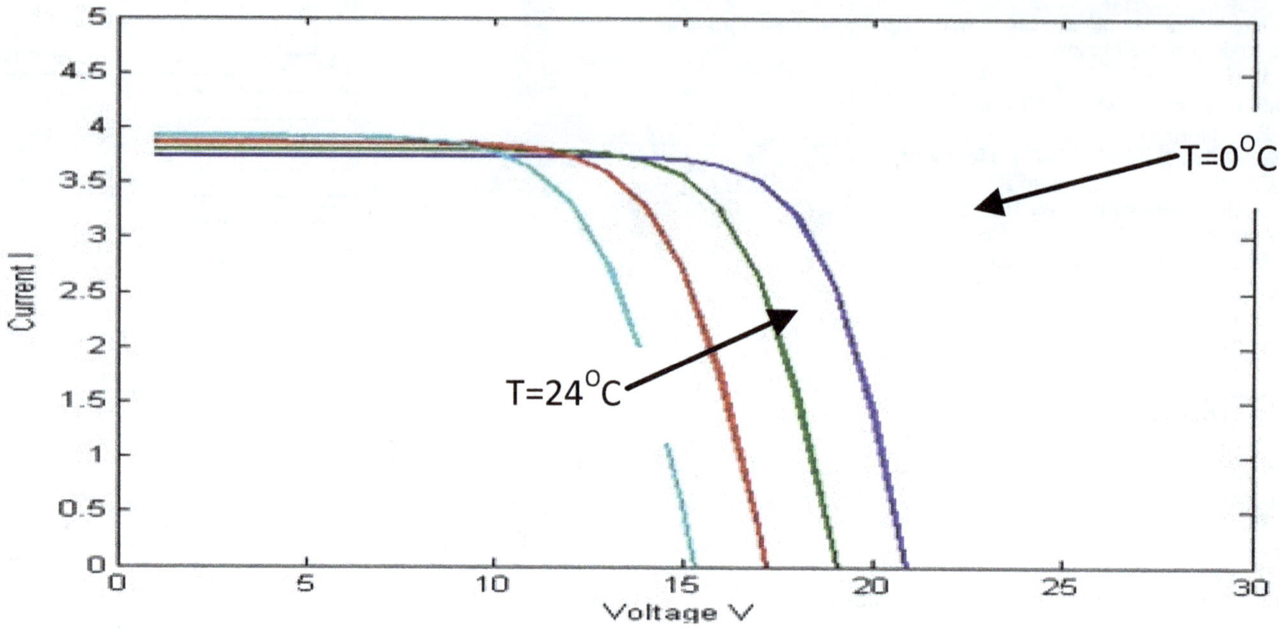

Figure 5. Module current against voltage at G= 1 Suns and at various temperature levels for zero bias.

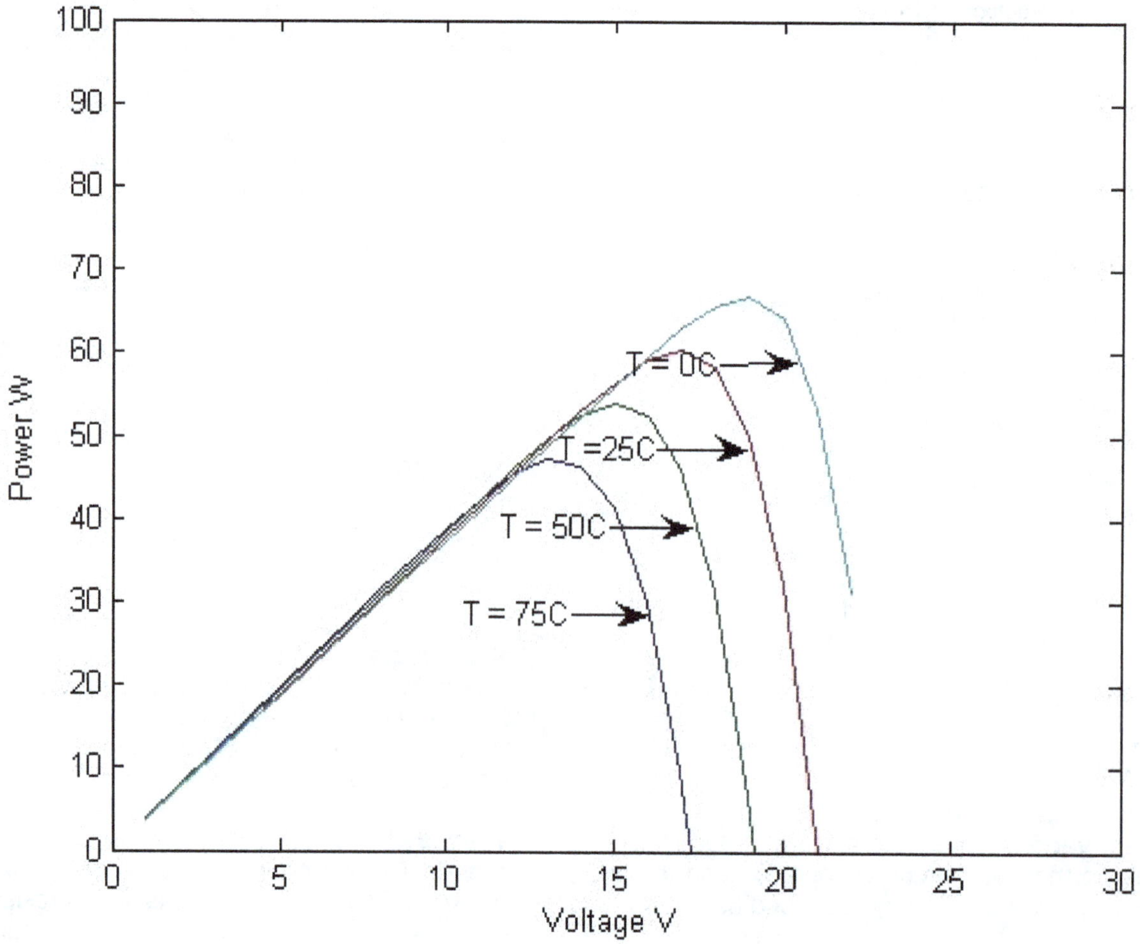

Figure 6. Power against module voltage at G = 1 Suns and at various temperature levels.

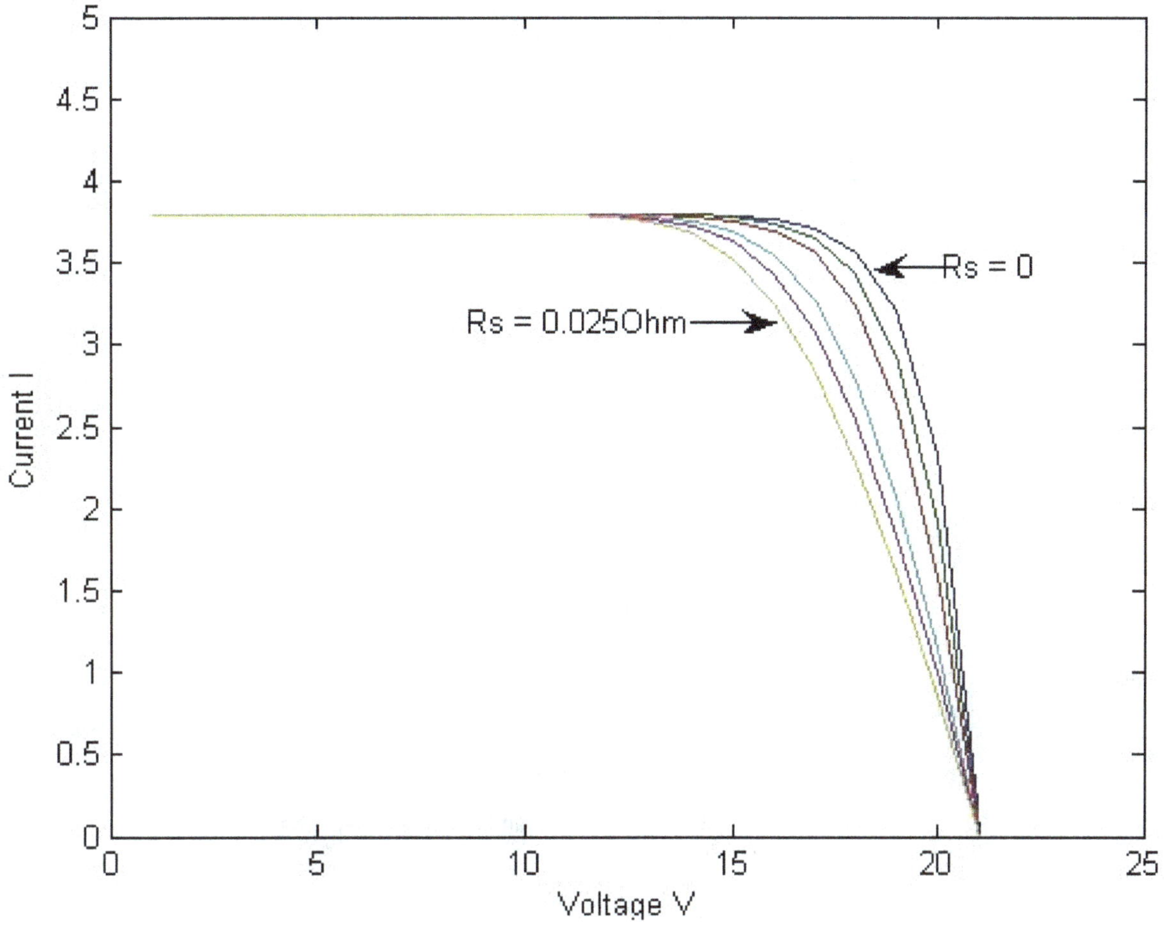

Figure 7. Graph of current against module voltage at G= 1 Suns and at various series resistance (Rs).

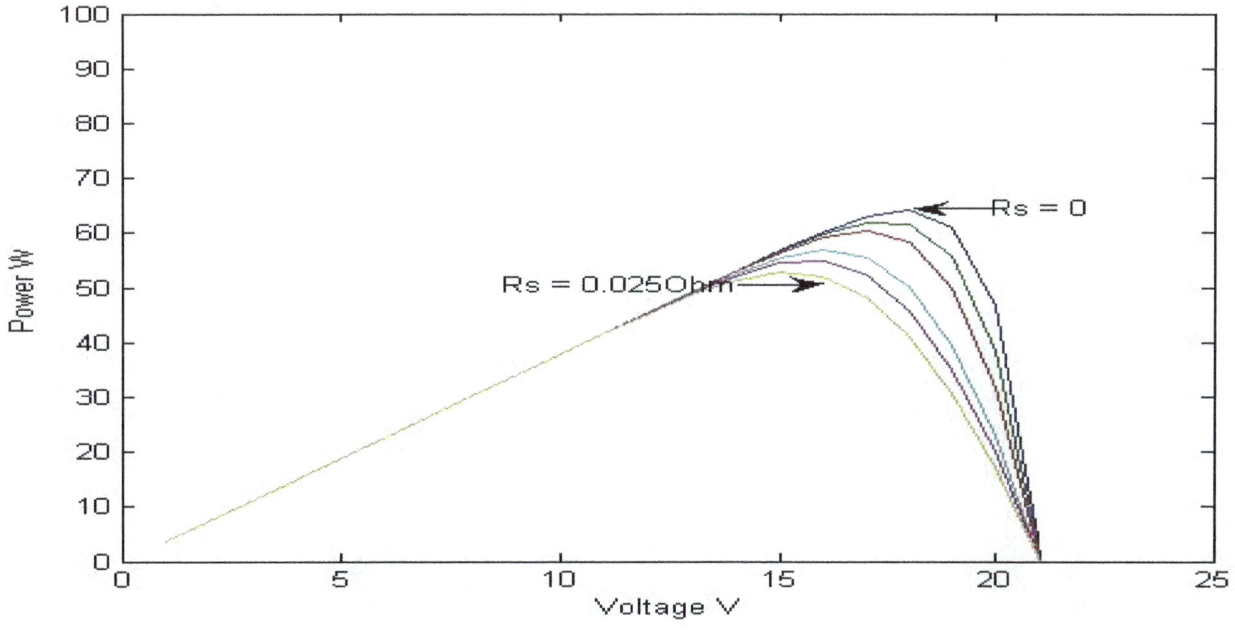

Figure 8. Power against module voltage at G = 1 Suns and at various series resistance, (Rs).

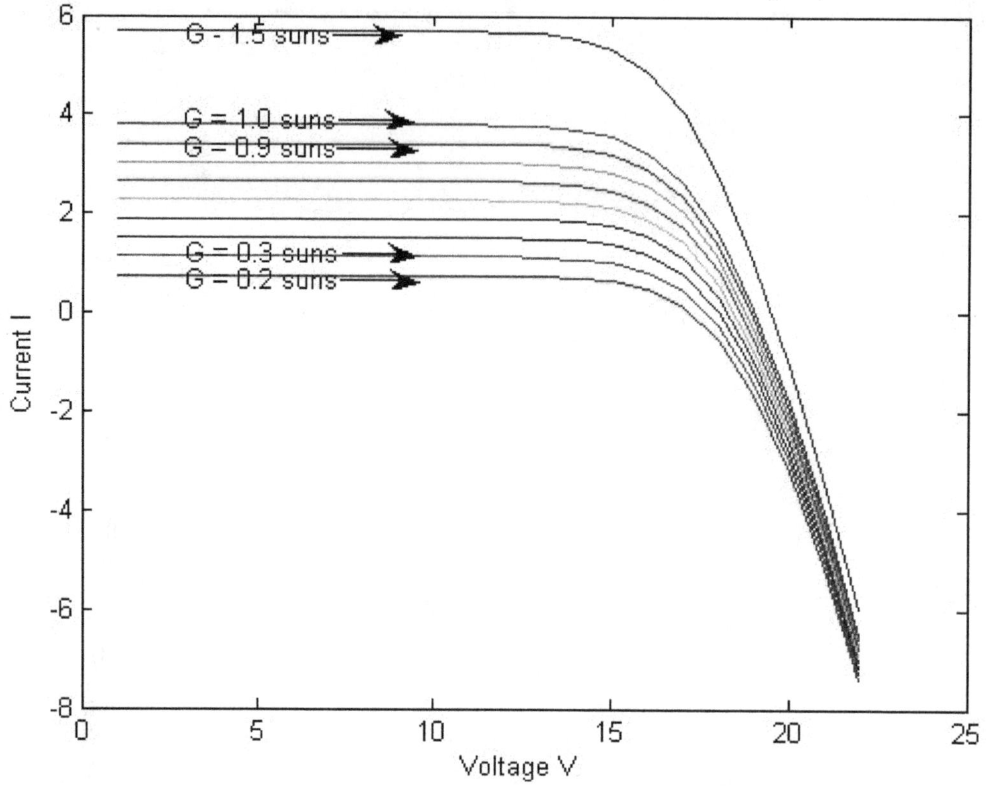

Figure 9. Module current against voltage at T = 25°C and at various irradiation levels.

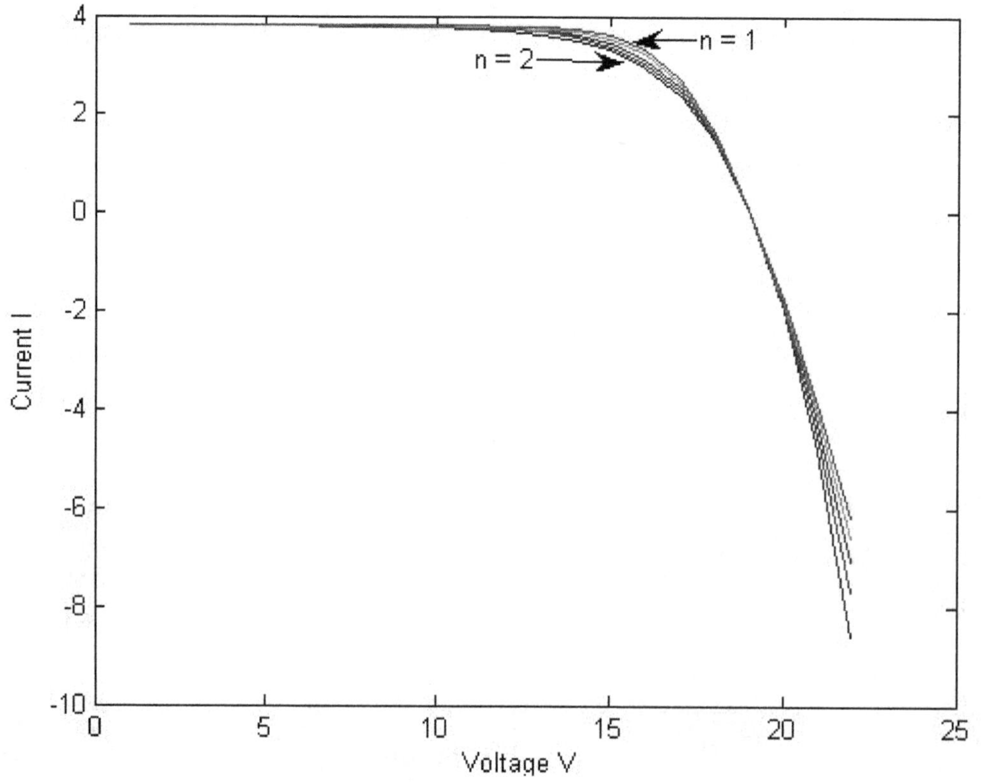

Figure 10. Module current against voltage at G = 1 Suns, 25°C with various diode quality factor values.

Figure 9 showed the effect of irradiance variation 0.2 – 1.5 Suns at constant temperature 25°C (Pandiarajan et al., 2011; Atlas and Sharaf, 2007; Yushaizad, 2004). The open circuit voltage, V_{oc} did not vary much as it is logarithmically dependent on the solar irradiance while the short circuit current is a linear function of the ambient irradiation since photocurrent depends on the irradiation. The higher the irradiation, the greater the current. It was also noted that reducing the irradiation will result in decreasing the output voltage of the panel.

As the diode quality factor is increased, the open circuit voltage of the cell increased. The ideal value of quality factor for solar cell is unity but its practical value for Silicon PV cell ranges between 1 and 2. Green (1992) states that diode quality factor takes a value between 1 and 2; being near 1 at high currents, rising towards 2 at low currents. This is demonstrated in Figure 10.

Conclusion

A Matlab simulation of PV solar cell using the Shockley diode circuit equations of a solar cell using the single diode equivalent circuit and taking into account the effects of physical and environmental parameters like cell temperature, series resistance, R_s , ambient irradiation and diode quality factor was carried out. The effect of shunt is however not considered. The results obtained followed the expected trends as reported by previous authors. This is the first step in the use of our results to model the I - V and P - V characteristics of a solar module deployed to take data in our location.

ACKNOWLEDGEMENT

One of the authors wishes to thank Mr Asonibare Oluwatosin for the MATLAB codes and implementation.

REFERENCES

Abdulkadir M, Samosir AS, Yatim AHM (2012). Modeling and simulation based approach of Photovoltaic system in Simulink Model APRN. J. Eng. Appl. Sci. 7(5):616-623.

Akihiro OI (2005). Doctoral Thesis. Design and simulation of photovoltaic water pumping system.

Ashish KS, Rakesh N (2011). PSIM and MATLAB based Simulation of PV Array for Enhance the Performance by using MPPT Algoirthm Int. J. Elect. Eng. 4(5):511-520.

Atlas IH, Sharaf AM (2007). A Photovoltaic Array Simulation Model for Matlab- Simulink GUI Environment. IEEE, Clean Electrical Power, International Conference on Clean Electrical Power (ICEP '07), June 14-16, Ischia, Italy.

Bourdoucen H, Gastli A (2007). Analytical Modeling and Simulation of Photovoltaic Panels and Arrays. J. Eng. Res. 4(1):75-81.

Gonzalez-Longatt FM (2005). "Model of Photovoltaic Module in MatlabTM", 2do congreso iberoamericano de estudiantes de ingeneria Electrica, electronica y Computacion (II CIBELEC).

Gonzalez-Longatt FM (2010). Model of Photovoltaic Module in MatlabTM", 2do congreso iberoamericano de estudiantes de ingeneria Electrica, electronica y Computacion (II CIBELEC).Spain. pp. 1-5.

Geoff W (2000). "Evaluating MPPT Converter Topologies using a MATLAB PV model" Australian Universities Power Engineering Conference, AUPEC '00, Brisbane.

Gow JA, Manning CD (1999). "Development of a photovoltaic array model for use in power electronic simulation studies," IEE Proc. Elect. Power Appl. 146(2):691-696.

Hansen A, Lars H, Bindner H (2000). "Models for a Stand-Alone PV System" Risø National Laboratory, Roskilde, December, ISBN 87 – 550-2776 – 8. [Online]. Available: http://www.risoe.dk/rispubl/VEA/ris-r-1219.htm

Kumari JS, Babu Ch. Sai H (2012). Mathematical Modeling and simulation of Photovoltaic Cell using Matlab_Simulink Environment. Int. J. Elect. Comput. Eng. 2(1):26-34.

Mohammed A (2009). Improved Circuit Model of Photovoltaic Array. Int. J. Elect. Power Energy Syst. Eng. 2:3.

Nishioka K, Nobuhiro S, Yakiharu U, Takashi F (2007). Analysis of multicrystalline silicon solar cells by modified 3-diode equivalent circuit model taking leakage current through periphery into consideration. Solar Energy Mater. Solar Cells. 91(13):1222-1227.

Pandiarajan N, Ramapragha R, Ranganath M (2011). Application of circuit model for Photovoltaic energy conversion system. Int. J. Adv. Eng. Technol. IJAPET. 2(4):118-127.

Ramos HJA, Campayo MJJ, Zamora BI, Larranaga LJ, Zulueta GE, Puelles PE (2010). Modelling of Photovoltaic Module In proceedings of the International Conference on Renewable Energies and Power Quality (ICREPQ'10) Granada Spain.

Salmi T, Bouzguenda M, Gastli A, Masmoudi A (2012). MATLAB/Simulink Based Modelling of Solar Photovoltaic Cell. Int. J. Renew. Energy Res. 2(2):213-217.

Savita N, Nema RK, Gayatri A (2010). "MATLAB/Simulink based study of photovolataic cells/modules/array and their experimental verification". Int. J. Energy Environ. 1(3):487-500.

Tsai HL, Tu CS, Su YJ (2008). Development of Generalized Photovoltaic Model Using MATLAB?SIMULINK Proceedings of the World Congress on Engineering and Computer Science 2008 WCECS, San Fransisco, USA. pp. 22-24.

Yushaizad Y (2004). Modeling and Simulation of maximum power point tracker for Photovoltaic System. National power and energy conference, proceedings, Kuala Lumpur, Malasia.

On three-dimensional quasi-Sasakian manifolds admitting semi-symmetric metric connection

Sunil Yadav and D. L. Suthar

Department of Applied Science and Humanities, Faculty of Mathematics, Alwar Institute of Engineering and Technology, Alwar North Ext. Alwar-301030, Rajasthan. India.

The object of the present paper is to study three-dimensional quasi-Sasakian manifold equipped with semi-symmetric metric connection. The geometrical properties of conformal curvature tensor and the conservative quasi-conformal curvature tensor are discussed with such connection. Among other we have deal the conservative properties of quasi-conformal curvature with respect to semi-symmetric metric connection.

Key words: Quasi-Sasakian manifold, conformal curvature tensor, quasi-conformal curvature tensor.

INTRODUCTION

Friedmann and Schouten (1924) introduced the notion of semi-symmetric linear connection on a differential manifold. Hayden (1932) introduced the idea of metric connection with torsion on Riemannian manifold. A systematic study of semi-symmetric metric connection on a Riemannian manifold has been given by Yano (1970) and later studied by Amur and Maralabhavi (1970), Bagewadi (1982), Bagewadi et al. (2007), and Prakasha et al. (2008). In the study Bagewadi and Venkatesh (2007), Friedmann and Schouten (1924) and Sharafuddin and Hussian, (1976), the others have obtained results on the conservativeness of projective, pseudo projective, conformal. Concircular, quasi-conformal curvature tensor on K-contact, Kenmotsu and Trans-Sasakian manifolds. Olszak (1986) considered the three-dimensional cases of normal almost contact metric manifold and Yadav et al. (2011) obtained some results on K-Contact and trans-Sasakian manifolds. Bagewadi et al. (2007), have been studied conservative projective curvature tensor on a trans-Sasakian manifold with respect to semi-symmetric metric connection. Prakasha et al. (2008) have studied the conservativeness of conformal and quasi–conformal curvature tensor on a trans-Sasakian manifold under the condition $\phi(grad\alpha) = (n-2)grad\beta$ that admitting the semi-symmetric connection.

In this study we stabilized some basic results and studied conformal curvature tensor the conservativeness of conformal quasi conformal curvature tensor under the condition $df(X) = g(gradf, X)$ on a three-dimensional qusi-Sasakian manifold admitting a semi –symmetric connection.

PRELIMINARIES

Let M be a $(2n+1)$–dimensional connected differentiable manifold endowed with an almost contact metric structure (ϕ, ξ, η, g) where ϕ is a tensor field of type $(1, 1)$, ξ is a vector field, η is an 1-form and g is the Riemannian metric on M such that (Amur and Maralabhavi, 1970; Blair, 2002).

$$\phi^2 X=-X+\eta(X)\xi \ \eta(\xi)=1 \tag{1}$$

$$g\ (X,Y)=g(X,Y)-(X)(Y)\ \ ,X,Y\in TM\ , \tag{2}$$

$$\phi\,\xi=0\,, \qquad \eta(\phi X)=0, \qquad \eta(X)=g(X,\ \xi)\,, \tag{3}$$

Let Φ be the fundamental $2-$form of the manifold M which is defined by $\Phi(X,Y)=g(X,\phi Y)\ \ ,X,Y\in TM$ then $\Phi(X,\xi)=0\,,\ \ X\in TM$ is said to be quasi –Sasakian if the almost contact structure $(\phi\ ,\ \xi\ ,\eta,\ g)$ is normal and the fundamental $2-$form Φ is closed $(d\Phi=0)$, which was first introduced by Blair (2002). The normal condition gives that the induced almost contact structure of $M\times R$ is integrable or equivalently, the torsion tensor field $N=[\phi\ ,\phi]+2\,\xi\otimes d\eta$ vanishes identically on M the rank of quasi Sasakian structure is always odd (Blair, 1967). It is equal to 1 if the structure is cosympletic and it is equal to $2n+1$ if the structure is Sasakian.

QUASI-SASAKIAN STRUCTURE OF DIMENSION THREE

An almost contact metric manifold of dimension three is called quasi-Sasakian manifold if and only if (De and Absos, 1997)

$$\nabla_X\xi=-\beta\phi X\ \ ,X\in TM\,, \tag{4}$$

for a function β defined on the manifold, ∇ being the operator of covariant differentiation with respect to Livi-civita connection of the manifold. Also we note that there is a function β on the manifold satisfying $\nabla_X\,\xi=-\beta\,\phi X$, then, $\xi\phi=0$ because from Equation (4) we find

$$\nabla_X(\nabla_Y\xi)=-(X\beta)\phi Y-\beta^2\{g(X,Y)\xi-\eta(Y)X\}-\beta\phi\nabla_X Y$$

which implies that

$$R(X,Y)\,\xi=-(X\beta)\phi Y+(Y\beta)\phi X+\beta^2\{\eta(Y)X-\eta(X)Y\}\,, \tag{5}$$

putting $X=\xi$,we obtain

$$R(\xi,Y,Z,\xi)=\beta^2\{g(Y,Z)-\eta(X)\eta(Y)\}+g(\phi Y,Z)\xi\beta\,, \tag{6}$$

Therefore, taking the skew properties, we can easily verify that $\xi\beta=0$. Clearly such a quasi-Sasakian manifold is cosympletic if and only if $\beta=0$ as the consequence of Equation (4), we have

$$(\nabla_X\phi)(Y)=\beta\big(g(X,Y)\xi-\eta(Y)X\big),\ \ X,Y\in TM\,, \tag{7}$$

$$(\nabla_X\eta)(Y)=g(\nabla_X\xi,Y)=-\beta\,g(\phi X,Y) \tag{8}$$

and

$$(\nabla_X\eta)\xi=-\beta\eta(\phi X)\,, \tag{9}$$

In three-dimensional Riemannian manifold the Weyl-conformal curvature tensor vanishes, that is

$$R(X,Y)Z=g(Y,Z)QX-g(X,Z)QY+S(Y,Z)X-S(X,Z)Y-\frac{r}{2}[g(Y,Z)X-g(X,Z)Y]\,, \tag{10}$$

where Q is the Ricci operator that is $g(QX.Y)=S(X,Y)$ and r is the scalar curvature of the manifold.

Let M^3 be a three-dimensional quasi-Sasakian manifold the Ricci tensor S of M^3 the is given by (Bagewadi and Venkatesh, 2007)

$$S(Y,Z)=\left(\frac{r}{2}-\beta^2\right)g(Y,Z)+\left(3\beta^2-\frac{r}{2}\right)\eta(Y)\eta(Z)-\eta(Y)d\beta(\phi Z)-\eta(Z)d\beta(\phi Y) \tag{11}$$

Now as the consequence of Equation (11), we get the Ricci operator Q .

$$QY=\left(\frac{r}{2}-\beta^2\right)Y+\left(3\beta^2-\frac{r}{2}\right)\eta(Y)\,\xi-\eta(Y)(\phi\ grad\,\beta)-d\beta\,(\phi Y)\,\xi\,, \tag{12}$$

where the gradient of the function f is related to the exterior derivative df by the formula $df(X)=g(gradf,X)$ from Equation (11), we find

$$S(Y,\xi)=2\beta^2\eta(Y)-\ d\beta(\phi Y)\,, \tag{13}$$

Moreover, as the consequence of Equations (10) and (13) we get

$$R(X,Y)\xi=\beta^2\,(\eta(Y)X-\eta(X)Y)+d\beta(\phi X)Y-d\beta(\phi Y)X\,,\ \text{for}\ \ X,Y\in TM\,, \tag{14}$$

Also from Equation (10), we get

$$\eta(R(X,Y)Z)=g(Y,Z)\eta(QX)-g(X,Z)\eta(QY)+S(Y,Z)\eta(X)-S(X,Z)\eta(Y)$$
$$-\frac{r}{2}[g(Y,Z)\eta(X)-g(X,Z)\eta(Y)]$$

for X,Y,Z orthogonal to ξ , we get from above

$$\eta\,(R(X,Y)Z)=g(Y,Z)\eta(QX)-g(X,Z)\,\eta(QY)\,, \tag{15}$$

Let $(M\ ,g)$ be a n-dimensional Riemannian manifold of class C^∞ with metric tensor g and let ∇ be the Levi -Civita connection on M. A linear connection $\tilde{\nabla}$ on (M,g) is said to be semi-symmetric if the torsion tensor T of the connection $\tilde{\nabla}$ satisfies.

$$T(X,Y)=\pi(Y)X-\pi(X)Y\,, \tag{16}$$

where π is 1-form on with ρ as the associated vector field that is $\pi(X)=g(X,\rho)$,for any differentiable vector field X on M.

A semi-symmetric metric connection $\tilde{\nabla}$ is called semi-symmetric metric connection if it further satisfies $\tilde{\nabla}_X g=0$. In an almost contact manifold semi-symmetric metric connection is defined by identifying a $1-$form π of Equation (16) with contact form η , that is

$$T(X,Y)=\eta(Y)X-\eta(X)Y\,, \tag{17}$$

with ξ is the associate vector field ,that is $g(X,\xi)=\eta(X)$. The relation between semi-symmetric connection $\tilde{\nabla}$ and Livi-Civita connection ∇ of (M,g) have been obtained by Yano (1970).

$$\tilde{\nabla}_X Y=\nabla_X Y+\eta(Y)X-g(X,Y)\xi\,, \tag{18}$$

further, relation between the curvature tensor R and $\tilde{R}$ of type $(1,3)$ of the connection ∇ and $\tilde{\nabla}$ respectively is given by

$$\tilde{R}(X,Y)Z=R(X,Y)Z-K(Y,Z)X+K(X,Z)Y-g(Y,Z)FX+g(X,Z)FY, \quad (19)$$

where K is a tensor of type $(0,\ 2)$ defined by

$$K(Y,Z) = g(FY,Z) = (\tilde{\nabla}_X \eta)(Z) - \frac{1}{2} g(Y,Z)' \quad (20)$$

From Equation (19) it follows that

$$\tilde{S}(Y,Z) = S(Y,Z) - (n-2)K(Y,Z) - A\,g(Y,Z), \quad (21)$$

where $\tilde{S}$ is the Ricci tensor of the connection $\tilde{\nabla}$, $A = Tr.K$

Differentiating Equation (21) conveniently with respect to X, we obtain

$$(\tilde{\nabla}_X \tilde{S})(Y,Z) = (\nabla_X S)(Y,Z) - (n-2)(\nabla_X K)(Y,Z) - \eta(Y)S(X,Z)$$
$$-\eta(Z)S(X,Y) + (n-2)\eta(Y)K(X,Z) + (n-2)\eta(Z)K(Y,X) +$$
$$g(X,Y)S(\xi,Z) + g(X,Z)S(Y,\xi) - (n-2)g(X,Z)K(Y,\xi) - (n-2)g(X,Y)K(Z,\xi), \quad (22)$$

Now let (e_i) be the orthonormal basis of the tangent space at each point of the manifold for $i = 1, 2, 3, \ldots, n$ putting $Y = Z = e_i$ in Equation (22) and taking summation over i, we get

$$\tilde{\nabla}_X \tilde{r} = \nabla_X r - (n-2)(\nabla_X A)\cdot \quad (23)$$

BASIC RESULTS

Theorem 1

On a three-dimensional quasi-Sasakian manifold (M^n, g), we have

$$\left[(\nabla_\xi S)(Y,Z) - (\nabla_Y S)(\xi,Z)\right] = \frac{1}{2}\xi r[\,g(\phi Y,\phi Z)] + \eta(Y)g(\nabla_\xi grad\beta,\ \phi Z)$$
$$-g(\nabla_\xi grad\beta\ ,\phi Y)\ \eta(Z) - 6\beta Y\ \eta(Z) + (3\beta^3 - \beta\frac{r}{2})\,g(\phi Y,Z) - g(\nabla_Y\ grad\beta\,,\phi Z)$$

where the gradient of a function f is related to the exterior derivative df by the formula $df\,(X) = g(gradf\ ,X)$

Proof

Differentiating conveniently Equation (11) along the vector field X, we get

$$(\nabla_X S)(Y,Z) = \left(\frac{1}{2}Xr - 2\beta X\beta\right)g(Y,Z) + (8\beta X\beta - \frac{1}{2}Xr)\eta(Y)\eta(Z) - \beta(3\beta^2 - \frac{r}{2}) \quad (24)$$
$$[\eta(Y)g(\phi X,Z) + \eta(Z)g(\phi X,Y)] + \eta\big(g(\phi X,Y)d\beta(\phi Z) + g(\phi X,Z)d\beta(\phi Y)$$
$$+\eta(Y)g(\nabla_X\ grad\ \beta,\phi Z) - \eta(Z)\ g(\nabla_X\ grad\ \beta,\ \phi Y)$$

Putting $X = \xi$ Equation (24) and using Equations (1) and (3), we get

$$(\nabla_\xi S)(Y,Z) = \frac{1}{2}\xi r[g(Y,Z) - \eta(Y)\eta(Z)] + \eta(Y)g(\nabla_\xi\ grad\beta,\phi Z) - \eta(Z)\ g(\nabla_\xi\ grad\beta,\ \phi Y), \quad (25)$$

Again from Equation (24), we get

$$(\nabla_X S)(\xi,Z) = \left(\frac{1}{2}Y r - 2\beta Y\beta\right)\eta(Z) + (8\beta Y\beta - \frac{1}{2}Y r)\eta(Z) - \beta(3\beta^2 - \frac{r}{2})g(\phi Y,Z), \quad (26)$$
$$+g(\nabla_Y\ grad\beta,\phi Z) - g(\nabla_Y\ grad\beta,\ \phi\xi)\eta(Z)$$

From Equations (24) and (25), we get the required result.

Theorem 2

A three-dimensional quasi-Sasakian manifold with semi-symmetric metric connection under the condition that $df(X) = g(gradf, X)$, we have

$$K(Y,Z) = -\beta\ g(\phi\ Y,Z) - \eta(Y)\eta(Z) - \frac{1}{2}g(Y,Z)' \quad \text{(i)}$$

$$K(Y,\xi) = -\frac{3}{2}\eta(Y)' \quad \text{(ii)}$$

$$K(\nabla_Y \xi, Z) = -\beta^2[g(Y,Z) - \eta(Y)\eta(Z)] + \frac{\beta}{2}g(\phi Y,Z)' \quad \text{(iii)}$$

$$K(Y,\nabla_Z \xi) = \beta^2[g(Y,Z) - \eta(Y)\ \eta(Z)] + \frac{\beta}{2}g(Y,\phi Z)' \quad \text{(iv)}$$

Proof

From Equation (20), we have

$$K(Y,Z) = g(FY,Z) = (\nabla_X \eta)(Z) - \eta(Y)\eta(Z) - \frac{1}{2}g(Y,Z)' \quad (27)$$

Using Equations (4) and (8) in (27), we get the first result

$$K(Y,Z) = -\beta\ g(\phi\ Y,Z) - \eta(Y)\eta(Z) - \frac{1}{2}g(Y,Z) \quad (28)$$

Taking $Z = \xi$ in the first result and using Equation (3) we get the second results that is

$$K(Y,\xi) = -\frac{3}{2}\eta(Y), \quad (29)$$

Next by contracting $Y = \nabla_Y \xi$ in result (i) and using Equations (1), (3) and (4), we get the third result.
Again put $Z = \nabla_Z \xi$ in result (i), we get the (iv) result.

Theorem 3

A three-dimensional quasi-Sasakian manifold with semi-symmetric metric connection under the condition that $df\,(X) = g(gradf\ ,X)$ we have following results.

$$\left[(\nabla_\xi K)(Y,Z) - (\nabla_Y K)(\xi,Z)\right] = -2\beta^2 g(\phi Y,Z) - \frac{\beta}{2}g(Y,\phi Z) -$$
$$\beta(1-\beta)\,[g(Y,Z) + \eta(Y)\ \eta(Z)] + \frac{3}{2}\,Y\eta(Z)$$

Proof

From Equation (20), we have

$$(L_\xi K)(Y,Z) = 2\beta K(Y,Z) + \beta\,\eta(Y)\eta(Z) \tag{30}$$

and

$$(\nabla_\xi K)(Y,Z) = \xi K(Y,Z) - K(\nabla_\xi Y, Z) - K(Y, \nabla_\xi Z), \tag{31}$$

Using the result of theorem 2,(i-iii-iv) in Equation (26), we get

$$\begin{aligned}(\nabla_\xi K)(Y,Z) &= (L_\xi K)(Y,Z) + \beta^2 g(Y,Z) + \eta(Y)\eta(Z) - \frac{\beta}{2} g(\phi Y, Z), \\ &- \beta^2 [g(Y,Z) - \eta(Y)\eta(Z)] - \frac{\beta}{2} g(Y,\phi Z)\end{aligned} \tag{32}$$

Using Equation (30) we get

$$(\nabla_\xi K)(Y,Z) = \left[\frac{-4\beta^2-\beta}{2}\right] g(QY,Z) - \frac{\beta}{2} g(Y,\phi Z) - \beta g(Y,Z) + \left[1+\beta^2-\beta\right]\eta(Y)\eta(Z), \tag{33}$$

and also

$$(\nabla_Y K)(\xi,Z) = YK(\xi,Z) - K(\nabla_Y \xi, Z) - K(\xi, \nabla_Y \xi)$$

Using Equation (4) and the result of theorem (2b) (i-ii), we get

$$(\nabla_Y K)(\xi,Z) = -\frac{3}{2} Y\eta(Z) + \beta^2 g(Y,Z) + \eta(Y)\eta(Z) - \frac{\beta}{2} g(\phi Y, Z), \tag{34}$$

Subtracting Equations (34) and (32) we get the required result.

THREE-DIMENSIONAL QUASI-SASAKIAN MANIFOLD ADMITTING SEMI-SYMMETRIC METRIC CONNECTION WITH $Div\tilde{W} = 0$

Theorem 4

A three-dimensional quasi-Sasakian manifold admitting a semi-symmetric metric connection whose quasi-conformal curvature tensor with respect to this connection is conservative then the scalar curvature of the manifold is given by

$$\begin{aligned}r &= \left(\frac{n(a+b)}{-b}\right)\left[\frac{1}{2}\xi r + 3\beta^2 - \frac{r}{2}\beta\right] - 2n(n-2)\beta^2 - \frac{n}{b}\beta - (n\beta(n-2)(1-\beta) - 2n\beta^2 + \frac{nb}{2}] \\ &- \frac{nr(a+b)}{2b}\left[(\nabla_\xi r) - (n-2)(\nabla_\xi A)\right]\end{aligned}$$

provided the vector field X, Y, Z orthogonal to ξ.

Proof

A quasi-conformal curvature tensor on a three-dimensional quasi-Sasakian manifold with respect to semi-symmetric metric connection is given by

$$\begin{aligned}\tilde{W}(X,Y)Z &= aR(X,Y)Z + b[g(Y,Z)QX - g(X,Z)QY + S(Y,Z)X - S(X,Z)Y], \\ &+ \frac{r}{6}(a+b)[g(Y,Z)X - g(X,Z)Y]\end{aligned} \tag{35}$$

Differentiating Equation (35) conveniently and contracting, we get

$$\begin{aligned}Di\tilde{W}(X,Y)Z &= a[(\nabla_X S)(Y,Z) - (\nabla_Y S)(X,Z)] + b[(\tilde{\nabla}_X S)(Y,Z) - (\tilde{\nabla}_Y S)(X,Z)], \\ &+ \frac{r(a+b)}{6}[g(Y,Z)(\tilde{\nabla}_X \tilde{r}) - g(X,Z)(\tilde{\nabla}_X \tilde{r})]\end{aligned} \tag{36}$$

Using Equation (22) and (23) in (36), we get

$$\begin{aligned}(a+b)[(\nabla_X S)(Y,Z) - (\nabla_Y S)(X,Z)] &= -b(n-2)[(\nabla_X K)(Y,Z) - (\nabla_Y K)(X,Z)] \\ &- b[\eta(X)S(Y,Z) - \eta(Y)S(X.Z)] - b(n-2)[\eta(Y)K(X,Z) - \eta(X)K(Y,Z)] \\ &- b[g(X,Z)S(Y,\xi) - g(Y,Z)S(X,\xi)] - b(n-2)[g(Y,Z)K(Y,\xi) - g(X,Z)K(Y,\xi)] \\ &- \frac{r(a+b)}{6}[g(Y,Z)\nabla_X r - (n-2)g(Y,Z)(\nabla_X A) - g(X,Z)\nabla_Y r - (n-2)g(X,Z)(\nabla_Y A)]\end{aligned}, \tag{37}$$

Putting $X = \xi$ in Equation (37) and using Equations (1),(3) and (13), result of theorem 2(i-ii), we get

$$\begin{aligned}(a+b)\left|(\nabla_\xi S)(Y,Z) - (\nabla_Y S)(\xi,Z)\right| &= -b(n-2)[(\nabla_X K)(\xi,Z) - (\nabla_\xi K)(Y,Z)] - bS(Y,Z) \\ &+ \eta(Y)\{2\beta^2(Z) - d\beta(\phi Z)\} + b(n-2)\eta(Y)\eta(Z) + \frac{b(n-2)}{2}\eta(Y)\eta(Z) - \beta g(Y,Z) - \eta(Y)\eta(Z) \\ (5.4)\ &- \frac{1}{2}g(Y,Z) + 2b\beta^2\eta(Y)\eta(Z) + b\eta(Z)d\beta(\phi Y) + 2\beta^2 bg(Y,Z) \\ &+ \frac{3b}{2}(n-2)g(Y,Z)\eta(Y) - \frac{3b(n-2)}{2}\eta(Y)\eta(Z) - \frac{r(a+b)}{6}\begin{bmatrix} g(Y,Z)(\nabla\xi r) - (n-2)g(Y,Z) \\ (\nabla\xi A) - \eta(Z)(\nabla_Y r) - (n-2)\eta(Z)(\nabla_Y A)\end{bmatrix}\end{aligned}$$

Using the results of theorem 1 and theorem 3 in Equation (38), we get

$$\begin{aligned}&(a+b)\begin{bmatrix}\frac{1}{2}\xi r\,(g(\phi Y,\phi Z) + \eta(Y)g(\nabla_\xi grad\beta,\phi Z) - g(\nabla_\xi grad\beta,\phi Y)\eta(Z) \\ -6\beta Y\beta\eta(Z) + (3\beta^2 - \beta\frac{r}{2})g(\phi Y,Z) - g(\nabla_Y, grad\beta,\phi Z\end{bmatrix} \\ &= -b(n-2)\left[\begin{Bmatrix}2\beta^2 g(\phi Y,Z) + \frac{\beta}{2}g(Y,\phi Z) + \beta(1-\beta)(g(Y,Z) + \eta(Y)\eta(Z)) \\ +\frac{3}{2}Y\eta(Z)\end{Bmatrix}\right] - bS(Y,Z) + 2\beta^2\eta(Y)\eta(Z), \\ &+ \eta(Y)d\beta(\phi Z) + b(n-2)\eta(Y)\eta(Z) + \frac{b(n-2)}{2}\eta(Y)\eta(Z) - \beta g(\phi Y,Z) - \eta(Y)\eta(Z) - \frac{1}{2}g(Y,Z) - 2b\beta^2\eta(Y)\eta(Z) \\ &+ b(Z)d\beta(\phi Y) + 2b\beta^2 g(Y,Z) + \frac{3b}{2}(n-2)g(Y,Z)\eta(Y) - \frac{3b(n-2)}{2}\eta(Y)\eta(Z) \\ &- \frac{r(a+b)}{6}\begin{bmatrix} g(Y,Z)(\nabla_\xi r) - (n-2)g(Y,Z)(\nabla_\xi A) \\ -\eta(Z)(\nabla_Y r) - (n-2)\eta(Z)(\nabla_Y A)\end{bmatrix}\end{aligned} \tag{38}$$

If the vector field X, Y, Z is orthonarmal to ξ, we have

$$\begin{aligned}r &= \left(\frac{n(a+b)}{-b}\right)\left[\frac{1}{2}\xi r + 3\beta^2 - \frac{r}{2}\beta\right] - 2n(n-2)\beta^2 - \frac{1}{n}\beta - (n\beta(n-2)(1-\beta) - 2n\beta^2 + \frac{nb}{2}] \\ &- \frac{nr(a+b)}{2b}2b\left[(\nabla_\xi r) - (n-2)(\nabla_\xi A)\right]\end{aligned}$$

THREE-DIMENSIONAL QUASI-SASAKIAN MANIFOLD ADMITTING SEMI-SYMMETRIC METRIC CONNECTION WITH $Div\tilde{R} = 0$

Theorem 5

A three-dimensional quasi-Sasakian manifold admitting a semi-symmetric metric connection then the scalar curvature of the manifold is given by

$$r = -6\beta(3+\beta) - 3\beta(1-\beta) - \left[\frac{1}{6} + 2\beta^2 - \frac{1}{2}(\nabla_\xi r + \nabla_\xi A)\right]$$

Proof

In a three-dimensional quasi-Sasakian manifold, weyl-conformal curvature tensor vanishes then with respect to semi symmetric metric connection is given by

$$\tilde{R}(X,Y)Z = [g(Y,Z)QX - g(X,Z)QY + \tilde{S}(Y,Z)X - \tilde{S}(X,Z)Y] - \frac{r}{2}[g(Y,Z)X - g(X,Z)Y], \quad (39)$$

Differentiating Equation (39) conveniently and contracting, we get

$$Div\tilde{R} = [(\tilde{\nabla}_X \tilde{S})(Y,Z) - (\tilde{\nabla}_Y \tilde{S})(X,Z)] - \frac{1}{2}[g(Y,Z)(\tilde{\nabla}_X \tilde{r}) - g(X,Z)(\tilde{\nabla}_Y \tilde{r})] \quad (40)$$

Using Equations (22) and (23) in (40), we get

$$\begin{aligned} Div\tilde{R} &= [(\nabla_X S)(Y,Z) - (\nabla_Y S)(X,Z)] + [(\nabla_Y K)(X,Z) - (\nabla_X K)(Y,Z)] - (n-2)\eta(Y)\eta(Z)K(X,Z) \\ &+ g(X,Z)S(Y,\xi) - (n-2)\eta(Y)g(Y,Z) + g(Y,Z)S(X,\xi) + (n-2)g(Y,Z)K(X,\xi) \\ &- \frac{1}{2}[g(Y,Z)\{(\nabla_X r) - (n-2)(\nabla_X A)\} - g(X,Z)\{\nabla_Y r) - (n-2)(\nabla_Y A)\}], \end{aligned} \quad (41)$$

Using the result of Theorem (1) (i-ii), we obtain

$$\begin{aligned} Div\tilde{R} &= [(\nabla_X S)(Y,Z) - (\nabla_Y S)(X,Z)] + (n-2)[(\nabla_X K)(X,Z) - (\nabla_X K)(Y,Z)] \\ &- \eta(Y)S(X,Z) - \beta(n-2)(Y)g(\phi X,Z)\eta(Y) - (n-2)\eta(Y)\eta(X)\eta(Z) \\ &- \frac{(n-2}{2}g(X,Z)\eta(Y) + g(X,Z)S(Y,\xi) + \frac{2(n-2)}{3}\eta(Y)g(X,Z) + \beta(n-2)g(\phi Y,Z)\eta(X) \\ &+ (n-2)\eta(X)\eta(Y)\eta(Z) - \frac{2(n-2)}{2}\eta(X)g(Y,Z) - \frac{1}{2}g(Y,Z)(\nabla_X r) - \frac{(n-2)}{2}g(Y,Z)(\nabla_X A) \end{aligned} \quad (42)$$

Taking $X = \xi$ in Equation (42) and using Equations (1), (3) and (13), we get

$$\begin{aligned} Div\ R &= [(\nabla_\xi S)(Y,Z) - (\nabla_Y S)(\xi,Z)] + (n-2)[(\nabla_y K)(\xi,Z) - (\nabla_\xi K)(Y,Z)] \\ &+ \frac{(n-2)}{3}\eta(Y)\eta(Z) + S(Y,Z) + \eta(Y)d\beta(\phi Z) - (Z)d\beta(\phi Y) + (n-2)\beta g(\phi Y,Z) \\ &+ \left[\frac{(2-n)}{6} + 2\beta^2 - \frac{1}{2}\nabla_\xi r - \frac{(n-2)}{2}\nabla_\xi A\right]g(Y,Z) \end{aligned} \quad (43)$$

Using the result of the theorem 1 and 3, we get

$$\begin{aligned} Div\ \tilde{R} &= \frac{1}{2}\xi_r[g(Y,Z) - \eta(Y)\eta(Z)] + \eta(Y)g(\nabla_\xi grad\ \beta, \phi Z) \\ &- g(\nabla_\xi grad\ \beta, \phi Y)\eta(Z) - 6\beta Y\eta(Z) + (3\beta^2 - \beta\frac{r}{2})g(\phi Y,Z) \\ &- g(\nabla_Y grad\ \beta, \phi Z) + (n-2)\begin{bmatrix} 2\beta^2 g(Y,Z) + \frac{\beta}{2}g(Y,\phi Z) \\ + (1-\beta)\beta[g(Y,Z) + \eta(Y)\eta(Z)] \\ + \frac{2}{3}Y\eta(Z) \end{bmatrix} \\ &+ \frac{n-2}{3}\eta(Y)\eta(Z) + S(Y,Z) + \{\frac{n-2}{6} + 2\beta^2 - \frac{1}{2}\nabla_\xi r - \frac{(n-2)}{2}\nabla_\xi A\}g(Y,Z) \\ &+ \eta(Y)d\beta(\phi Z) - \eta(Z)d\beta(\phi Y) + (n-2)g(\phi Y,Z) \end{aligned} \quad (44)$$

Using the results Theorem 1, and simplifying, we get

$$r = -6\beta(3+\beta) - 3\beta(1-\beta) - \left[\frac{1}{6} + 2\beta^2 - \frac{1}{2}(\nabla_\xi r + \nabla_\xi A)\right]. \quad (45)$$

Conclusion

The layout of the work is to characterize the geometrical properties of conformal curvature tensor and the conservative quasi-conformal curvature tensor are on three-dimensional quasi-Sasakian manifold with respect to semi-symmetric metric connection.

ACKNOWLEDGEMENTS

The authors are thankful to Prof. U. C. De, Department of Pure Mathematics, University of Calcutta (W.B.) and the reviewers for their available suggestions in improving the paper.

REFERENCES

Amur K, Maralabhavi YB (1970). On quasi-conformally flat space, Tensor, N.S. 31:194.
Bagewadi CS (1982). On totally real submanifolds of a Kahlerian manifold admitting semi-symmetric metric F - connection, Indian J.
Bagewadi CS, Prakasha DG, Venkatesha (2007). Conservative projective curvature tensor on a trans-Sasakian manifold with respect to semi-symmetric metric connection. An. St. Univ. Ovidus Constanta 15(2):5-18.
Bagewadi CS, Venkatesh (2007). Some curvature tenor tensor on trans-sasakian manifolds. Turk. J. Math. pp. 301-311.
Blair DE (1967). The geometry of quasi-sasakian structure. J. Differ. Geom. 1:331-345.
Blair DE (2002). Riemannian geometry of contact and symmetric manifold, progress mathe. Vol. 203, Birkhauser, Bostn-Basel-Berlin
De UC, Absos AS (1997). K -contact and Sasakian manifold with conservative quasi-conformal curvature tensor. Bull. Cal. Math. Soc. 89:349-354.
Friedmann A, Schouten JA (1924). Uberdie geometric der holbsymmetricschen Ubertragurgen. Math. Zeitschr. 21:211-239.
Hayden HA (1932). Subspace of space with torsion. Proc. Lon. Math. Soc. 34:27-50.
Olszak Z (1986). Normal almost contact metric manifolds of dimension three. Ann. Polon. Math. 47:41-50.
Prakasha DG, Bagewadi CS, Venkatesha (2008). Conformaly and quasi-conformaly conservative curvature tensor on a trans-Sasakian manifolds with respect to semi-symmetric metric connection, Differential Geometry-Dynamical system. 10:263-274.
Pure. Appl. Math. 13(15):528-536.
Sharafuddin A, Hussian SI (1976). Semi-symmetric metric connection in almost contact manifold, Tensor, N.S. 30:133-139.
Yadav S, Dwivedi PK, Suthar DL (2011). Some results on K-contact and trans-Sasakian manifolds. Int. J. Math. Arch. 2(7):1202-1207.
Yano K (1970). On semi-symmetric metric connections, Revue Roumaine Ae Math. Pure Appl. 15:1579-1586.

13

Temperature dependent current-voltage characteristics of electrodeposited p-ZnO/n-Si heterojunction

Hatice Asil[1], Kübra Çinar[2] , Emre Gür[2], Cevdet Coşkun[3] and Sebahattin Tüzemen[2]

[1]Faculty of Education, Kilis 7 Aralik University, 79000 Kilis/Turkey
[2]Department of Physics, Faculty of Sciences, Atatürk University, 25240 Erzurum/Turkey.
[3]Department of Physics, Faculty of Arts and Sciences, Giresun University, 28100 Giresun/Turkey.

p-ZnO thin films were grown by electrochemical deposition (ECD) technique on n-Si substrate in order to form the p-ZnO/n-Si heterojunction. Hall measurement and hot probe techniques were used to determine the conductivity type of the ECD grown ZnO thin film. X-ray diffraction measurements revealed the peaks corresponding to the ZnO crystal directions of (002), (101) and (200) confirmed by the Joint Committee on Powder Diffraction Standards (JCPDS) files, indicating the polycrystalline nature of the films. The electrical characterization of p-ZnO/n-Si heterojunction was carried out in the temperature range of 80-300 K. The ideality factor and barrier height of the structure exhibited a variation between 2.49 to 5.36 and between 0.574 and 0.173 eV for this temperature ranges, respectively. The variation with temperature observed on the electrical parameters of the p-ZnO/n-Si heterojunction was explained by the introduction of a spatial distribution of barrier heights due to barrier height in homogeneities that prevail at the p-ZnO/n-Si heterojunction interface.

Key words: ZnO thin films, p-n heterojunction, electrodeposition.

INTRODUCTION

ZnO is a member of II-VI compound semiconductor family which has a wide and direct band-gap of about 3.4 eV. Its physical properties such as high transparency, large excitonic binding energy of 60 meV, wurtzite crystal structure (almost the same lattice parameters with GaN), resistance to high energy radiation and temperature makes it very attractive for both optical and electrical devices. Many ZnO based devices have been already reported such as lasers (LDs) (Buzás and Geretovszy, 2007) and light emitting diodes (LEDs) (Guo et al., 2008) working in ultraviolet region and thin film transistors (TFTs) (Fortunato et al., 2005; Kim et al., 2009) for solid-state electronics. Also, it can be easily grown with any growth technique since Zn is one of the most easily oxidized elements. Variety of growth methods have been used to grow ZnO thin films so far, such as chemical vapor deposition (Sen Chien et al., 2010), ionized cluster-beam deposition (Whangbo et al., 2000), pulsed laser deposition (Zhu et al., 2010), DC sputtering (Ye et al., 2003) and magnetron sputtering (Kim et al., 2003), metal organic chemical vapour deposition (MOCVD) (Mohanta et al., 2008), sol-gel method (Li et al., 2010), chemical spray pyrolysis (Krunks et al., 2009) and electrochemicaldeposition (ECD) (Sharma et al., 2010; Cembrero and Busquets-Mataix, 2009; Dalchiele et al., 2001; Inguanta et al., 2013; Mouet et al., 2010). ECD growth technique has a number of advantages that makes it ideal for some specific applications, which are being low cost, allowing growth of thin films with large area at low temperature, controllable thickness and deposition rate which is relatively higher than other techniques (Izaki and Omi, 1997). The technique is also less hazardous and more environmentally friendly (Izaki and Katayama, 2000). One of the most important challenges of ZnO based technologies which is preventing the progress in device arena is difficulty in obtaining p-type conductivity. Although first p-type conductivity in ZnO reported in 1997, it has still been difficult to deposit p-type ZnO with high homogeneity, quality and purity for optoelectronic device applications. However, a lot of different doping elements have been reported as a source of p-type conductivity such as N, Li, Mg, K, Cu, Na, As, P, Sb, Ag, etc. (Liu et al., 2008; Xiao et al., 2006; Fang and Kang, 2010; Jun and Yintang, 2008; McCluskey and Jokela, 2009; Janotti and Van de Wall, 2009; Kim et al., 2009; Lin et al., 2008; Fan et al., 2007; Doggett et al., 2007; Pan et al., 2007; Kim et al., 2009). The most promising, reliable and preferable one is doping with only N and co-doping N with group III elements (Tüzemen and Gür, 2007).

Zn_3N_2 powder was used to obtain p-type ZnO in the present study, which was shown previously that annealing Zn_3N_2 powder in oxygen media helps to obtain p-ZnO (Zou et al., 2009; Kaminska et al., 2005). Also, Wang et al. (2003) and Erie et al. (2008) reported that ZnO thin films doped with Zn_3N_2 deposited by DC magnetron sputtering and pulsed laser deposition techniques are resulted in p-type conductivity.

In this study, we performed the p-type growth of ZnO thin film on n-type Si substrate using ECD technique in order to investigate the electrical characteristics of p-ZnO/n-Si heterojunction. The electrical parameters of p-ZnO/n-Si heterojunction were determined with current-voltage (I-V) measurements in the temperature range of 80-300 K.

EXPERIMENTAL

Before forming the p-ZnO/n-Si heterojunction, n-Si (100) substrate were cleaned using the root cause analysis (RCA) procedure (waited in boiling NH_3+H_2O_2+6H_2O for 10 min and then, HCl+H_2O_2+6H_2O_2 at 60°C for 10 min) as reported in literature (Aydoğan et al., 2009). Then, Zn_3N_2 powder was spread out on n-Si wafer and annealed at 450°C for one hour to obtain a homogeneous film on the Si substrate. The electrochemical growth process was carried out by using an electrochemical cell having three electrodes which are counter (Zn), reference (Ag/AgCl) and working electrodes (n-Si). The growth process was controlled by a Gamry Reference 600 Potentiostat-Galvanostat/ZRA (Zero Resistance Ammeter). For p-ZnO growth, we prepared a solution of 0.05 M $Zn(ClO_4)_2$ and 0.1 M $Li(ClO_4)$ in dimethyl sulfoxide [DMSO-$(CH_3)_2SO$] as a solution. The ECD growth was carried out under the cathodic potential of -1 V and lasted one hour at solution temperature of 130°C. After deposition, sample was cleaned with de-ionized water.

Indium is used as an ohmic contact material on both sides of the structure which can be seen in Figure 1. The conductivity type of the ZnO thin films was determined using the Hall and resistivity measurements and hot probe technique. The Hall measurements were performed by a home-made Hall kit using a Varian 2901 regulated magnet power supply. The carrier concentration, resistivity and mobility parameters were calculated from Hall measurements. The I-V characteristics of p-ZnO/n-Si heterojunction were obtained by using a Keithley 487 picoammeter equipped with a closed-cycle helium cryostat. The samples were also characterized by X-ray diffraction (XRD) technique structurally using Rigaku D/Max-IIIC diffractometer, with Cu Kα radiation of 1.54 Å, within the 2θ angle ranging from 20-80°.

RESULTS AND DISCUSSION

Michael Faraday reported that the thickness of the film can be calculated with the help Faraday's laws of electrolysis focused on the electrochemical researches in 1834 as follows:

$$F=mQ/(\rho.n.A.t) \tag{1}$$

$$d=mQ/(nFA\rho) \tag{2}$$

where are d is the thin film tickness, A; the surface area, F; Faraday's number (96485 C/mol), n; the charge number (2 for ZnO), Q; the charge passed, ρ; density (5.6 g/cm^3), m; the molecular weight (81.4 g/mol). We have calculated about 1.5 µm to the thickness of the thin film with the help of this formula.

Figure 2 shows XRD pattern of ECD grown p-type ZnO thin film on n-Si substrate. The ZnO crystal planes of (002), (101) and (200) confirmed by the Joint Committee on Powder Diffraction Standards (JCPDS) files (36-1451 and 65-0682) can be clearly seen which indicates the polycrystalline structure of the grown thin films. The full width at half maximum (FWHM) values of the peaks corresponding to the (002), (101) and (200) planes are 0.094, 0.312 and 0.086, respectively. In addition, grain sizes of the (002) and (101) peaks were calculated as 88.54, 26.7 and 98.8 nm, respectively.

The p-type conductivity of the grown ZnO thin film was confirmed by both hot probe and Hall measurements with the Van der Pauw configuration. The carrier concentration and Hall voltage of p-ZnO thin film were determined as 3×10^{15} cm^{-3} and 0.986 mV using Hall measurement technique, respectively.

Figure 3 demonstrates both forward and reverse bias log (I-V) plot of p-ZnO/n-Si hetero junction. The current flow through a p-n junction can be defined by thermionic emission (TE) theory as follows (Rhoderick and Williams, 1998):

$$I = I_0 \exp\left(\frac{qV}{nkT}\right)\left[1-\exp\left(-\frac{qV}{kT}\right)\right] \tag{3}$$

where q is the electronic charge, k Boltzmann constant, T temperature, V the applied voltage, n ideality factor which is given by:

$$n = \frac{q}{kT}\left(\frac{dV}{d(\ln I)}\right) \tag{4}$$

I_0 is the reverse saturation current which is obtained

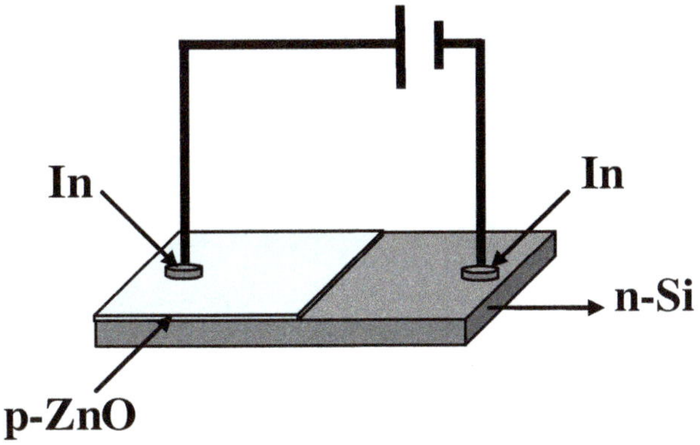

Figure 1. The schematic diagram of the p-ZnO/n-Si heterojunction

from the straight line intercept of $\ln I$ at $V = 0$ in Equation (3) and is given:

$$I_0 = A^* A T^2 \exp\left(-\frac{q\Phi_{b0}}{kT}\right) \quad (5)$$

where A^* is the effective Richardson constant, A the diode area and the zero-bias barrier height.

The experimental values of n and Φ_{b0} can be determined from intercepts and slopes of the forward bias I-V plot at each temperature (Figure 3) using Equations (4) and (5), respectively. n and Φ_{b0} values of the p-ZnO/n-Si heterojunction exhibit a variation between 2.49 and 0.574 eV at 300 K and 5.36 and 0.173 eV at 80 K, respectively. There may be a mixture of different metallic phases with different barrier heights (BHs) at heterojunction interface due to incomplete interfacial reaction. Image force-lower, generation-recombination, interface states and TFE are the mechanisms caused the large values of the ideality factor. The absolute value of ideality factor can be determined by the help of these calculated mechanisms. Also, the contamination at a interface is often present at the interfaces of junction prepared by the routine processing methods used in the semiconductor electronics industries. These contaminants may act directly to introduce inhomogeneity or they may simply promote inhomogeneity, through the generation of defects, additional interfacial chemical phases and etc. Even if the absence of chemical contaminants, BH inhomogeneity may be present. Thus, interface roughness may contribute to the presence of BH inhomogeneity due to effectively increasing or decreasing the low-BH patches. Finally, there are numerous structural defects, grain boundaries, dislocations, stacking faults, at interfaces, and these may contribute to BH inhomogeneity (Sullivan et al., 1991; Soylu and Yakuphanoglu, 2010; Tung, 1992).

Figure 4 shows the variation of ideality factor and barrier height with temperature. As can be seen in Figure 4, the values of n are indicated by the open triangles (experimental) and the closed triangles show estimated value of ideality factor using Equation (12) with coefficients ρ_2 and ρ_3. The values of Φ_{b0} by the open squares (experimental) and the closed squares represent estimated values of Φ_{b0} using Equation (11) with σ_{s0}. While Φ_{b0} increases with increasing temperature, n decreases. This behavior may be explained by current flow through the patches of lower barrier height and larger ideality factor and has been successfully explained on the basis of TE mechanism which takes into consideration spatial distribution of the barrier heights due to the inhomogeneities prevailing at the p-n junction interface. Since current transport across the interface is a temperature activated process, electrons at low temperatures are able to surmount the lower barriers and therefore, the current transport will be dominated by current flow through the patches of lower barrier height (BH) and a larger ideality factor. As the temperature increases, more and more electrons have sufficient energy to surmount the higher barrier (Sullivan et al., 1991; Tung, 1992; Biber et al., 2002; Karataş et al., 2003; Güllü et al., 2007).

A conventional activation $\ln(I_0/T^2)$ energy $1/kT$ versus $1/nkT$ or plot is shown in Figure 5. $\ln(I_0/T^2)$ versus $1/kT$ plot is found to be non-linear in the measurement temperature range. The non-linearity of $\ln(I_0/T^2)$ versus $1/kT$ plot is caused by the temperature dependence of the barrier height and ideality factor. According to Equation (5), one obtains:

$$\ln\left(\frac{I_0}{T^2}\right) = \ln(AA^*) - \frac{q\Phi_{b0}}{kT} \quad (6)$$

According to Equation (6), the plot yields a straight line with a slope given by a barrier height at 0 K, $\Phi_{b0}(T = 0)$ and intercept is the Richardson constant (A*). This straight line yields activation energy of 0.257 eV, as shown in Figure 5. The deviation in the Richardson constant may be due to the spatial distribution of inhomogeneous barrier heights and potential fluctuations at the interface that consist of low and high barrier areas (Tung, 1992; Karataş et al., 2003; Güllü et al., 2007).

The Richardson constant obtained from $\ln(I_0/T^2)$ versus $1/kT$ plot was determined as 1.53 × 10^{-7} A/cm^{2}K^2. This value obtained from the temperature dependence of the $I-V$ characteristics may also be affected by lateral in homogeneities of the barrier heights. This discrepancy can be explained by supposing that the distribution of the barrier heights is a Gaussian distribution of barrier heights with a mean value $\overline{\Phi}_b$ and standard deviation σ_s, in the form of:

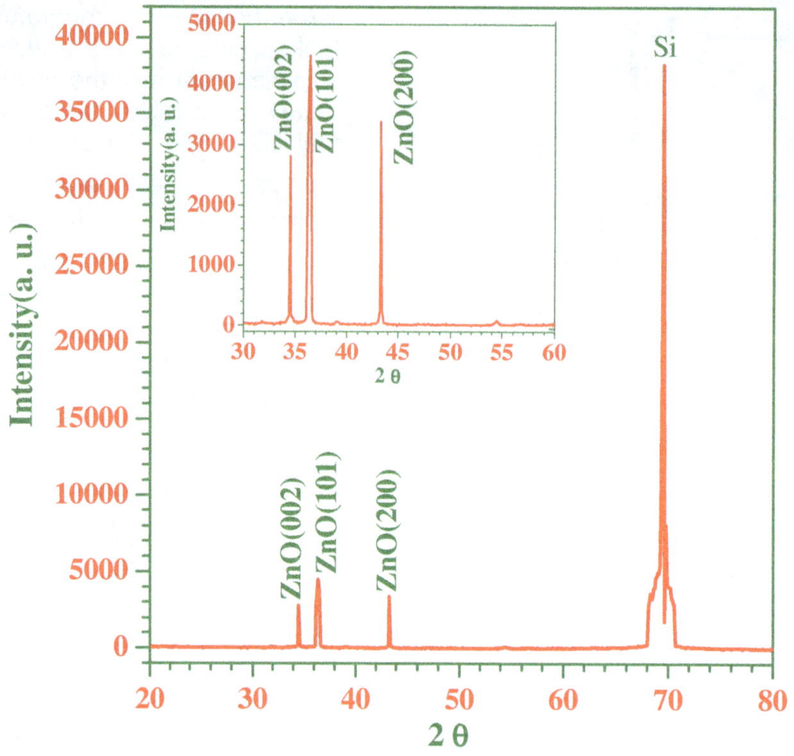

Figure 2. XRD measurement result of the p-type ZnO thin film deposited on n-Si substrate.

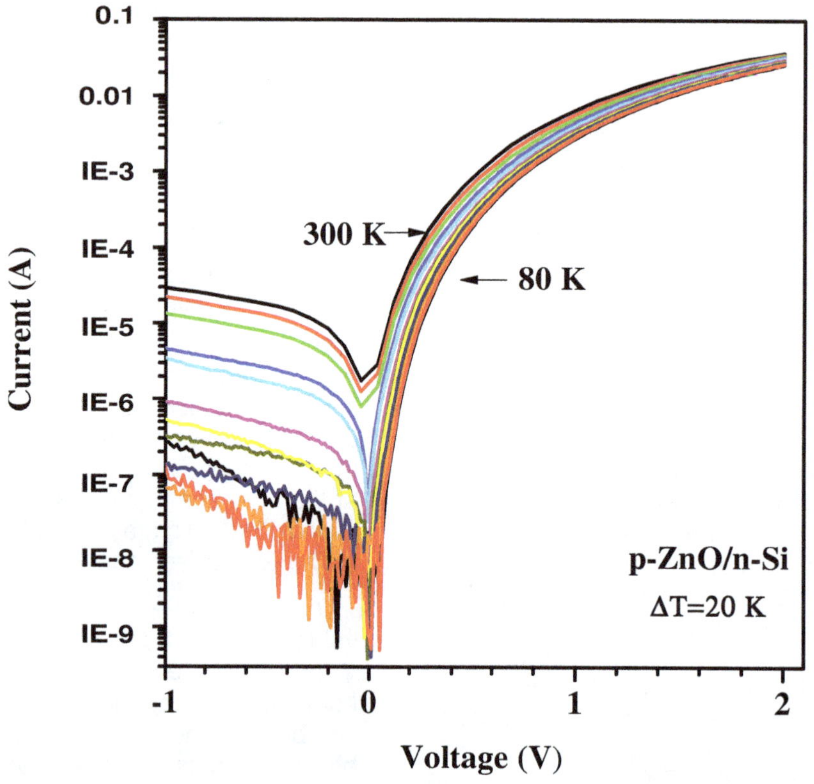

Figure 3. The current-voltage characteristics of the p-ZnO/n-Si heterojunction at temperature between 80-300K.

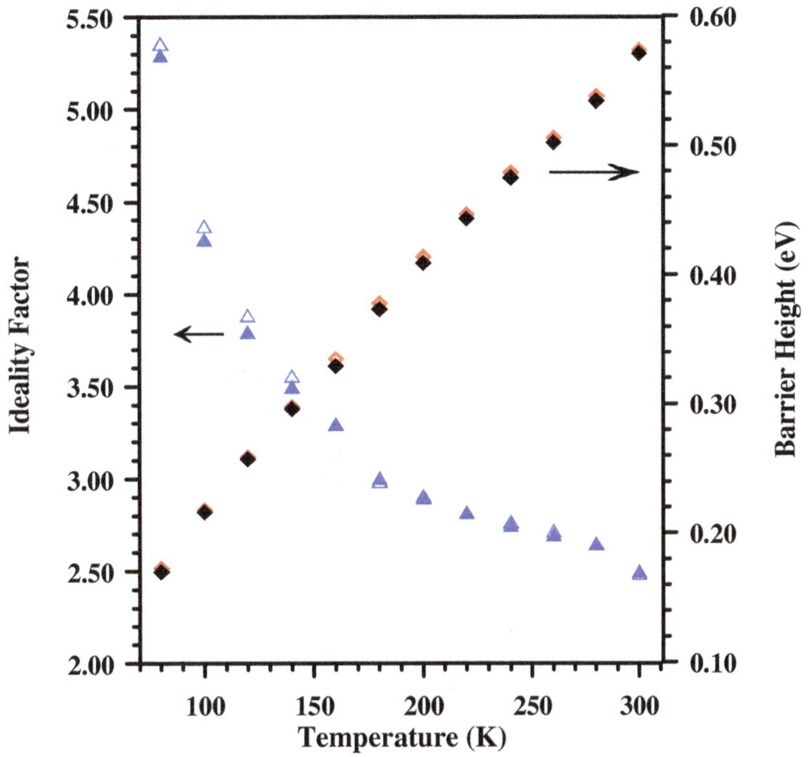

Figure 4. The ideality and barrier height versus temperature plots for p-ZnO/n-Si heterojunction.

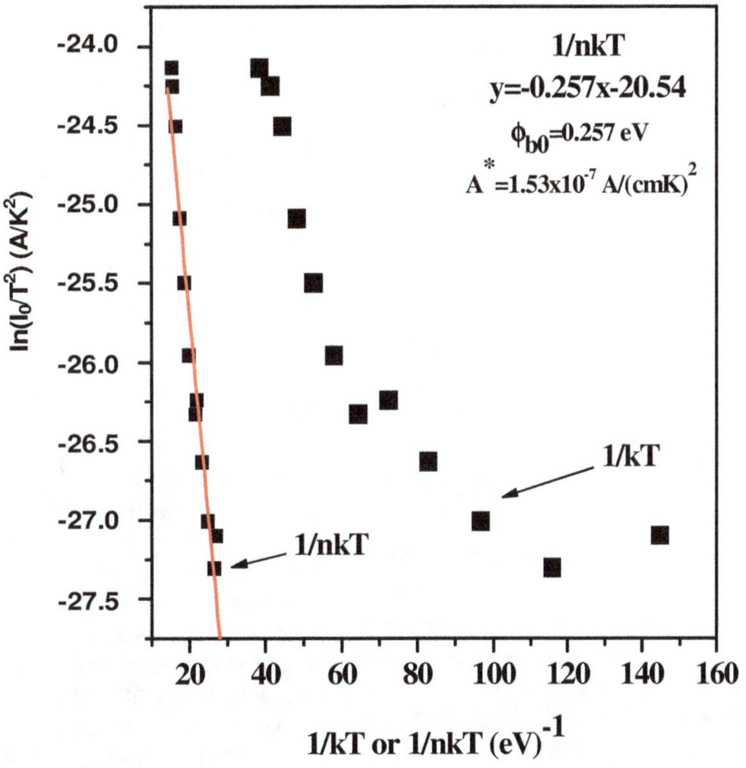

Figure 5. The $\ln\left(I_0/T^2\right) - 1/kT$ or $1/nkT$ plot for heterojunction.

$$P(\Phi_b)=\frac{1}{\sigma_S\sqrt{2\pi}}\exp\left(-\frac{\left(\Phi_b-\bar{\Phi}_b\right)^2}{2\sigma_S^2}\right) \quad (7)$$

where $1/\sigma_s\sqrt{2\pi}$ is the normalization constant of the Gaussian barrier height distribution. The total current under the forward bias of V can be expressed as:

$$I(V)=\int_{-\infty}^{+\infty} I(\Phi_b,V)P(\Phi_b)d\Phi \quad (8)$$

where $1/\sigma_s\sqrt{2\pi}$ is the current at the bias for a barrier height based on the ideal thermionic emission-diffusion theory and $P(\Phi_b)$ is the normalized distribution function giving the probability of accuracy for barrier height. Upon integration,

$$I(V)=A^*T^2\exp\left[-\frac{q}{kT}\left(\bar{\Phi}_b-\frac{q\sigma_S^2}{2kT}\right)\right]\exp\left(\frac{qV}{n_{ap}kT}\right)x\left[1-\exp\left(\frac{qV}{kT}\right)\right] \quad (9)$$

with

$$I_0 = AA^*T^2\exp\left(-\frac{q\Phi_{ap}}{kT}\right), \quad (10)$$

where Φ_{ap} and n_{ap} are the apparent barrier height and apparent ideality factor , respectively, and are given by:

$$\Phi_{ap}=\bar{\Phi}_{b0}-\frac{q\sigma_{s0}^2}{2kT} \quad (11)$$

$$\left(\frac{1}{n_{ap}}-1\right)=\rho_2+\frac{q\rho_3}{2kT} \quad (12)$$

In Equation (12), ρ_2 and ρ_3 values are the coefficients of voltage deformation of the barrier distribution. That is, the voltage dependencies of the mean barrier height and the barrier distribution width are given by coefficients and ρ_2 and ρ_3, respectively. Then, σ_{so} is the zero bias standard deviation of barrier height distribution. The standard deviation is a measure of the barrier homogenity. Song et al. (1986) and Werner and Güttler (1991) were used in the above expression for the apparent barrier homogeneity construction. If standard deviation is small it can be neglected. The temperature dependence of the ideality factor can be understood on the basis of Equation (12). The plot of n_{ap} versus $1/2kT$ should give a straight line that gives voltage coefficients ρ_2 and ρ_3 from intercept and slope respectively, as shown in Figure 6. The ρ_2 values from the slope of $(n^{-1}-1)$-$(1/2kT)$ plot were obtained as -0.515 at a temperature range between 300-160K and determined as -0.586 for 160-80K, respectively. Also, the ρ_3 values from the same plot were determined as -0.0049 for 300-160 K range and then measured as -0.0031 at for the range of 160-80K. $\bar{\Phi}_{bo}$ and σ_{s0} values were calculated from the slope of barrier height vs. (1/2kT) plot (Figure 7). $\bar{\Phi}_{bo}$ and σ_{s0} were measured as 0.818 eV and 117 mV at the range 300-160K while they were determined as 0.481 eV and 54 mV between 160-80K, respectively.

The continuous solid line in Figure 7 represents data estimated with these parameters using Equation (11). Combining the Equations (10) and (11), we get:

$$\ln\left(\frac{I_0}{T^2}\right)-\left(\frac{q^2\sigma_0^2}{2k^2T^2}\right)=\ln(AA^*)-\frac{q\bar{\Phi}_{b0}}{kT} \quad (13)$$

According to Equation 13, a modified $\ln(I_0/T^2)-(q^2\sigma_0^2/2k^2T^2)$ versus $1/T$ plot should give a straight line with the slope directly performed by the mean and the intercept (=lnAA*) at the ordinate, determining A* for a given diode area A. Figure 8 shows $\ln(I_0/T^2)-(q^2\sigma_0^2/2k^2T^2)$ versus $1/T$ plot. The $\ln(I_0/T^2)-(q^2\sigma_0^2/2k^2T^2)$ versus $1/T$ plot yielded $\bar{\Phi}_{bo}$ of 0.794 eV in the range of 300-160K and 0.329 eV in the range of 160-80K, respectively. These values are convenient with what it is obtained from the mean BHs as shown in Figure 7. Richardson constant is obtained as 0.011 A/cm^2K^2 in the range of 160-80 K and 6.235 A/cm^2K^2 for 300-160K from the modified activation energy plot, respectively.

Conclusion

p-ZnO thin films were deposited by ECD technique on n-Si substrate to obtain p-ZnO/n-Si heterojunction. The I-V characteristics of the electrochemically fabricated p-ZnO/n-Si heterojunction were measured at a temperature range of 80-300 K. It is observed that the ideality factor of p-ZnO/n-Si heterojunction increases while the barrier height decreases within this temperature range. The I-V characteristics of this structure have been interpreted on the basis of TE mechanism with Gaussian distribution of the barrier heights. Richardson constant is calculated between 0.011-6.235 A/cm^2K^2 for the temperature range of 300-80 K from $\ln(I_0/T^2)-q^2\sigma^2_{s0}/2k^2T^2-1/kT$ plot.

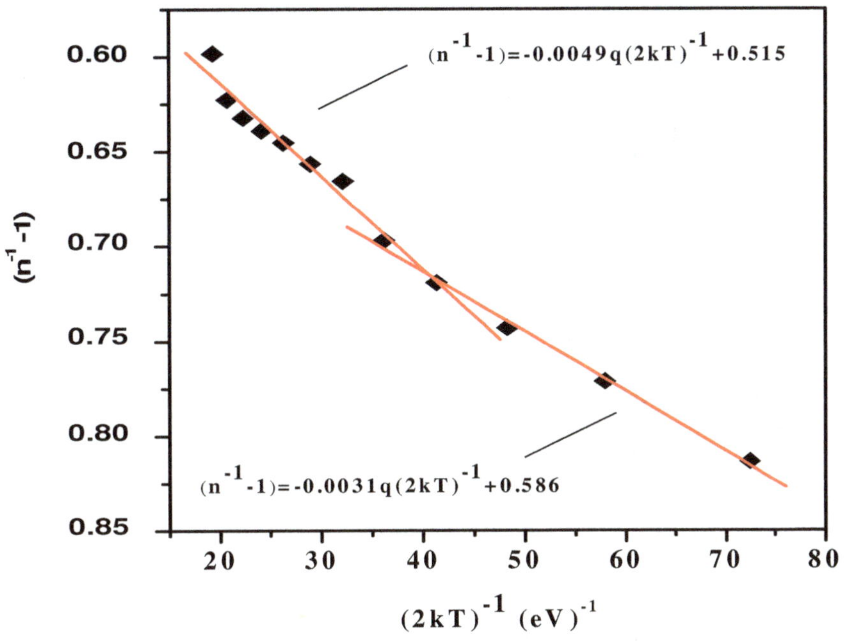

Figure 6. $(n^{-1}-1)$ versus $(1/2kT)$ plot of the p-ZnO/n-Si heterojunction. The continuous curves show estimated values of ideality factor using Eq. (10) for Gaussian distributions of barrier heights with $\rho_2 = -0.515$, $\rho_3 = -0.0049$ V in 300-160 K and $\rho_2 = -0.586$, $\rho_3 = -0.0031$ V in 160-80 K.).

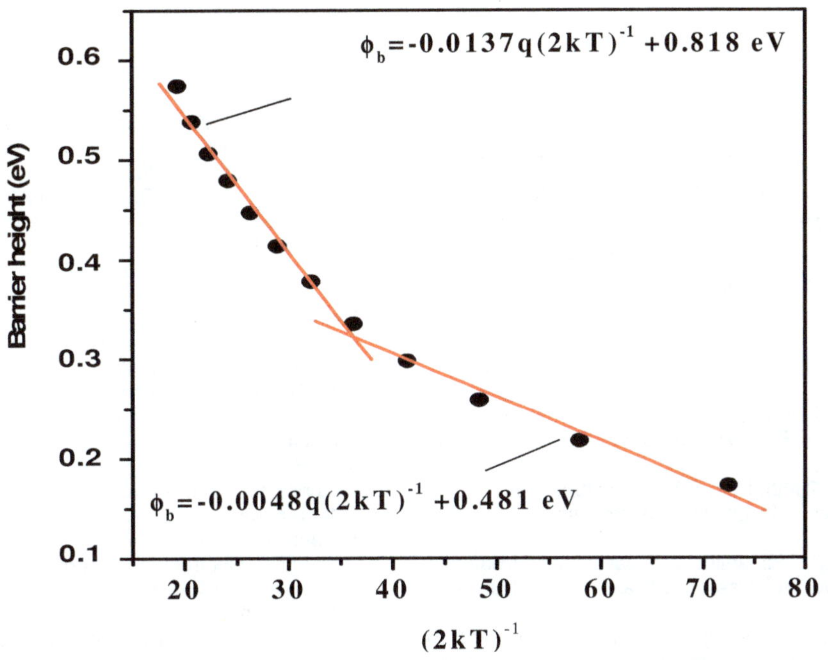

Figure 7. The barrier height-1/(2kT) curve of the p-ZnO/n-Si hetrojunction. The continuous curve related to the filled circles represents estimated values of Φ_{ap} using Equation (9) for Gaussian distributions of barrier heights with $\bar{\Phi}_{b0} = 0.818$ eV and $\sigma_{s0} = 117$ mV in 300-160 K and $\bar{\Phi}_{b0} = 0.481$ eV and $\sigma_{s0} = 54$ mV in 160-80K.

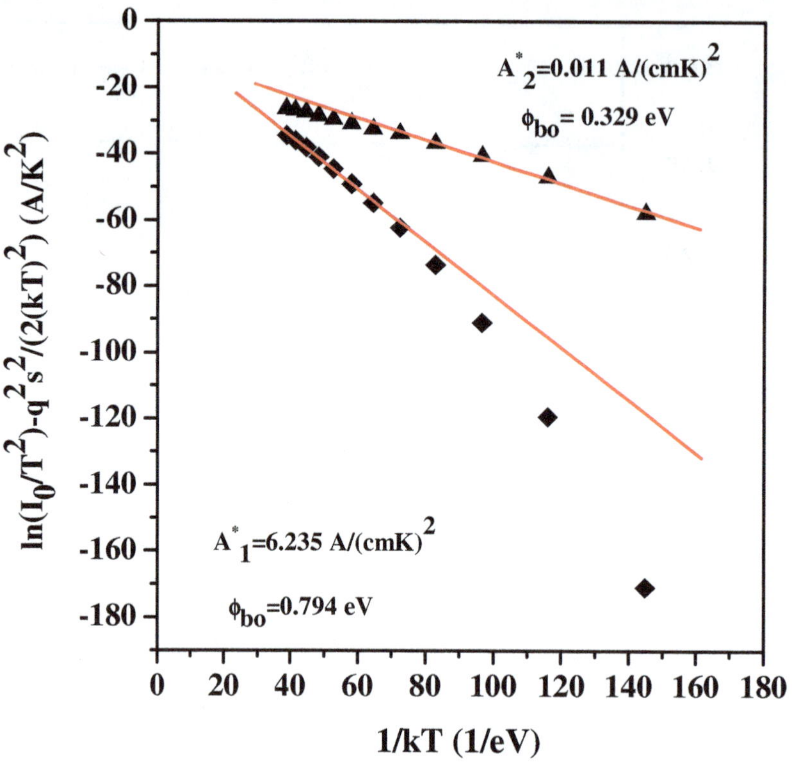

Figure 8. The $\ln(I_0/T^2) - q^2\sigma^2_{s0}/2k^2T^2 - 1/kT$ plot for the heterojunction according to Gaussian distributions of the barrier heights. The filled triangels represent the plot calculated for $\sigma_{s0} = 117$ mV in 300-160 K and the filled squares represent the plot calculated for $\sigma_{s0} = 54$ mV in 160-80 K.

ACKNOWLEDGEMENTS

We would like to thank Dr. Şakir Aydoğan for his help in the preparation of the paper. This work was carried out as part of Atatürk University Research Fund project (PN: 2009/89), and Scientific and Technical Research Council of Turkey (TUBITAK) (PN:107T822).

REFERENCES

Aydoğan Ş, Çinar K, Asil H, Coşkun C, Türüt A (2009). Electrical characterization of Au/ZnO Schottky contacts on n-Si. J. Alloy. Compd. 476:913-918.

Biber M, Temirci C, Turut A (2002). Barrier height enhancement in the Au/n-GaAs Schottky diodes with anodization process. J. Vac. Sci. Technol. B 20(1):10-13.

Buzás A, Geretovszy Z (2007). Patterning ZnO layers with frequency doubled and quadrupled Nd:YAG laser for PV application. Thin Solid Films 515:8495-8499.

Cembrero J, Busquets-Mataix D (2009). ZnO crystals obtained by electrodeposition:Statistical analysis of most important process variables. Thin Solid Films 517:2859-2864.

Dalchiele EA, Giorgi P, Marotti RE, Martin F, Barrado JRR, Ayouci R, Leinen D (2001). Electrodeposition of ZnO thin films on n-Si(100). Solar Energy Mater. Solar Cells 70:245-254.

Doggett B, Chakrabarti S, O'Haire R, Meaney A, McGlynn E, Henry MO, Mosnier JP (2007). Electrical characterisation of phosphorus-doped ZnO thin films grown by pulsed laser deposition. Superlattice. Microst. 42:74-78.

Erie JM, Li Y, Ivill M, Kim HS, Pearton SJ, Gila B, Norton DP, Ren F (2008). Properties of Zn3N2-doped ZnO thin films deposited by pulsed laser deposition. Appl. Surf. Sci. 254:5941-5945.

Fan JC, Xie Z, Wan Q, Wang YG (2007). As-doped p-type ZnO films prepared by cosputtering ZnO and Zn3As2 targets. J. Cryst. Growth 307:66-69.

Fang TH, Kang SH (2010). Preparation and characterization of Mg-doped ZnO nanorods. J. Alloy. Compd 492:536-542.

Fortunato E, Barquinha P, Pimentel A, Gonçalves A, Marques A, Pereira L, Martins R (2005). Recent advences in ZnO transparent thin film transistor. Thin Solid Films 487:205-211.

Güllü Ö, Biber M, Duman S, Türüt A (2007). Electrical characteristics of the hydrogen pre-annealed Au/n-GaAs Schottky barrier diodes as a function of temperature. Appl. Surf. Sci. 253:7246-7253.

Guo H, Zhou J, Lin Z (2008). ZnO nanorod light-emitting diodes fabricated by electrochemical approaches Electrochem. Commun. 10:146-150.

Inguanta R, Garlisi C, Spano T, Piazza S, Sunseri C (2013). Growth and photoelectrochemical behavior of electrodeposited ZnO thin films for solar cells. J. Electrochem 43:199-208.

Izaki M, Katayama J (2000). Characterization of boron incorporated zinc oxide film chemically prepared from an aqueous. J. Electrochem. Soc. 147:210-213.

Izaki M, Omi T (1997). Characterization of transparent zinc oxide films prepared by electrochemical reaction. J. Electrochem. Soc. 144:1949-1952.

Janotti A, Van de Wall CG (2009). Fundamentals of zinc oxide as a semiconductor Rep. Prog. Phys. 72:126501-1265029.

Jun W, Yintang Y (2008). Deposition of K-doped p type ZnO thin films on (0001) Al2O3 substrates. Mater. Lett. 62:1899-1901.

Kaminska E, Piotrowska A, Kossut J, Barcz A, Butkute R, Dobrowolski W, Dynowska E (2005). Transparent p-type ZnO films obtained by oxidation of sputter-deposited Zn3N2. Solid State Commun. 135:11-15.

Karataş Ş, Altindal Ş, Türüt A, Özmen A (2003). Temperature dependence of characteristic parameters of the H-terminated Sn/p-Si(100) Schottky contacts. Appl. Surf. Sci. 217:250-260.

Kim GH, Kim DL, Ahn BD, Lee SY, Kim HJ (2009). Investigation on doping behavior of copper in ZnO thin film. Microelectron. J. 40:272-275.

Kim IS, Jeong EK, Kim DY, Kumar M, Choi SY (2009). Investigation of p-type behavior in Ag-doped ZnO thin films by E-beam evaporation. Appl. Surf. Sci. 255:4011-4014.

Kim K, Lee K, Oh MS, Park CH, Im S (2009). Surface-induced time-dependent instability of ZnO based thin-film transistors. Thin Solid Films 517:6345-6348.

Kim KS, Kim HW, Kim NH (2003). Structral characterization of ZnO films grown on SiO2 by the RF magnetron sputtering. Physica B 334:343-346.

Krunks M, Dedova T, Karber E, Mikli V, Acik IO, Grossberg M, Mere A (2009). Growth and electrical properties of ZnO nanorod arroys prepared by chemical spray pyrolysis. Physica B 404:4422-4425.

Li Y, Xu L, Li X, Shen X, Wang A (2010). Effect of aging time of ZnO sol on the structural and optical properties of ZnO thin films prepared by sol-gel method. Appl. Surf. Sci. 256:4543-4547.

Lin SS, Lu JG, Ye ZZ, He HP, Gu XQ, Chen LX, Huang JY, Zhao BH (2008). p-type behavior in Na-doped ZnO films and ZnO homojunction light emitting diodes. Solid State Commun. 148:25-28.

Liu W, Gu SL, Ye JD, Zhu SM , Wu YX, Shan ZP, Zhang R, Zheng YD, Choy SF (2008). High temperature dehydrogenation for realization of nitrogen-doped p-type ZnO. J. Cryst. Growth 310:3448-3452.

McCluskey MD, Jokela SJ (2009). Defects in ZnO. J. Appl. Phys. 106:071101-0711013.

Mohanta SK, Kim DC, Cho HK, Chua SJ, Tripathy S (2008). Structral and optical properties of ZnO nanorods grown by metal organic chemical vapor deposition.s J. Cryst. Growth 310:3208-3213.

Mouet T, Devers T, Telia A, Mesai Z, Harel V, Konstantinov K, Kante I, Ta MT (2010). Growth and characterization of thin ZnO films deposited on glass substrates by electrodeposition technique. Appl. Surf. Sci. 256:4114-4120.

Pan X, Ye Z, Li J, Gu X, Zeng Y, He H, Zhu L, Che Y (2007). Fabrication of Sb-doped p-type ZnO thin films by pulsed laser deposition. Appl. Surf. Sci. 253:5067-5069.

Rhoderick EH, Williams RH (1998). Metal-Semiconductor Contacs, second ed., Clarendon Press, Oxford.

Sen Chien FS, Wang CR, Chan YL, Lin HL, Chen MH, Wu RJ (2010). Fast-response ozone sensor with ZnO nanorods grown by chemical vapor deposition. Sensor. Actuat. Phys. B 144:120-125.

Sharma SK, Rammohan A, Sharma A (2010). Templated one step electrodeposition of high aspect ratio n-type ZnO nanowire arrays. J. Colloid Interf. Sci. 344:1-9.

Song JH, Van Meirhaeghe RL, Laflére WH, Cordon F (1986). On the difference in apparent barrier heights as obtained from capacitance-voltage and current-voltage-temperature measurements on Al/p-InP Schottky barriers. Solid-State Electron. 29 633-638.

Soylu M, Yakuphanoglu F (2010). Analysis in of barrier height homogeneity in Au/n-GaAs Schottky barrier diodes by Tung model. J. Alloys Compd. 506:418-422.

Sullivan JP, Tung RT, Pinto MR, Graham WR (1991). Electron transport of inhomogeneous Schottky barriers:A numerical study. J. Appl. Phys. 70:7403-7424.

Tung RT (1992). Electron transport at metal-semiconductor interfaces: General theory. Phys. Review B 45:13509-13523.

Tüzemen S, Gür E (2007). Principal in issues producing new ultraviolet light emmiters based on transparent semiconductor zinc oxide. Opt. Mater. 30:292-310.

Wang C, Ji Z, Liu K, Xiang Y, Ye Z (2003). P-Type ZnO thin films prepared by oxidation of Zn3N2 thin films deposited by DC magnetron sputtering. J. Cryst. Growth 259:279-281.

Werner JH, Güttler HH (1991). Barrier inhomogeneities at Schottky contacts. J. Appl. Phys. 69:1522-1533.

Whangbo SW, Jang HB, Kim SG, Cho MH, Jeong KH, Whang CN (2000). Properties of ZnO thin films grown at room temperature by using ionized cluster beam deposition. Korean J. Phys. Soc 37(4):456-460.

Xiao B, Ye Z, Zhang Y, Zeng Y, Zhu L, Zhao B (2006). Facrication of p-type Li-doped ZnO films by pulsed laser deposition. Appl. Surf. Sci. 253:895-897.

Ye ZZ, Lu JG, Chen HH, Zhang YZ, Wang L, Zhao BH, Huang JY (2003). Preparation and characteristics of p-type ZnO films by DC reactive magnetron sputtering. J. Cryst. Growth 253:258-264.

Zhu BL, Zhao XZ, Su FH, Li GH, Wu XG, Wu J, Wu R(2010). Low temperature annealing effects on the structure and optical properties of ZnO films grown by pulsed laser deposition. Vacuum 84:1280-1286.

Zou CW, Chen RQ, Gao W (2009). The microstructures and electrical and optical properties of ZnO:N films prepared by thermal oxidation of Zn3N2 precursor. Solid State Commun. 149:2085-2089.

14

Temperature dependence of wall-plug efficiency of high power laser diodes

Muzahim I. Azawe

Department of Physics, College of Education, University of Mosul, Mosul, Iraq.

Temperature dependence of wall-plug efficiency of high power laser diodes is theoretically analyzed in terms of experimentally measured parameters of the device. This issue has an influence on the available power output and on the degradation mechanism. High power laser diodes have become increasingly important as the output power of such devices continues to rise. The present calculations revealed that the decrease of wall-plug efficiency with temperature could be explained, presumably in terms of excess voltage loss and internal waveguide optical loss. Understanding the dynamic mechanism of the decrease of efficiency can lead to optimum operating conditions of the device. Thermally rolling power runaway is found to be an important factor in the high power laser diodes degradation and eventually to device failure

Key words: Laser diodes, high power, wall-plug efficiency, internal loss, waveguide loss.

INTRODUCTION

High power laser diodes are appropriate light sources for enormous applications due to their high power conversion efficiency. The primary issue limiting the maximum power of these devices is self-heating (thermal rolling) at high drive currents (Li et al., 2007). For low-series-resistance and thermal-resistance device, it is found that the junction-temperature rise in CW operation is a strong function of both the characteristic temperature for external differential efficiency as well as of the heatsink thermal resistance (Botez, 1999) thermal resistance effect will be shown in this article.

Thermally affected high power laser (HPL) parameters will be studied in this work. Based on our experimental results, the temperature dependence of the crucial parameters of high power laser diodes will be investigated numerically. Among these are; the slope efficiency, wall-plug efficiency, and the internal loss. These critical device parameters, in addition to the hetero-barrier leakage and series resistance have to be optimized to yield maximum light powers (Schmidt et al., 2005). Thermal resistance will be shown in this article to be increased with optical power output, which degrades the efficiency of the device. Recently, it has been showed that low thermal resistance can provide high slope efficiency by the lithographic approach (Demir et al., 2010). Thermal properties of laser diode chips can be described by a thermal time constant associated with the heat spreading volume in the substrate (Shan et al., 2010). The efficient heat extraction from the substrate by excellent cooling system will be the most crucial issue for high power conversion efficiency. The temperature dependence has to be taken into account in any laser design for high power conversion efficiency.

Theory

The wall-plug efficiency of a high power laser diode is given by:

$$\eta_C = \frac{P_{opt}}{P_{elect}} \quad (1)$$

where P_{opt} is the optical power output, P_{elect} is the electrical input power. The dependence of P_{opt} on the pumping current above threshold can be described as (Diehl et al., 2006):

$$P_{opt} = \eta_{slope}.(I_{op} - I_{th}) \quad (2)$$

with η_{slope} is the slope efficiency. The slope efficiency is related to the external differential quantum efficiency η_{ext} as (Behringer, 2007):

$$\eta_{slope} = \eta_{ext}.\frac{\hbar\omega}{q} \quad (3)$$

$\frac{\hbar\omega}{q}$ is the necessary voltage needed to get the required injection current, I_{op} is the operating current, and I_{th} is the threshold current. The electrical input power can be written as:

$$P_{elect} = V_{op}.I_{op} + R_s.I_{op}^2 \quad (4)$$

with V_{op} is the applied voltage at the operating current, and R_s is the series resistance. In this paper, the applied voltage will depend on the input power. The dissipated power in the laser is:

$$P_{dis} = I_{th}.\eta_{ext}.\frac{\hbar\omega}{q} + I_{op}(V_{op} - \eta_{ext}.\frac{\hbar\omega}{q}) + R_s I_{op}^2 \quad (5)$$

The temperature rise of the active region is connected to the dissipated power through the thermal resistance, R_{th} as (Laikhtman et al., 2005):

$$\Delta T = R_{th}.P_{dis} \quad (6)$$

The temperature dependence of threshold current and slope efficiency can be described as exponential functions with two specific constants T_o and T_l as (Bai et al., 2010; Botez et al., 1996):

$$I_{th}(T_{j2}) = I_{th}(T_{j1})\exp(\frac{T_{j2} - T_{j1}}{T_o}) \quad (7)$$

$$\eta_{slope}(T_{j2}) = \eta_{slope}(T_{j2})\exp(\frac{T_{j1} - T_{j2}}{T_l}) \quad (8)$$

Here, I_{th} and η_{slope} are measured at the different junction temperatures T_{j1} and T_{j2}. It is very difficult to get values for T_o and T_l from principles, but relatively simple to extract them from measurements. Higher barriers and lower threshold current will increase the values of T_o and T_l giving laser with better temperature stability.

Device parameters

Non-coated anti-reflection (AR) bar laser diodes (BLD's) with reduced linewidth, increased spatial brightness, and low thermal resistance was used in the experiments. Low thermal resistance was achieved in our setup by minimizing the intra-package and package to heat sink interface which will enable the device to work with high power and external efficiency. On the other hand, low thermal resistance enable the BLD to be operated CW without sacrificing operating lifetime in addition to the superior performance.

Depending on the thermal resistance, the heat dissipation in laser diodes results in an active region temperature rise. This temperature rise limits the output power (through so-called thermal rollover) and enhances the device degradation mechanisms. It is therefore critical to minimize the thermal resistance in high-power laser devices.

Narrow linewidth and stable wavelength emission were maintained during the operation by the influence of high accurate current driving circuitry and temperature controller. The mount of the BLD had the ability to control the temperature of the active region within the required stability with both the thermo-electric cooler (TEC) and the liquid nitrogen (the cryostat). A slope efficiency η_{slope} of (1.3024 W/A), threshold current I_{th} of (6.2 A), series resistance R_s (7.6 mΩ), thermal resistance R_{th} (1.2 K/W), and wavelength (803.08 nm) at a temperature (19°C) were the device parameters found experimentally. Other parameters, which can be affected by temperature, were also obtained for the sake of numerical simulation, $\partial\lambda/\partial T = 0.43 nm/^o C$ and $\partial V_{be}/\partial T = -2.7 mV/^o C$. The characteristic temperature coefficient of threshold current density was extracted from the measurements, as differ for every type of laser diode (T_o = 113°C).

The maximum power conversion efficiency (PCF) is expressed as (Ma and Zhong, 2007; Kanskar et al., 2003):

$$\eta_{PCF_{max}} \cong \eta_{ext}.\frac{V_F}{V_{be}}(1 - 2\sqrt{R_s I_{th}/V_{be}}) \quad (9)$$

where V_F is the quasi-Fermi level difference, and V_{be} is the overall built- in voltage. In order to calculate the slope efficiency of the device, light power output versus current at different temperatures were recorded. The slope of $(dP/dI)_T$ above threshold (stimulated emission), converted to the external differential quantum efficiency η_{ext}, was calculated and drawn at each temperature. The dependence of η_{ext} on the temperature from this plot, for the device under test, was found to be of the form:

$$\eta_{ext}(T) = (1.77_{T=16^o C})\exp\left[-.(\frac{T-16}{24.63})^2\right] \quad (10)$$

Provided that T is in °C. This formula is numerically obtained for the experimental variation of the external differential quantum efficiency with temperature with curve fitting tool in Matlab provided that R-square of the fitting is nearly equal one. This equation differs slightly with the equation given by (Suchalkin et al., 2002).

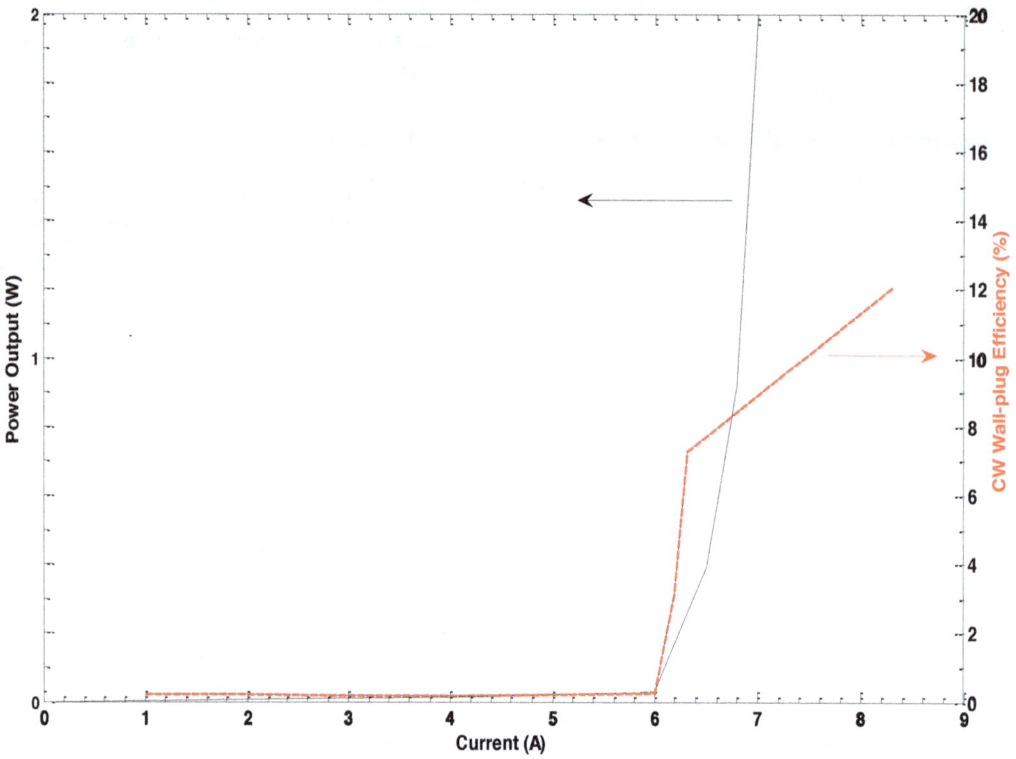

Figure 1. CW power output and wall-plug efficiency at a controlled temperature of 16°C.

Power conversion efficiency is temperature dependent through the dependence of all parameters given in Equation (9) on temperature, in other words, η_{PCF} is temperature dependent as will be shown later.
The next step is to find the modal gain from the threshold condition which can be stated as:

$$R.\exp\{(\Gamma G-\alpha_{tot})L\}=1 \tag{11}$$

where R is the facet power reflectivity, Γ the confinement factor, G the material gain, α_{tot} is the total loss, and L is the laser cavity length. The material gain G is sufficiently small when the total loss α_{tot} is to be estimated from the low energy side of the modal gain spectra (Suchalkin et al., 2002), that is:

$$g=\Gamma G-\alpha_{tot} \tag{12}$$

$$g\approx-\alpha_{tot} \tag{13}$$

Subtracting the mirror loss α_m, $\left\{\alpha_m=\left(\frac{1}{L}\right)\ln(R)\right\}$ from the total loss gives the internal loss. The temperature increase of α_{tot} is the issue of this work. The coefficient of gain saturation had been neglected in the calculations, since it has a minor effect on the high power output of the device.

RESULTS AND DISCUSSION

In order to get a full understanding of the HPL characteristics, the power and wall-plug efficiency versus the input current are shown in Figure 1, provided that the temperature was controlled at 16°C. This experimental plot was recorded while maintaining the driving in control circuit with accuracy around ±0.1A. The power was obtained just over 2W without any kink or thermal runaway to ensure linear operation and without mode hopping. At this power level, wall-plug efficiency can be estimated with good accuracy.

Wall-plug efficiency increases with the input current and reaches its maximum value at a current value of (8A). This available power output and high efficiency were all due the coolant instrument applied to the HPL mount.

The parameters that have an impact on the power conversion efficiency (Equation (2) is V_{be}, and the quasi-Fermi separation V_F (it can be determined from the lasing wavelength), in addition to their dependence on temperature, they are affected by the input power as shown in Figure 2. The wavelength had been recorded with a high stability in driving current and temperature. The wavelength was recorded by the monochromator with a wavelength step of 0.1 nm. The assumption of that the quasi-Fermi separation V_F to be consistent with the

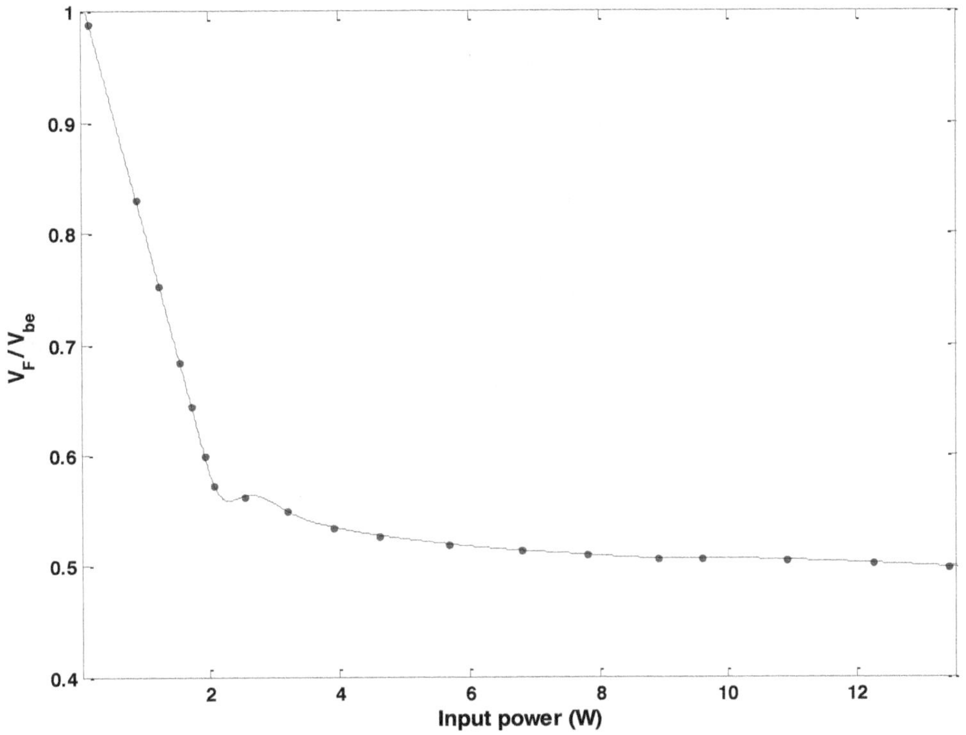

Figure 2. The ratio of quasi-Fermi level difference to the overall built- in voltage as a function of input power of HPL.

clamping of the carrier density to its threshold value when the stimulated emission was reached (gain = losses). The usual measurements are to record the spectra of laser diode above threshold and driving with the laser diode with pulses of low duty cycle to reduce the heating effect. Mechanical chopper is used to chop the laser beam. The detector was integrated with preamplifier and lock-in amplifier with reference frequency from the chopper.

As can be seen from the plot that, the quasi-Fermi separation and the built-in potential equals each other before applying any current to the HPL device, and the ratio decreases rapidly and reaches its maximum decrease after threshold current had been reached. Figure 3 shows the influence of temperature on the power dissipation in the laser given by the Equation 4.

The evaluation of power dissipation in the laser was calculated by taking into account the dependence of the terms given in Equation 4 on the temperature, especially threshold current, slope efficiency, and series resistance, as stated in the above. As can be seem from the graph, the power dissipation is high at elevated temperature.

The wall-plug efficiency, as stated above, is mainly dependent on the internal loss mechanism of the high power laser diode. The input power that is not converted to light output power is considered as wasted power (heat). This heat has to be removed from the laser mount, otherwise, more power must be added to overcome the losses, and hence the junction temperature will rise and ultimately shorten lifetime of the laser diode (Williamson and Kanskar, 2004).

The temperature rise in the active region of the HPL when the temperature control was stabilized only by the TEC is found to be very high as given in the following Figure 4. This temperature rise was calculated from power dissipation and thermal resistance as a function of temperature at different operating currents, Equations 4 to 7, and taking into account Fourier's law of heat conduction for the heat sink temperature calculations. This rise can be very crucial on the lifetime and power conversion efficiency of the HPL and hence on the wall-plug efficiency. The figure also illustrates the need for more efficient temperature control, for example cascaded TEC, good thermal interface materials.

The variation of thermal resistance with optical power output is shown in Figure 5. Light output above threshold is expressed as in Equation 2 and it is temperature dependent and a self-consistent numerical simulation was applied to obtain the two characteristics T_o and T_l. The thermal resistance values for our device which was cooled actively were found and plotted as a function of power output, taking into account the variation of slope efficiency and threshold current with temperature. The line was drawn as a theoretical fit for the experimental

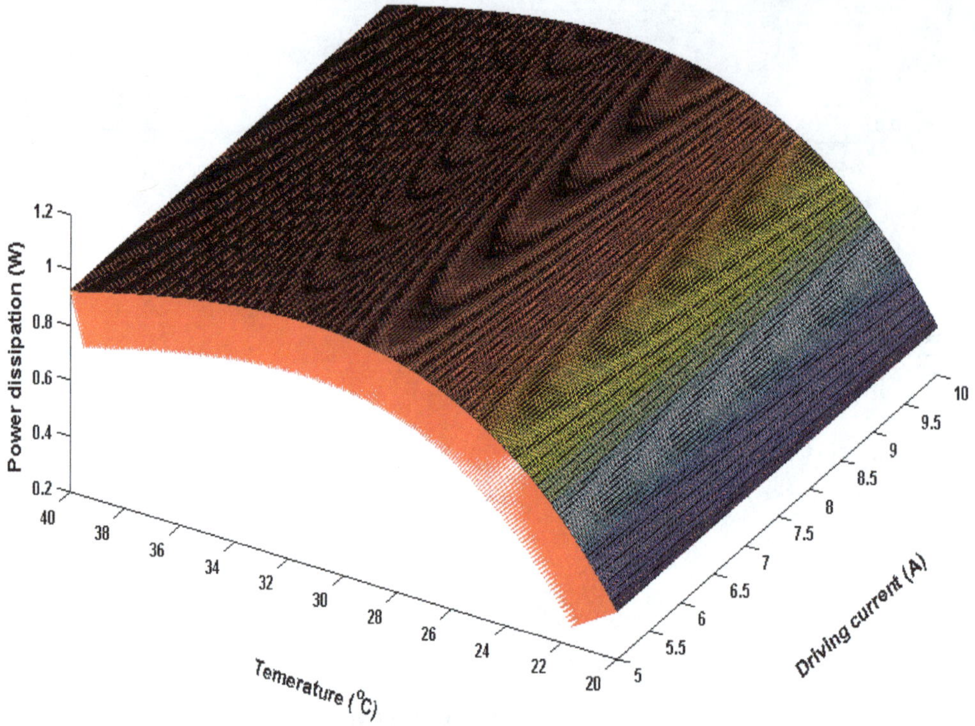

Figure 3. The power dissipation dependence on both the driving current and the temperature of the heat sink.

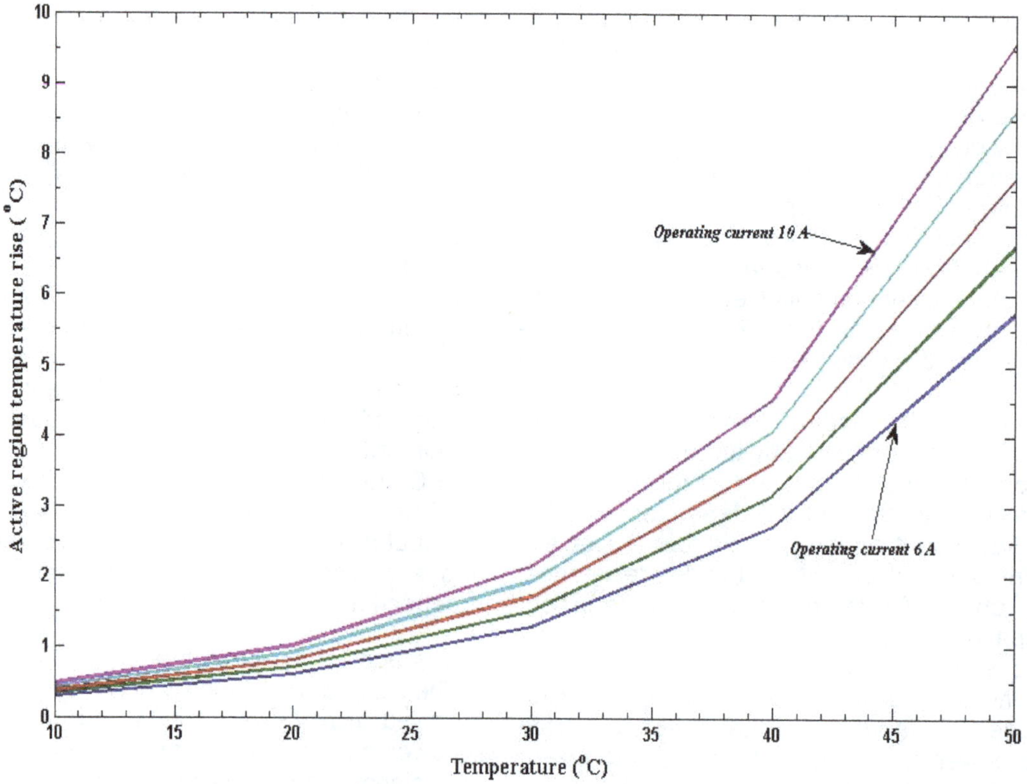

Figure 4. Temperature rise of the HPL active region as a function of heat sink temperature control with TEC at different operating currents.

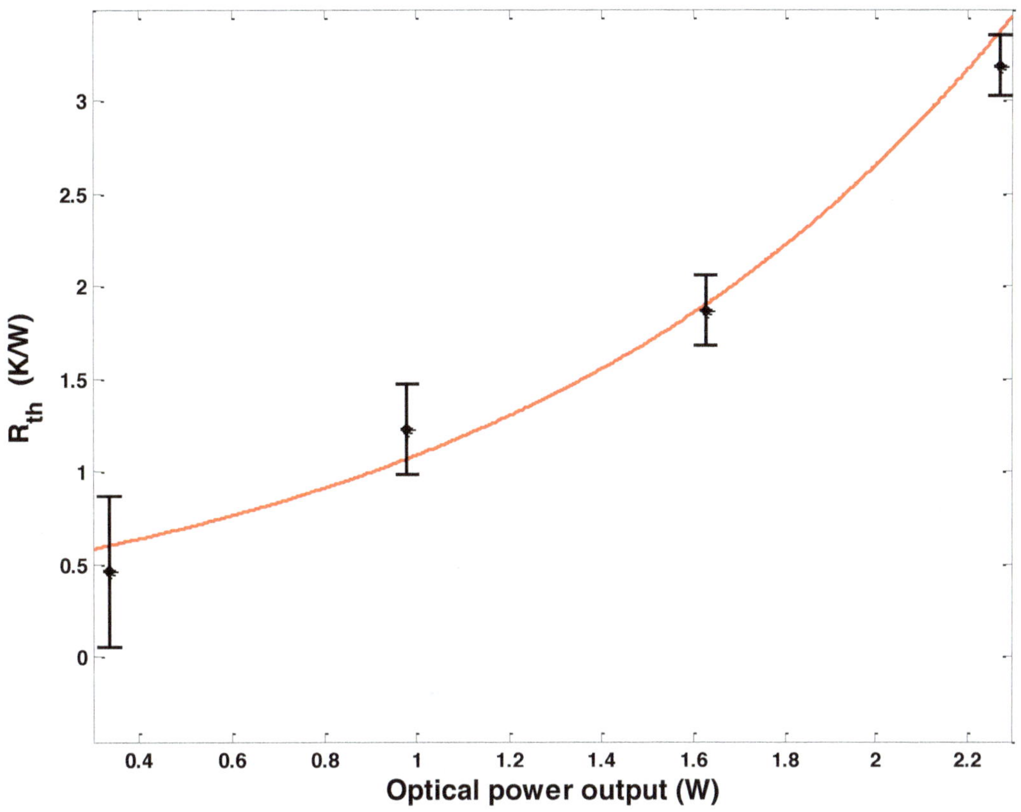

Figure 5. Thermal resistance of the HPL with optical power output. Solid line represents the theoretical simulation.

data. Thermal resistance high increase will have an impact on the wall-plug efficiency of the device, as will be shown.

The band gap energy of the semiconductor lasers variation with temperature follows the Varshni relation (Sze, 1985), which indicates quadratic temperature dependence, and this dependence has to be included in our calculations as well. The other effect that has to be included is the wavelength-shift with temperature.

To study the power conversion efficiency of the HPL as a function of temperature, Equation 8, we have to consider the terms given by this equation as temperature variables, that is, threshold current, external differential quantum efficiency, overall built- in voltage, and series resistance. Figure 6 shows the power conversion efficiency as a function of temperature. The plot shows that overall built- in voltage, series resistance and external differential quantum efficiency are the most effective parameters on this efficiency when they are considered to be as temperature dependent or not. The efficiency drops out to 50% when these parameters were taken as temperature dependence, provided that the BLD was stabilized efficiently to overcome the active temperature rise.

The reliable operation of HPL with improvement power conversion efficiency requires high T_o and T_l, in order to suppress the dependence on the temperature of the series resistance, built-in voltage, and external differential quantum efficiency. In addition, the decrease of series resistance leads the increase in the electrical-to-optical conversion. The influence of series resistance has to be considered in any design of the HPL.

Now, the effect of temperature on the characteristics of the HPL especially that can influence the wall-plug efficiency variation with temperature will be studied thoroughly. The slope efficiency in conjunction with η_{ext}, and the internal loss had to be evaluated with temperature variations in order that the wall-plug efficiency and its variation with temperature need to be asserted. The slope efficiency as a function of temperature is shown in Figure 7.

The decrease of η_{slope} with the temperature is due to the increase of internal loss, loss of carriers over the heterobarriers, and nonradiative recombination (C.F. Equations 4 and 5). The thermal carrier loss mechanism can be combined with gain measurement and calculation (Jian and Summers, 2010).

As stated above that the internal loss increases with temperature due to many mechanism influenced by the

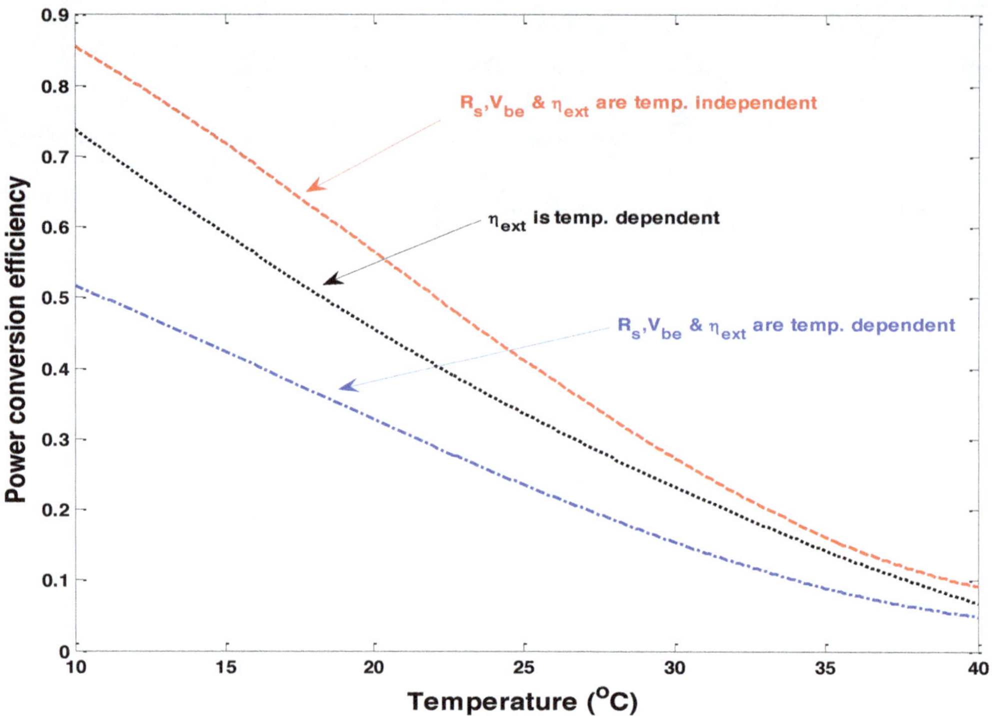

Figure 6. Illustrates the influence of R_s, V_{be}, and η_{ext} on the power conversion efficiency as a function of temperature.

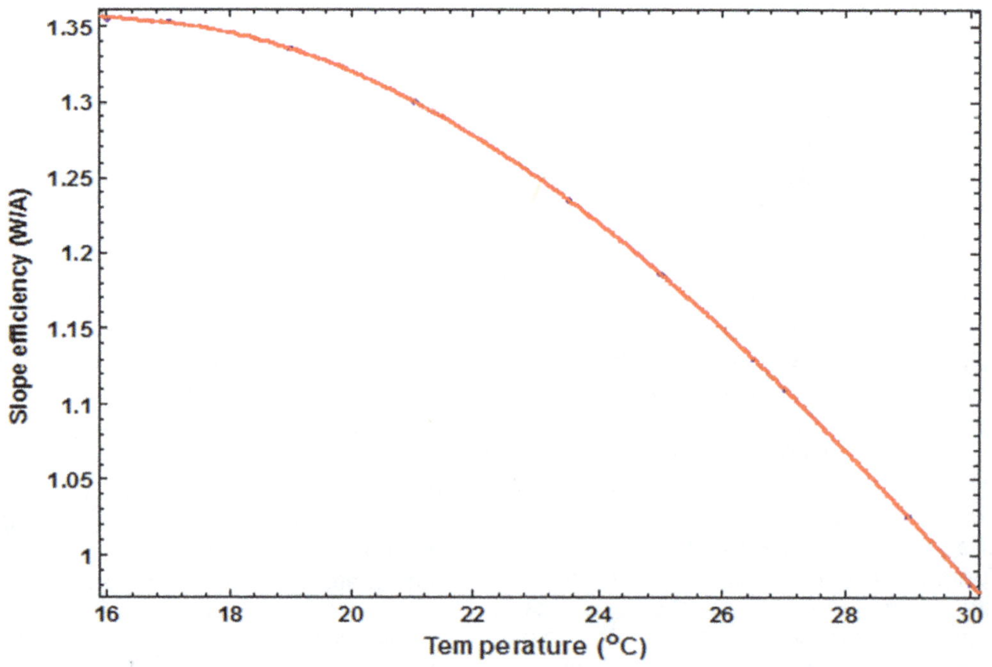

Figure 7. Slope efficiency of HPL variation with temperature.

temperature rise and their effects became pronounced. The internal loss increment with temperature is illustrated in Figure 8. Low internal loss increases external quantum efficiency, it is the low series resistance R_s, and hence,

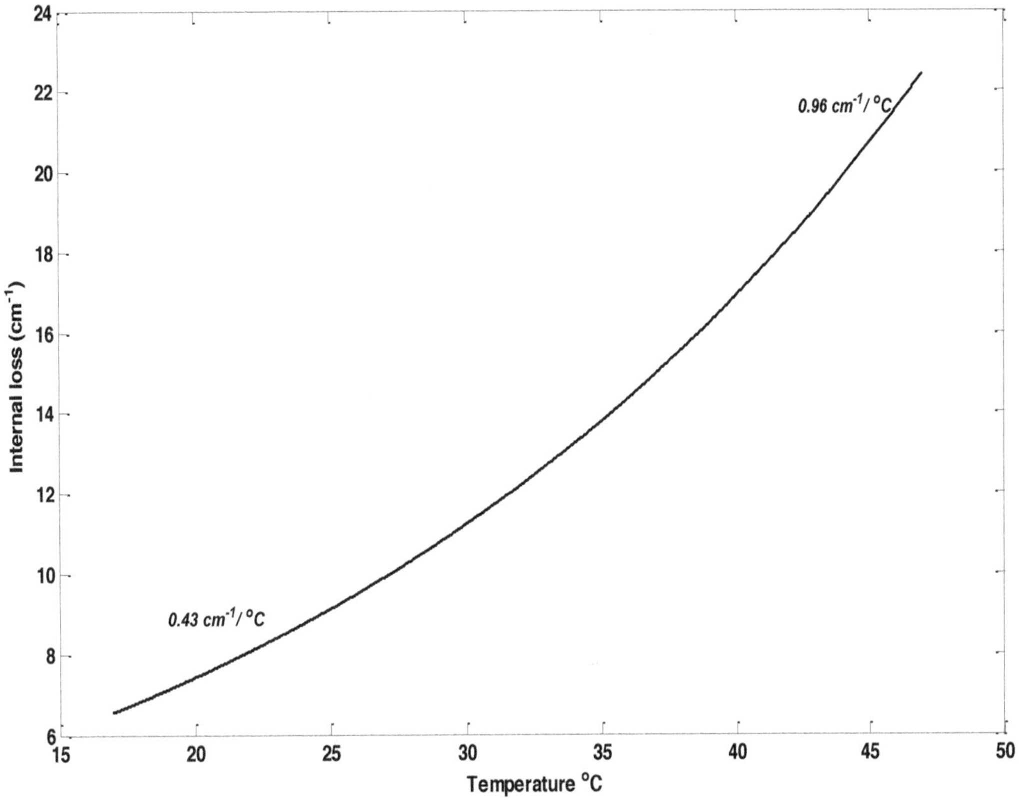

Figure 8. Dependence of internal loss on the temperature of the HPL.

the low forward voltage of these devices, that enables the relatively high CW output powers (Equation 1).

The important parameter of the HPL is the wall-plug efficiency, and hence its dependence on the operating temperature. Wall-plug efficiency calculation had been found to decrease with temperature, as can be seen from Figure 9. This efficiency of the high power laser diodes is a collection of over-all efficiency of the device and gives the exact picture of the characteristics parameters. The efficiency η_{PCF} can be written as:

$$\eta_{PCF} = \eta_1 . \eta_2 . \eta_3 . \eta_4 \qquad (13)$$

where η_1 accounts for laser diode efficiency, η_2 for the excess voltage ratio, η_3 accounts for pumping carrier loss, and η_4 for total loss (internal loss + mirror loss). The impact of temperature on the wall-plug efficiency can be interpreted by many factors, carrier leakage and nonradiative recombination for lowering the value of η_1. The second stands for the excess voltage loss. The third term accounts the material quality (series resistance, structure). The last term is the internal waveguide optical loss.

To focus on η_{PCF} to be dependent on the HPL device parameters, we can draw the following assumptions because, all the efficiency terms are interrelated; the laser diode efficiency η_1 =1, the η_{PCF} can be further elaborated as:

$$\eta_{PCF} = \frac{V_{op}.I_{op} + R_s.I_{op}^2}{I_{th}.\eta_{ext}.\dfrac{\hbar\omega}{q} + I_{op}(V_{op} - \eta_{ext}.\dfrac{\hbar\omega}{q}) + R_s I_{op}^2} x \frac{\ln\left(\dfrac{1}{R}\right)}{2\alpha_{\text{int}} L + \ln\left(\dfrac{1}{R}\right)} \qquad (14)$$

The maximum obtainable wall-plug efficiency is formulated as in Equation 9. The maximum value of η_{PCF} is mainly dependent on the operating current and the parasitic device parameters. The series resistance can be decreased by increasing the length of the cavity and in turn lowering the threshold gain. It also implies the reduction in the forward operating voltage for higher electrical power efficiency. The conventional approach to determine the current injection efficiency is in the above threshold regime, using a cavity length analysis (Coldren and Corzine, 1995).

Conclusions

The wall-plug efficiency (WPE) of HPL had been

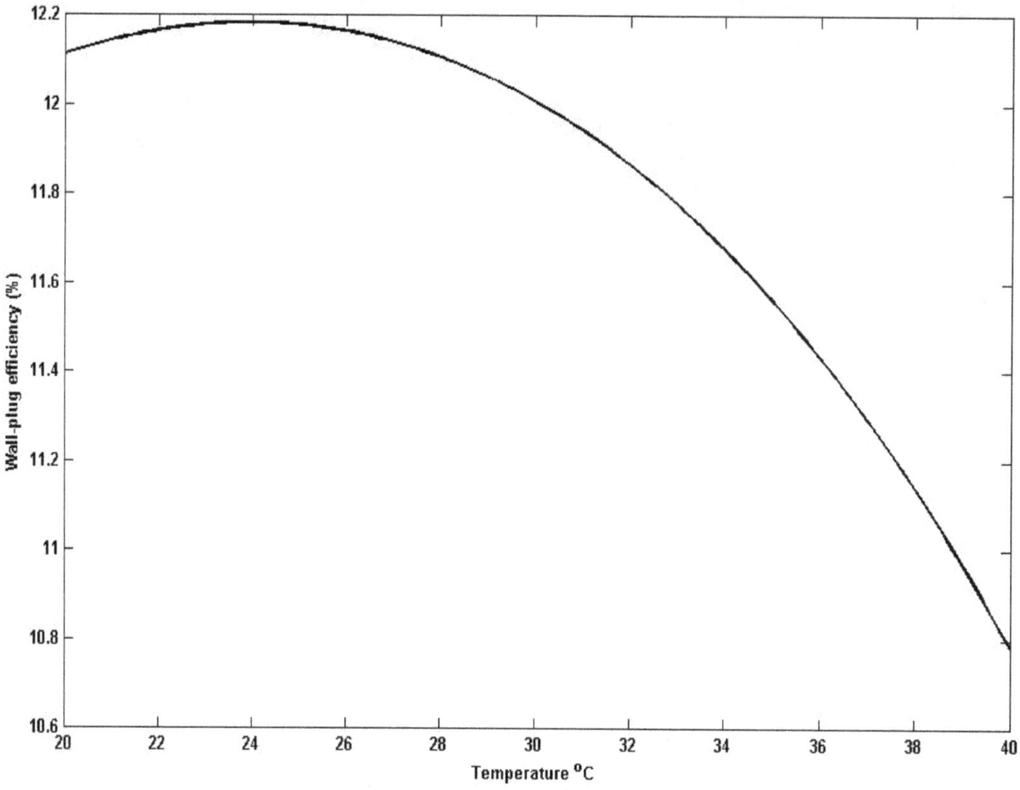

Figure 9. Variation of wall-plug efficiency with temperature.

studied in association with temperature rise and interpreted in terms of the parametric dependence of the device for the optimization for high power output. The main reasons for the decrease of the WPE are the excess voltage loss and internal waveguide optical loss. These effects may help for optimization of the most efficient device for high power output even though there is a trade-off between these factors.

REFERENCES

Bai Y, Bandyopadhyay N, Tsao S, Selcuk E, Slivken S, Razeghi M (2010). Highly temperature insensitive quantum cascade lasers. Appl. Phys. Lett. 97(25):251104-3.

Behringer M (2007). High Power Diode Lasers, Technology and Applications, edited by F. Bachmann, P. Loosen, and R. Poprawe, Berlin Springer. ISBN. 978-0-387-34453-9.

Botez D (1999). Design consideration and analytical approximations for high continuous-wave power, broad-waveguide diode lasers. Appl. Phys. Lett. 74(21):3102-3104.

Botez D, Mawst LJ, Bhattacharya A, Lopez J, Li J, Kuech TF, Iakovlev VP, Suruceanu GI, Caliman N, Syrbu AV (1996). 66% CW wallplug efficiency from Al-free 0.98 µm-emitting diode lasers. Electron. Lett. 32(21):2012-2013.

Coldren L A, Corzine SW (1995). Diode Lasers and Photonic Integrated Circuits. Wiley, New York. pp. 52-55.

Demir A, Zhao G, Deppe DG (2010). Lithographic lasers with low thermal resistance. Elect. Lett. 46(16):1147-1149.

Diehl L, Bour D, Corzine S, Zhu J, Hofler G, Loncar, Troccoli M, Capasso F (2006). High-power quantum cascade lasers grown by low-pressure metal organic vapor-phase epitaxy operating in continuous wave above 400 K. Appl. Phys. Lett. 88(20):201115-3.

Jian W, Summers HD (2010). An efficient approach to characterizing and calculating carrier loss due to heating and barrier height variation in vertical-cavity surface-emitting lasers. Chin. Phys. B. 19(1):014213-0142135.

Kanskar M, Nesnidal M, Meassick S, Goulakov A, Stiers E, Dai Z, Earles T, Forbes D, Hansen D, Corbett P, Zhang L, Goodnough T, LeClair L, Holehouse N, Botez D, Mawst LJ (2003). Performance and reliability of ARROW single mode and 100 µm laser diode and the use of NAM in al-free lasers. Proc. SPIE. 4995:196-208.

Laikhtman B, Gourevitch A, Westerfeld D, Donetsky D, Belenky G (2005). Thermal resistance and optimal fill factor of a high power diode laser bar. Semicond. Sci. Technol. 20(10):1087-1095.

Li H, Towe T, Chyr I, Brown D, Nguyen T, Reinhardt F, Jin Y, Srinivasn R, Berube M, Truchan T, Bullock R, Harrison J (2007). Near 1 kW of continuous-wave power from a single high-efficiency diode-laser bar. IEEE Photon. Techn. Lett. 19(13):960-962.

Ma X, Zhong L (2007). Advance in high power semiconductor diode lasers. Proc. SPIE. 6824:682402-1.

Schmidt B, Sverdlov B, Pawlik S, Lichtenstein N, Muller J, Valk B, Baettig R, Mayer B, Harder C (2005). 9xx High-power broad area laser diode. Proc. SPIE. 5711:201-208.

Shan Q, Dai Q, Chhhajed S, Cho J, Schubert F (2010). Analysis of thermal properties of GaInN light-emitting diodes and laser diodes. J. Appl. Phys. 108(8):08504-8.

Suchalkin S, Westerfeld D, Donetski D, Luryi S, Belenky G (2002). Appl. Phys. Lett. 80(16):2833-2835.

Sze SM (1985). Semiconductor Devices. Wiley, New York.

Williamson R, Kanskar M (2004). Improving the efficiency of high-power diode lasers. Compd. Semiconductors Rev. 2:2.

Effect of dielectric constant on energy losses in lead sulphide thin films grown by solution method at room temperature

Mosiori, Cliff Orori

Department of Physics, School of Pure and Applied Sciences, Kenyatta University, Kenya.

Thin films of lead sulphide (PbS) were deposited using chemical bath deposition (CBD) at different lead ion concentrations. A mixture of sodium hydroxide, varied concentrations of lead nitrate, triethanolamine (TEA), ammonia solution, thiourea, di-ionized and distilled water were used. A dip time of 120 min and pH of 9 at room temperature were maintained. It was found out that dielectric constants of the films varied from a maximum value of 12 to a minimum value of 2.3 in the photon energy range of 1.0 to 4.8 eV. Energy losses in the thin films were also found to be dependent on the concentration of lead ions in the bath and also this energy losses decreased as dielectric constants increased. It was concluded that the films could be used in photoconductivity, capacitance and solar cell absorber applications.

Key words: dielectric constant, thin films, lead sulphide, temperature.

INTRODUCTION

Semiconducting thin films and particularly lead chalcogenides have been widely studied owing to their interesting switching property (Prakash and Ashokan, 2004). These materials are used to fabricate a variety of electronic devices, which arises when the material is cast into thin film form. It is also observed that most physical properties reported on chalcogenides have been investigated using polycrystalline pellets or electrodeposits (Bresser et al., 1996; Lade et al., 1994). A good amount of work on DC conduction (Sagbo et al., 1994), contact capacitance (Simashkevin et al., 1994), spectral properties (Vidourek et al., 1995), AC conduction (Giuntini et al., 1995), and structural and magnetic properties (Dauoudi and Ekpunobi, 1996) has also been reported. However, dielectric constant behaviour as a function of energy loss and photon energy (frequency) has been over looked to some extent. In semiconductor thin film IR detector integrated circuits (for which high capacitance in small area is required), capacitors are grown by either evaporation or sputter techniques. When large-area capacitors are required, then the appropriate method of choice of growing them could be chemical bath method (CBD). To use lead sulphide thin film circuits, it is necessary that their dielectric and energy losses be understood so as they have appropriate value ranges. The dielectric coefficient of a thin film (capacitance) is an important practical parameter for assessing the expected behaviour of any thin film device. This makes it necessary to study the effect of energy, dielectric losses and (photon energy) frequency on thin films for any device fabrication. The dielectric behavior of thin film devices depends not only on their material properties but also on the method and conditions of preparation. Fringing effects at the edges of thin film dielectrics is usually negligible because the thickness of the dielectric is usually very small and uniform compared to its lateral

Table 1. Parameters for depositing PbS thin films.

Conc. Pb^{+2} (mol.)	Vol. Pb^{+2} (cm^3)	Vol. 1M NaOH (cm^3)	Vol. 1M TU (cm^3)	Vol TEA (cm^3)
0.3	5	5	6	2
0.4	5	5	6	2
0.5	5	5	6	2
0.6	5	5	6	2
0.7	5	5	6	2

dimensions. The magnitude of its geometrics and measured influence of dielectric capacitance gains or losses may differ if the electric field at the thin film/metal insulator interface varies with the insulator over a certain region. For given material the film thickness alone establishes the capacitance density which in turn can be used to determine the area needed for a particular capacitance value. Since most capacitors utilize constant and uniform thickness then much concern is not only based on thickness but also on dielectric and energy losses in capacitor performance. The dielectric loss as part of the energy of an electric field is dissipated without recovery as heat in the dielectric material and is comprised of two parts; the first part arises due to lead resistance and electrode resistance. This part is frequency dependent. It is effective and very influential at higher frequencies or photon energies. It is minimized using high conducting metal electrodes. The other part arises due to material property and is also frequency dependent (Maissel and Glang, 1970). Dielectric strength is found to reduce rapidly below about 100 nm wavelength as the presence of pinholes or discrete defects in thin films increases. To the best of our knowledge, little or no reports is available on the study of dielectric properties of lead sulphide (PbS) thin films deposited by chemical bath at room temperature. In this paper, an attempt is made to report some of the dielectric properties of solution deposited PbS thin films for solar cell, charge storage/detection applications. In all the thin films used in this study, thickness thin film was kept at about 103 nm to minimize the inherent defects and pinholes.

METHODOLOGY

The chemical bath deposition technique was used. Glass substrate (microscope slides) which had been previously degreased in concentrated nitric acid for 24 h, cleaned in cold water with detergent, rinsed with di-ionized water and dried in clean dry air to provide better surface nucleation for growth of the films were used. A mixture of 1 M sodium hydroxide, 0.3 – 0.7 M lead nitrate solutions, 7.4 M triethanolamine (TEA), 14 M ammonia, 1 M thiourea, di-ionized and distilled water, microscopic glass slides and beaker were used. 5 ml of lead nitrate was poured into a 100 ml beaker followed by 5 ml of 1 M sodium hydroxide and then thoroughly stirred to obtain a milky solution. 6 ml of 1 M thiourea and 2 ml of 1 M tri-ethanolamine were immediately added, the mixture thoroughly stirred with a glass rod before glass slides was vertically introduced into the beaker. The dip time was kept at 120 min, and pH at 9 and at room temperature. Lead nitrate concentration was varied for the subsequent films from 0.4-0.7 M PbS as illustrated in Table 1. The grown samples were removed, rinsed with distilled water and allowed to dry before they were analyzed and characterized.

RESULTS

During deposition, cations and anions in the solution react to become neutral atoms, which precipitate slowly under the control of a complexing agent (TEA). With the addition of TEA, the reaction proceeded slowly so that PbS thin films of neutral atoms were formed on the substrate. The complexing agent slowed down precipitation action while ammonia served as a pH stabilizer. Sulphide ions were released by hydrolysis from thiourea and complexes formed were adsorbed onto the substrate as heterogeneous nucleation. Growth took place by ionic exchange reaction of S^{2-} with Pb^{2+} ions and by this process of ion-by-ion exchange, PbS thin films were deposited. The films were transparent, uniform and adherent though specularly reflecting. In discussing the dielectric properties of PbS thin films, a model to represent the dielectric property was used. A quantum–mechanical model to outline the characteristics of dielectrics of PbS thin films was used. The model predicts on the basis of classical mechanics. Based on the wave mechanical theory of matter, a dielectric is a material which is so constructed such that the lower bands of its allowed energy levels are completely full at the absolute zero of temperature (Exclusion Principle) and at the same time isolated from higher unoccupied bands by a large zone of forbidden energy levels. This is observed in PbS thin films (they have a dielectric constant of above 8). This shows that conduction in the lower fully occupied bands is impossible since there are no un-occupied energy levels to take care of the additional energy which would be acquired by the electrons from the applied field caused photons. The zone of forbidden energy levels is so wide that there is only a negligible probability that an electron in the lower band of allowed levels will acquire enough energy to make the transition to the unoccupied upper band where it could take part in conduction. That is why PbS thin films are poor solar cell absorber layers. When photons fall on PbS thin film layer, an electric field due to photons is impressed upon a PbS, positive and

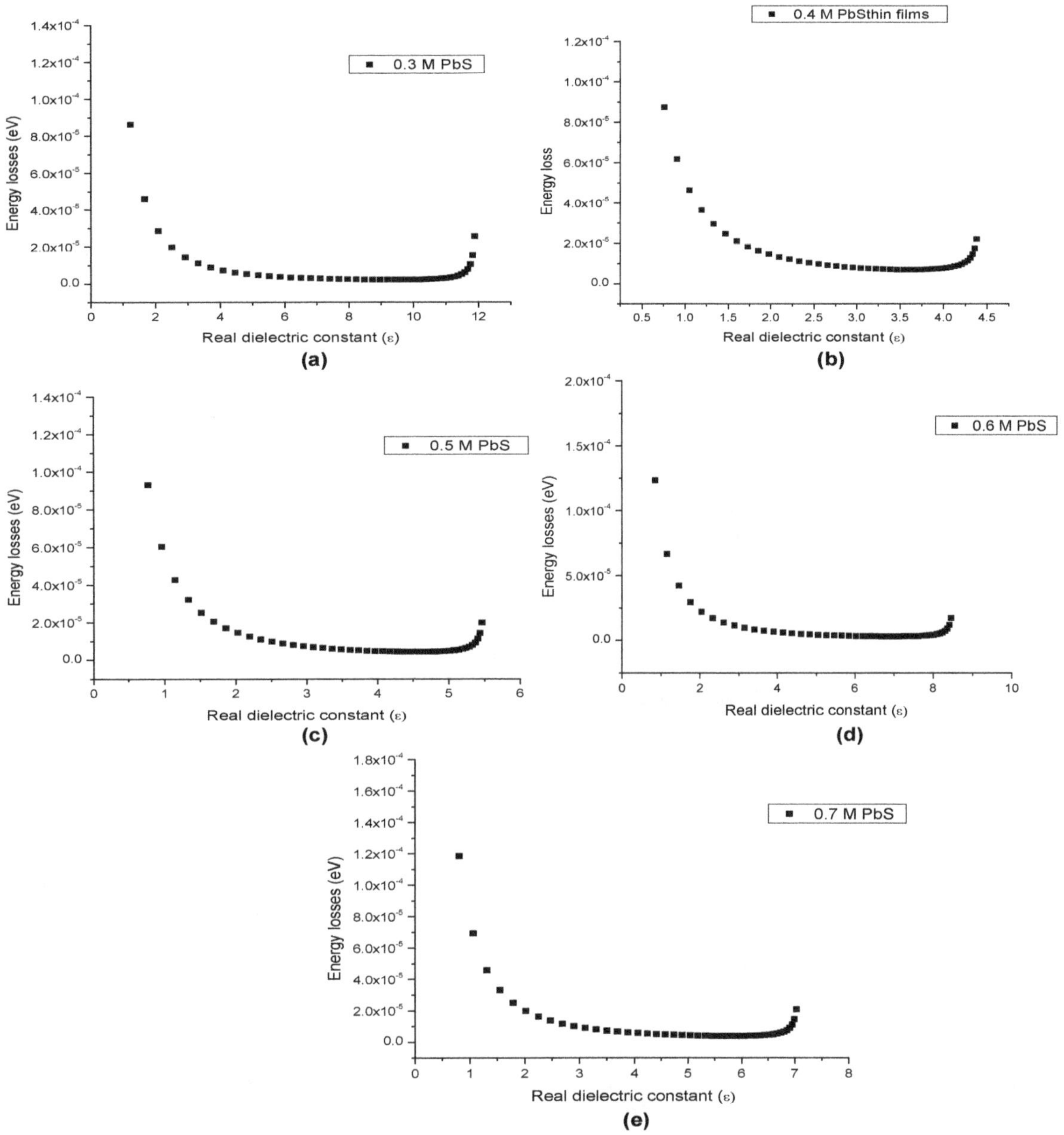

Figure 1. (a) Variation of real dielectric constant versus energy losses in 0.3 M PbS films; (b) Variation of dielectric constant against energy losses for 0.4 M PbS films; (c) Variation of real dielectric constant versus energy losses in 0.5 M PbS thin films; (d) Variation of real dielectric constant versus energy losses in 0.6 M PbS thin films; (e) Variation of real dielectric constant versus energy losses in 0.7 M PbS films.

negative charges in their atoms and molecules are displaced in opposite directions causing polarization. Since the motion of charges of opposite sign move in opposite directions, they constitute an electric current called a polarization current. This polarization causes energy losses in PbS thin films (Figure 1a to d). A charge accumulating in PbS thin film layer in an un-measurably short time referred to as the instantaneous dielectric constant or geometric dielectric constant describes the property of the medium giving rise to the effect called dielectric states (Figure 1 a-d). This is recognized by the modern theory that identifies two distinct types of charges and charging currents that rapidly results into forming instantaneous polarizations and absorptive polarizations. PbS thin films depends on this mechanism as shown in Figure 2. The effect of chemical and physical structures in PbS dielectric constant depend on two quantities; magnitude and relaxation-time which in turn determine many of the properties of dielectric polarizations of the absorptive type. The magnitude of the polarizability (*k*) of PbS can be expressed in terms of a directly measurable quantity or by simulation in relation to dielectric constant

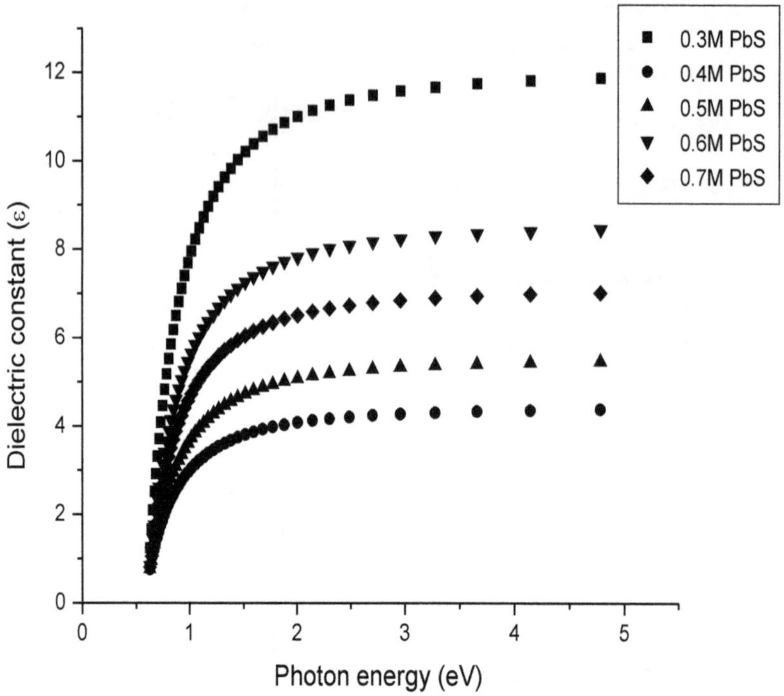

Figure 2. Variation of dielectric constants versus photon energy (eV) for PbS.

by the relation:

$$k = \frac{3(\varepsilon - 1)}{4\pi(\varepsilon + 2)} \quad (1)$$

A classification of dielectric polarization into absorptive polarization has been made based on this expression. Electronic polarizations are due to the displacement of charge within the atoms and are the most important of the instantaneous polarizations that determine the dielectric constant of PbS thin films. Polarizability per unit volume due to the electronic polarizations is considered to be a quantity which is proportional to the number of bound electrons in a unit volume and inversely proportional to the forces binding them to the nuclei of the atoms. This occurs in PbS molecules. This is depicted by the small band gap of about 0.4 eV for PbS thin films. It is known that the time required for any applied field to displace electrons within an atom to a new position with respect to their nuclei is so short that there is no observable effect of time or frequency upon the value of the dielectric constant. This is true only until frequencies corresponding to absorption lines in the visible or ultra-violet spectrum are attained (called the optical frequency range). The frequencies in the optical range corresponding to the absorption lines in the spectrum, the dielectric constant or better the refractive index changes rapidly with frequencies and till absorption starts to appears. This is why this report is justified to use refractive index (*n*) and dielectric constant (*ε*) interchangeably for the qualitative discussion of the properties of dielectric polarization that follows the relation $\varepsilon = n^2$, which is known as Maxwell's rule based on electromagnetic theory and only applicable whenever 'ε' and '*n*' are measured at the same frequency no matter how high or low they may be. The refractive index (*n*) in the visible spectrum provides the means of determining the magnitude of electronic polarizations since other types of polarizations have negligible magnitude when frequency of the impressed field lies in the visible spectrum. For materials like PbS having only electronic polarizations, dielectric properties are simply independent of frequency in the electrical frequency range and are characterized by complex dielectric behaviours. PbS thin films also experience atomic polarizations. Atomic polarizations form part of the polarization of a molecule which can be attributed to the relative motion of the atoms in a molecule which it is composed. They are attributed to perturbation by any applied field causing vibrations on atoms due to energy gained from photons. Ions having their characteristics or resonance in the infra-red range experience them and thus PbS thin films that are sensitive to infra-red and atomic polarization begin to contribute to ε or n^2 at frequencies below approximately 10^{14} seconds; that is, in the near infrared and about 10^{10} cycles per second frequency where optical and electrical frequencies ranges merge. Atomic polarizations contribute a constant amount to ε and n^2 for any given material. It is taken as the difference between polarization (measured at some low infrared or high electric frequency) and the

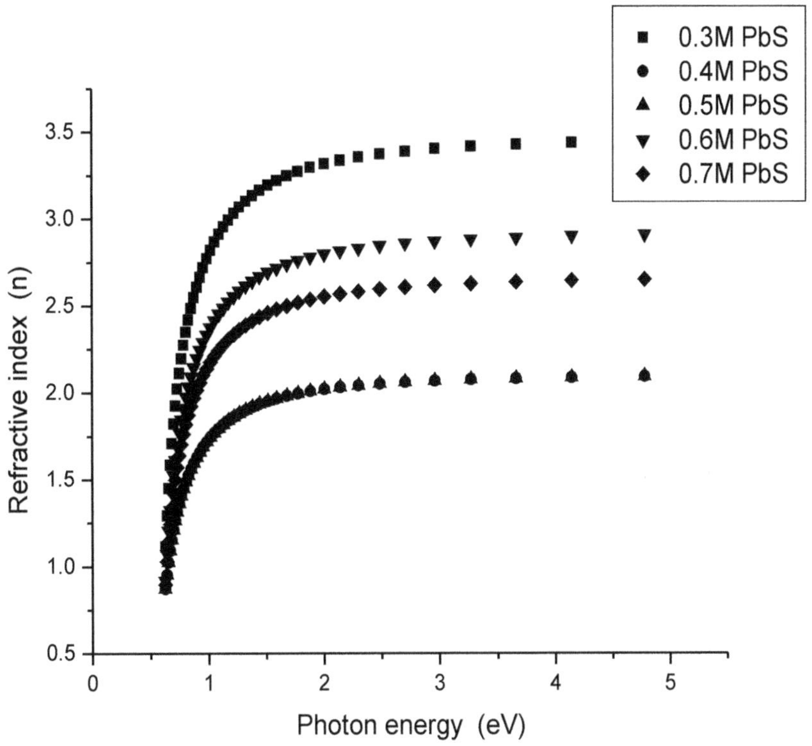

Figure 3. Variation of refractive index versus photon energy for PbS thin films.

electronic polarization as determined from refractive index measurements in the visible spectrum. When photoelectric effect causes a field to be applied, there is a tendency for molecules in a material to align themselves with their dipole axes in the direction of that filed. This dipole polarization is superimposed upon electronic and atomic polarizations induced by the field. Since dipole polarization depends upon the displacement of charges within an atom, rather than upon the displacement of charges within a molecule, time required for this type of polarization to form depends on internal frictions of the material. Debye expressed time of relaxation of the dipole polarization in terms of the internal frictional force by the equation:

$$\tau = \frac{\xi}{2kT} = \frac{8\pi\eta r^2}{2kT} \quad (2)$$

where ξ is the internal friction coefficient, η is the coefficient of viscosity, *r* is the radius of the molecule and T is the absolute temperature. Applying this equation to the calculation of the relaxation time for PbS thin films at room temperature, τ is found to be 10^{-10} seconds assuming that PbS molecular radius is 2 x 10^{-8} cm and taking viscosity as 0.01 poises as they form in the bath, the result agrees with the experimental studies extending from 10^{9} to10^{11} cycles. This is what causes the dielectric constant to decrease from its highest value to a value approximately equal to the square of their refractive index as seen in Figure 3.

Conclusion

Thin films of PbS were successfully deposited at room temperature using CBD at different lead ion concentrations. Their dielectric constant varied from a maximum value of 12 to a minimum value of 2.3 in the photon energy range of 1.0 to 4.8 eV. Energy loss was found to decrease as the dielectric constant of the thin film increases and it was concluded that the films could be used in photoconductivity, capacitor fabrication and solar cell absorber applications.

ACKNOWLEDGEMENTS

We acknowledge the assistance offered by the Department of Physics, Kenyatta University where this study was carried out and the Department of Material Science, University of Nairobi (Chiromo campus) for kindly providing UV-VIS-NIR spectrophotometer for spectral analysis.

REFERENCES

Bresser J, Fajinmi G, Sasusi Y (1996). Deposition time dependence on absorptivity of chemically deposited lead sulphide thin films. Res. J.

Appl. Sci. 9:931-937.
Chouldrury N, Sarma K (2008). Structural characterisation of nanocrystralline PbS thin films: synthesised by CBD method. Indian J. Pure. Appl. Phys. 46:261-265.
Dauoudi I, Ekpunobi A (1996). Optical properties and band offsets of CdS/PbS superlattice. The Pacific J. Sci. Technol, 11:404-407.
Glang M, Maisset A (1970). The influence of hydrazine hydrate in the preparation of lead sulphide thin film. Iranian J. Sci. Technol. 29:151-162.
Giuntini R, Pawar S, More P, Pawar A, Patil V (2010). Nanocrystalline PbS thin films: Synthesis, microstructural and optoelectronic properties. Scholars Res. Libr. 2:1-6.
Osherov A, makai J, Balasz J, Horvath Z., Gutman, N, Amir S, Golan Y (2010). Tenability of optical band edge in thin PbS films chemically deposited on GaAs(100). J. Condensed Matter. 22:1-7.
Patil R, Lisca M, Stancu V, Buda M, Pentia E, Botila A (2006). Crystalline size effect in PbS thin films grown on glass substrates by chemical bath deposition. J. Optoelectr. Advan. Mater. 1:43-45.
Pentia E, Pintilie L, Matei I, Botila T, Ozbay E (2001). Chemical prepared nanocrysalline PbS thin films. J. Optoelectr. Advan. Mater. 3:525-530.
Prakash V, Jumate N, Popescu G, Moldovan M, Prejmeriean C (2004). Studies of some electrical and photoelectrical properties of PbS films obtained by sonochemical methods. Chalcogenide Lett. 7:95-100.
Simashkevin S, Kamoun N, Brini R. Amara A (1994). Structural and optical properties of PbS thin films deposited by chemical bath deposition. Elsevier Mater. Chem. Phys. 97:71-80.
Sagbo A, Junghare A, Wadibhasme N, Darypurkar A, Mankar R, Sangawar V (1994). Thickness Dependent Structural, Electrical and Optical Properties of Chemically Deposited Nanopartical PbS Thin Films. Turk. J. Phys. 31:279-286.
Vidourek J, J´auregui R, Ram´ırez-Bon A, Mendoza-Galva´n, M, Sotelo-Lerma A (1995). Optical properties of PbS thin films chemically deposited at different temperatures. Thin Solid Films, 441:104–110.

16

Capacitor bushing optimization via electrostatic finite element analysis

Amin Mahmoudi[1], Seyed Mahdi Moosavian[2*], Solmaz Kahourzade[1] and Seyed Nabi Hashemi Ghiri[3]

[1]Department of Electrical Engineering, University of Malaya, Kuala Lumpur, Malaysia.
[2]Department of Engineering, Shahrood Branch, Islamic Azad University, Shahrood, Iran.
[3]Department of Electrical Engineering, Shiraz University, Shiraz, Iran.

In high-voltage capacitor bushings, there is maximum field intensity at the edge of the aluminum foil and where foils overlap. Any technique decreasing much field intensity in those regions will significantly optimize the bushings' electrostatic characteristics. This paper presents the study of the effect of edge diversion in the overlapped region of an aluminum foil and of the shape of aluminum foil edge in improving the maximum field intensity of the overlapped region. Electrostatic analysis was used to minimize the field intensity. 2D and 3D simulations and analyses were run, respectively via finite element method magnetic (FEMM) and Vector Field Opera. Techniques improving field intensity distribution of capacitor bushing are presented. Results show that foils that are not folded and have the inner edge placed lower than the outer edge better distribute the field around the edges.

Key words: Capacitor bushing, cavities, electric field, electrostatic simulation, finite element analysis.

INTRODUCTION

As voltage increases, so does the importance of electrical insulation in high-voltage equipment (Guoqiang et al., 1999). The importance of insulation materials lies in their resistance against voltage or electric field intensity (Ueda et al., 1985). High voltage and high electric field intensity cause partial discharge, which, through the cavities formed during fabrication (Seghir et al., 2006; Ghourab and El-Makkawy, 1994; Hsu et al., 2011) leads to failure of insulation materials and deterioration and breakdown of dielectric insulation (Pompili and Mazzetti, 2002). Field analysis method allows analysis of electric field intensity in various parts of an insulator (Arora and Mosch, 2006). It allows calculation of maximum electric field and its distribution in all parts of a bushing and for various insulating materials. Various numerical techniques let designers solve problems insolvable by analytical methods (Faiz and Ojaghi, 2002). Finite element analysis is a powerful and precise method in complex geometrical modeling, providing solutions to many complicated problems in various fields of engineering (Monfared, 2011; Kumar Singh and Sharma, 2011; Jumaat and Ashraful Alam, 2011). Its capability in numerical solving also makes it the main analyzing method for calculation of the electric field in high-voltage bushing (Monga et al., 2006; Lesniewska, 2002).

Bushing is used in high-voltage equipment to insulate the high-voltage conductor from the external earthed body; 25 to 30% of large power transformer failures have been reported as due to bushing (Smith et al., 2010). At high voltage, aluminum plates uniformly distribute the high voltage potential between conductor and body. The plates, referred to as foils, make a series of cylindrical capacitors that uniformly distributes electric field. They are called capacitor bushings. In their core, typically, are thick papers that fill the distance between two capacitor plates that usually are conductors, semiconductor layers (graphite) or metallic (in which case they are mostly aluminum). The capacitor plates are placed as coaxial cylinders. The thick papers distancing the plates provide the required mechanical stability to the cylindrical capacitors. Figure 1 is a general view of a capacitor bushing (Ellis, 2004). When voltage increases, more aluminum foil

*Corresponding author. E-mail: sm.moosavian@gmail.com.

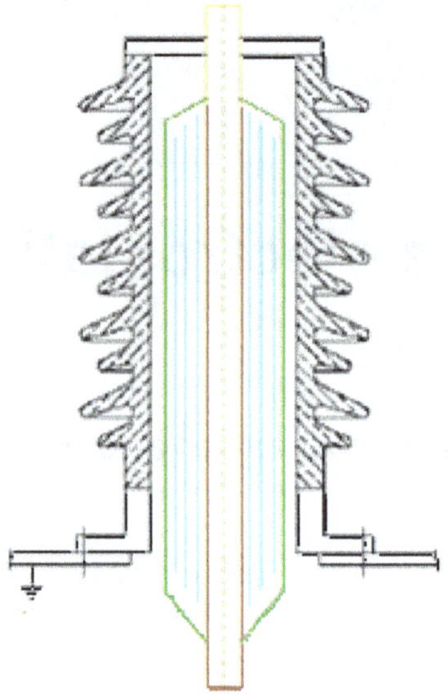

Figure 1. Typical configuration of a capacitor bushing.

is used, so, foil placement and the shape of the foil edge become crucial.

The part of a capacitive foil most sensitive to electric field intensity is its edge, which, when overlapping, increases the sensitivity. The electric field intensity is higher around that region than in the other regions. The relative position of two edges and their shape affect the field intensity of the surrounding and the overlapping regions, and so, partial discharges are expected to start there. This paper presents the calculation of the electric field intensity at the edges of a high-voltage bushing, and proposes a new technique and a new foil shape, both improving the field intensity of the affected regions. It first presents a mathematical model for the insulation system of a capacitive insulator, then compares the simulations of capacitor bushing electric field, via finite element method (FEM) software (2D simulation) and Vector Field Opera (3D simulation). Next it describes how folded and non-folded edges enhance electric field distribution of capacitive bushing, concluding with suggestions for improvement to electric field distribution.

THE MATHEMATICAL MODEL OF AN INSULATION SYSTEM

Considering the relation between electric potential and electric field intensity, the electric field in an insulation system is described by Laplace equation;

$$\nabla^2 V = o \qquad (1)$$

The equation is described with Neuman and Dirichlet boundary conditions. Solving Laplace's equation with specific boundary conditions gives:

$$E = -\nabla V \qquad (2)$$

To calculate the potential of capacitor plates and the cylindrical capacitor value between them, the capacitance dispersion between the plates should be considered. Capacitance dispersion is usually due to the foil's sharp edge. The value of each capacitor, even with a complex shape, can be derived by calculating the amount of electrical energy the field stores;

$$C = \frac{2W_e}{V^2} \qquad (3)$$

The electrical energy of the electric field between the foils is:

$$W_e = \int \frac{D.E}{2} dv \qquad (4)$$

How foil edge affects capacitor value can be calculated by finite element method (FEM). Cylindrical symmetry causes the electric field vector in the capacitor bushing to have two axial and radial components. In the region of two foil edges overlapping, the field intensity is higher owing to the resultant fields of the two edges' plates. Figure 2 shows the relative positions of foils 1, 2 and 3. The resultant field near the edges can be calculated.

Assuming the fields due to the edge of the second foil in the overlap region to be E21 and E22, and the fields due to the adjacent foils to be E1 and E3, the combined field of the two edges of the second foil and the fields of the other adjacent foils (foils 1 and 3) will determine the field intensity in the overlap region of foil 2. The radial component of the field there equals:

$$E_r = E_{r21} + E_{r22} + E_{r1} + E_{r3} \qquad (5)$$

Similarly, the axial component of the field equals:

$$E_a = E_{a21} + E_{a22} + E_{a1} + E_{a3} \qquad (6)$$

The electric field intensity can be calculated through Equation 7:

$$E = \sqrt{(E_r^{\,2} + E_a^2)} \qquad (7)$$

Figure 3 shows in 3D the overlap region of the foil edges. Figure 4 is the resultant field in an overlap region of foil edges. It shows that owing to the position of both foils 1 and 3 relative to 2, the radial component of foil 1 is higher than its axial component, but the axial component in foil 3

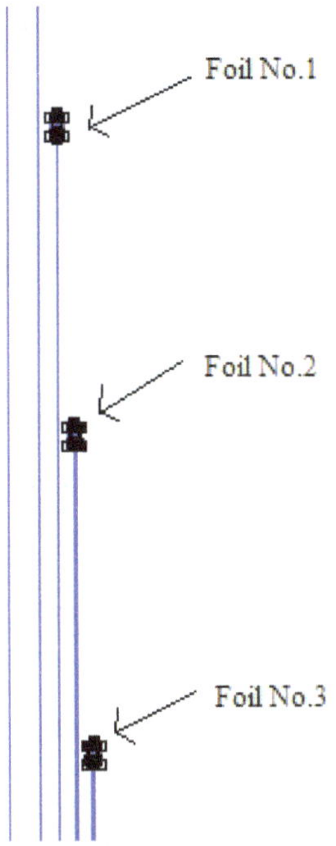

Figure 2. The proposed relative position of the foil edges.

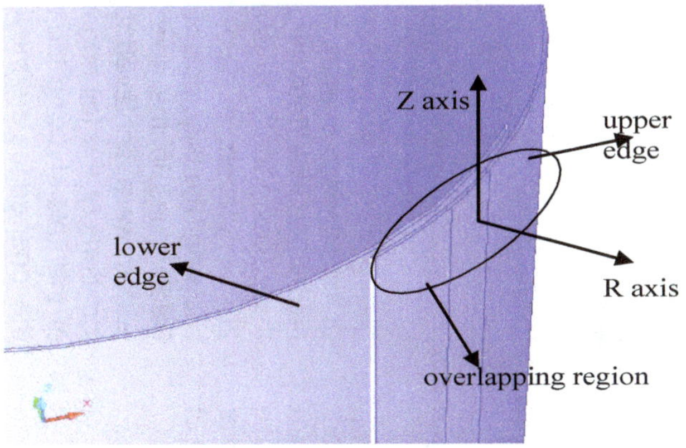

Figure 3. Position of a foil-edge in an overlap region.

is higher than its radial component.

SIMULATION

Finite element was used to investigate the electric field distribution of a capacitive bushing. All available finite element analysis (FEA) softwares were first compared before the one suitable for distribution was selected. FEA yielded the voltage for each foil of the selected test bushings. The field intensity of the overlap regions was obtained.

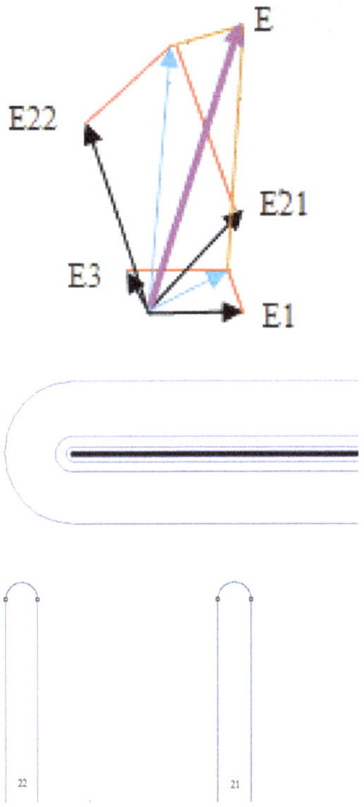

Figure 4. The resultant field in the overlap region of the foil edges.

Electrostatic simulation

The cylindrical symmetry in capacitor bushing allows 2D simulations. The capabilities of three simulation softwares (Table 1) were compared; finite element method magnetic (FEMM) was chosen for 2D simulation and Vector Field Opera 14.0 for 3D simulation (Meeker, 2006).

HVB 1 to 145 kV capacitor bushing was selected for 2D simulation (Nirou Company, 2011). Its main insulator comprises several compressed layers of oil-impregnated paper. There are 47 foils, of which 23 are used in the air section of the bushing and 23 in the oil section (the resin part). The last foil is used as grounding foil, connected to ground through a flange, its potential zero. The oil-impregnated paper isolates the foils. Table 2 lists the bushing materials and their characteristics (Gremmel and Kopatsch, 2001; ABB company, 2011). Figure 5 is a 2D figure of the bushing.

SIMULATION RESULTS

Figure 6 is an auto-mesh generated on Vector Field Opera 14.0 software, a tetrahedral element with 6 nodes that fit foil-edge curvature. Figure 7 gives the simulation results of all parts of the bushing: good voltage distribution

Table 1. Comparing three simulation softwares.

	Software	Vector field	ANSYS	FEMM
Pre-Processing	Possibility of 3D simulation	No	Yes	Yes
	Geometry problem definition	Easy	Difficult	Partly difficult
	Possibility of importing files from AutoCAD and other softwares	Yes	Yes	No
	2D mesh classification	Easy	Partly easy	-
	3D mesh classification	-	Difficult for huge amount of meshes	Difficult
	Solution for speed	Good	Low	Good
Post-Processing	Possibility of displaying the electric field	Yes	Yes	Yes
	Possibility of displaying the field components	Yes	Yes	Yes
	Display of the equipotential lines	Yes	No	Yes
	Possibility of displaying the fields in 3D	No	Yes	Yes
	Possibility of displaying quantities in specific pages	-	Yes	Yes
	Possibility of using software's output for other processing softwares	Yes	Yes	Yes

Table 2. Material properties of the bushing components (ABB).

Material	Relative permittivity	Conductivity σ
Oil impregnated paper	2.3	6.3×10^{-11}
Mineral oil	2.2	1×10^{-10}
Porcelain	6	1.9×10^{-9}
Resin	4.2	2.2×10^{-9}
Air	1	8.0×10^{-15}

distribution throughout the bushing and non-uniform distribution in the distal region. Next described is the simulation performed to optimize foil-edge field. Table 3 lists the numeric voltage of each foil. Figures 8 and 9 give the voltage value of each side of the bushing.

Simulations of the overlap region

Simulations were performed for various situations of foil edges in overlap regions, and the maximum field value was determined for the regions around the edges. Figure 10 is a visual presentation of a 2D simulation of two overlapping foil edges, showing the equipotential lines. The maximum field values in the specified interval around the edges were achieved by changing the relative locations of the foil edges. Figure 11 is a contour of the field showing the equipotential lines where the inner edges are moved in Z-axis direction and displaced lower. Parameter "M" indicates the displacement of the inner edge from the outer edge. Figure 12 compares the maximum intensities of various situations.

The horizontal axis represents the deviation of the inner edge (the left edge in Figure 10) from the outer edge. Units are in millimeter. Field value clearly varies with changes to the position of the inner edge. Table 4 lists the results for when the edges are far from each other. The negative sign shows the inner edge is lower than the outer edge.

Figure 12 shows maximum voltage intensity

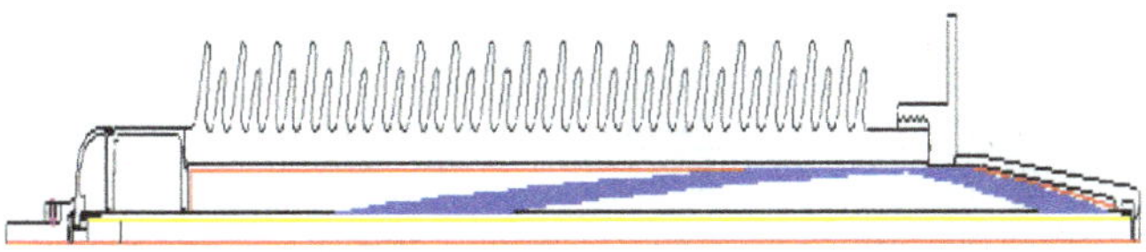

Figure 5. 2D view of the HVB 1 to 145 kV capacitor bushing selected for simulation.

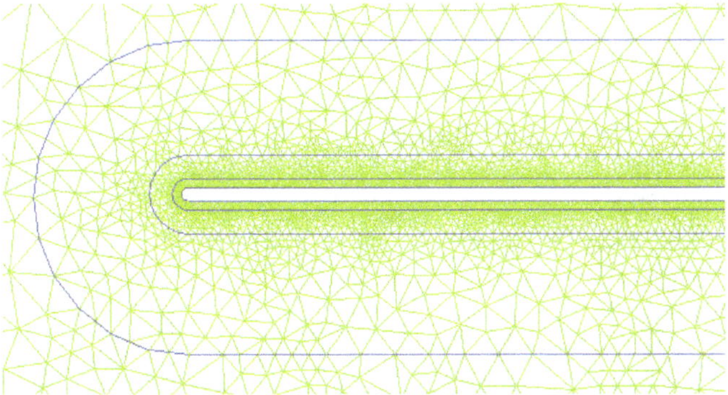

Figure 6. 3D auto-mesh generation in Vector Field Opera 14.0 (Opera Version 14.0 User Guide, 2011).

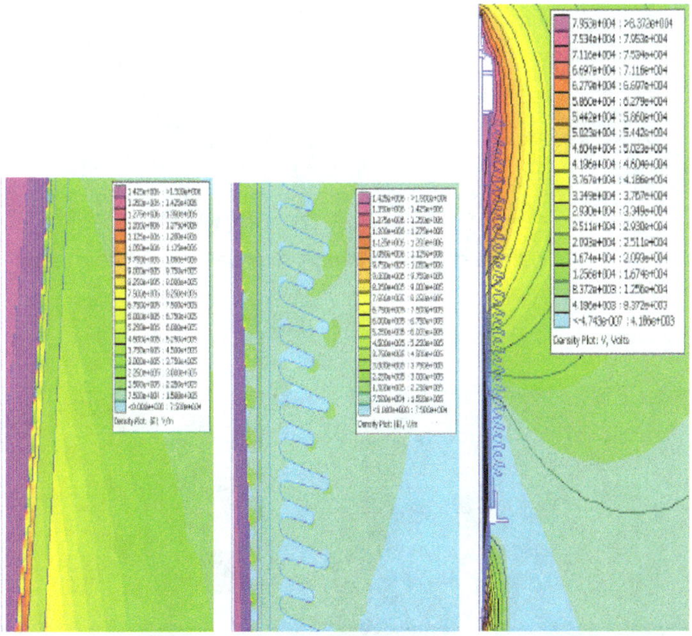

a b c

Figure 7. (a) Field intensity on oil-side of bushing; (b) Field intensity on air-side of bushing; (c) Voltage contours in various regions of bushing.

varying with movement of the edges relative to each other. The positive value of the X-axis shows that the outer edge is higher than the inner edge.

Techniques to improve field intensity in overlap areas

The best way to find the optimum configuration improving field intensity in overlap region of foil edges is 3D simulation (the non-symmetrical edges make 2D simulation impossible). Vector Field Opera 14.0 3D was used to simulate the electric field in the region of edges (Opera Version 14.0 User Guide 2011).

To simplify simulation, the cylindrical edges are considered curvature-less, so modeling them becomes easy. This study simulated various conditions of the edges: their relative positions and their shapes. It also compared the field values of the regions near the edges. Simulation results of the conditions are plotted in field contours. The overall results for the conditions are compared, and solutions for optimization were presented.

Folded edge

Folded edge is rectangular, so is sharp. Field intensity in sharp regions is high, so the shape of an edge should be changed. The foil edge is thus folded onto itself. Figure 13 shows the inner edge folded. Its field contour, for a plane near the edge overlap, is as shown in Figure 14.

Non-folded edges

The foil edges were, for this simulation, considered non-folded. Their corners are as shown in Figure 15. Figure 16 is the simulation results for a plated contour.

Non-folded and displaced

The edges were not folded and the inner edge was lower than the outer edge. The placing of the top of the inner edge at a level lower than the top of the outer edge is based on the 2D simulation described previously. Figure 17 shows edge positions and Figure 18 their field contours. Table 5 lists the maximum electric field for folded, non-folded and non-folded and displaced-low edges.

CONCLUSION AND SUGGESTION

The present study has been a finite-element analysis of the electric field of an HVB 1 to 145 kV capacitor bushing. The electric field in various regions of the bushing was calculated through a numeric solution of Laplace's equation with Neumann and Dirichlet boundary conditions. The high field-intensity at the edges of the aluminum foil and where the foils overlap motivated the technique for making it uniform at that point. Simulation results show that placing the inner edge at a position

Table 3. Voltage of each foil.

Foil number	Voltage (V)	Foil number	Voltage (V)	Foil number	Voltage (V)
1	79532.5	17	22680.7	33	43228.7
2	75720.3	18	19329.9	34	39967.2
3	71790.7	19	15862	35	36642.4
4	68147.1	20	12775.2	36	33250.4
5	64546.5	21	9561.2	37	29786.9
6	60991	22	6039.04	38	26515.8
7	57471.7	23	3172.94	39	23301.8
8	53869.7	24	78648	40	20004.4
9	50413.5	25	74125.8	41	16933.1
10	46972.7	26	69579.3	42	13763.2
11	43675.7	27	65476	43	11005.6
12	40271	28	61505	44	8182.66
13	36756.4	29	57670.8	45	5132.29
14	33129.6	30	53951.4	46	2681.35
15	29661.2	31	50214.2	-	-
16	26229.2	32	46683.5	-	-

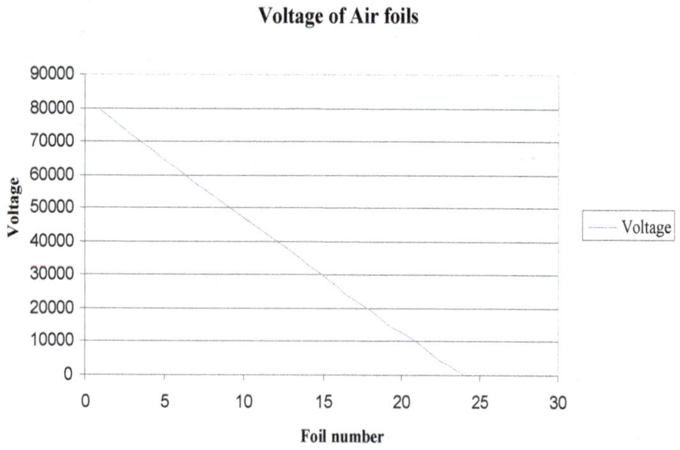

Figure 8. Voltage on air-side of bushing.

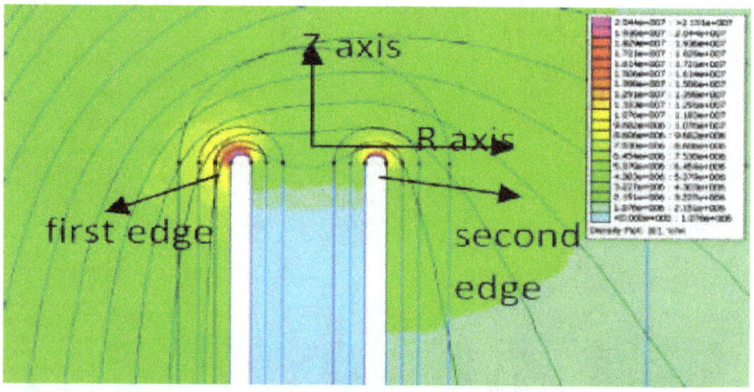

Figure 10. Voltage contour and equipotential lines in the overlap region.

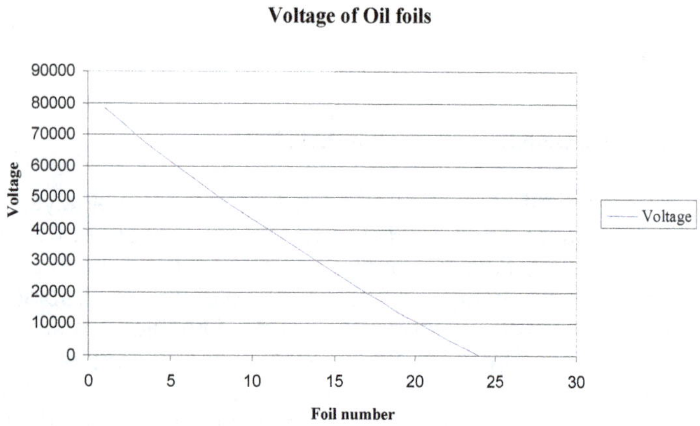

Figure 9. Voltage on oil-side of bushing.

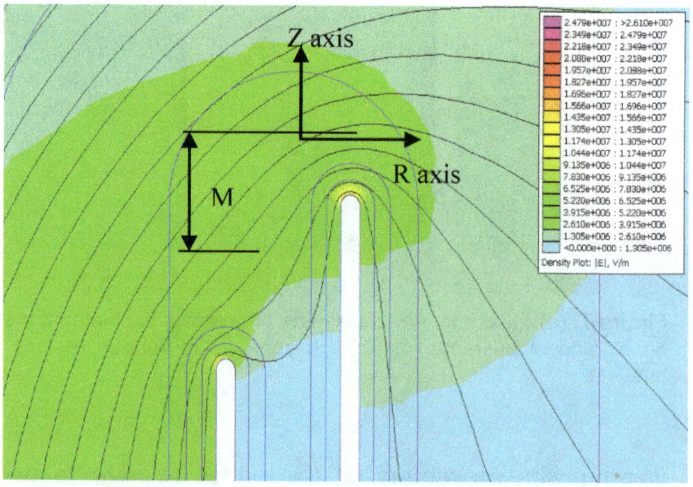

Figure 11. Field contour and the displacement value of the inner edge from the outer edge.

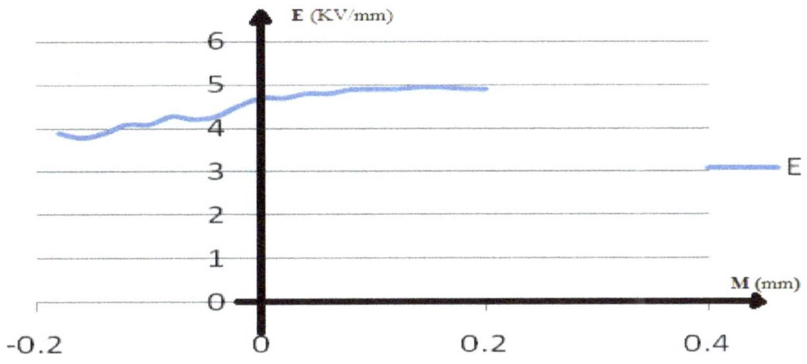

Figure 12. Variation of the maximum field intensity around edge against edge position.

Table 4. Electric field values with increased distance between edges.

M (distance between inner edge and outer edge) (mm)	E_{max} (kV/mm)
-1.12	2.9
1.0	4.74

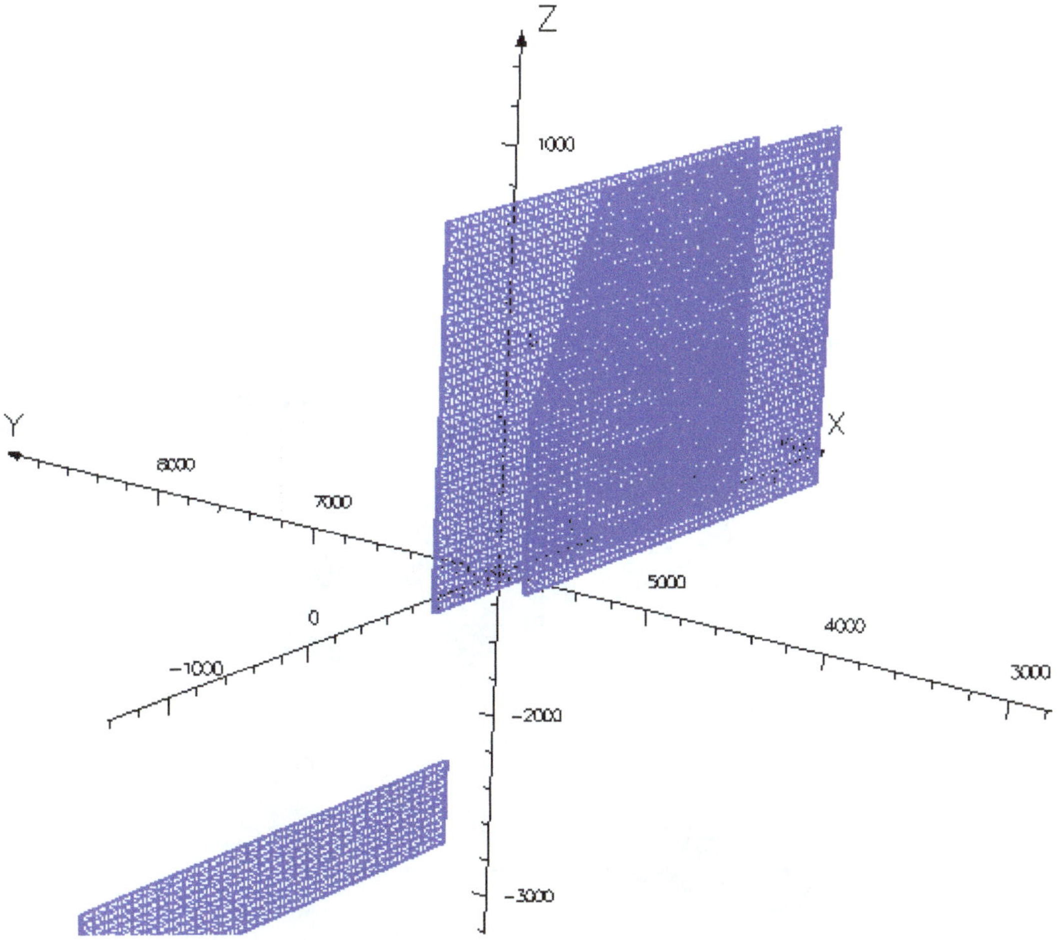

Figure 13. Position of edge against each other (folded edges).

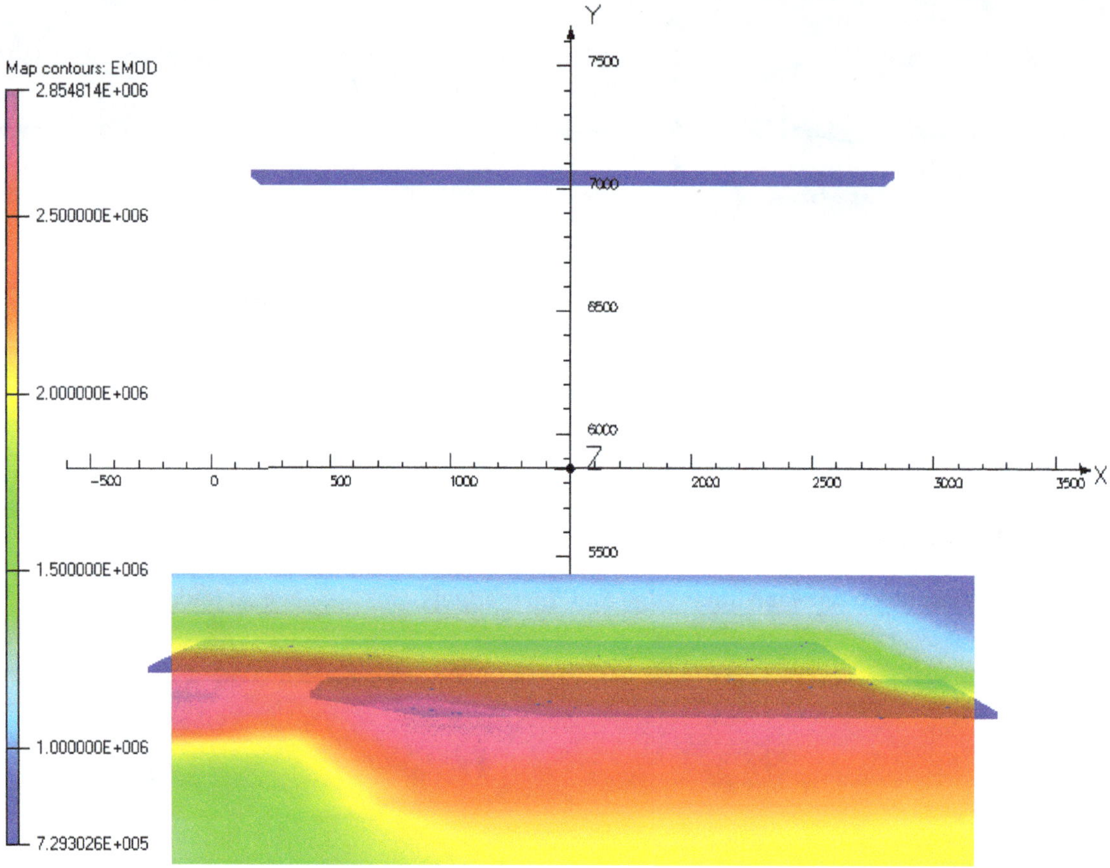

Figure 14. Electric field contour (folded edges).

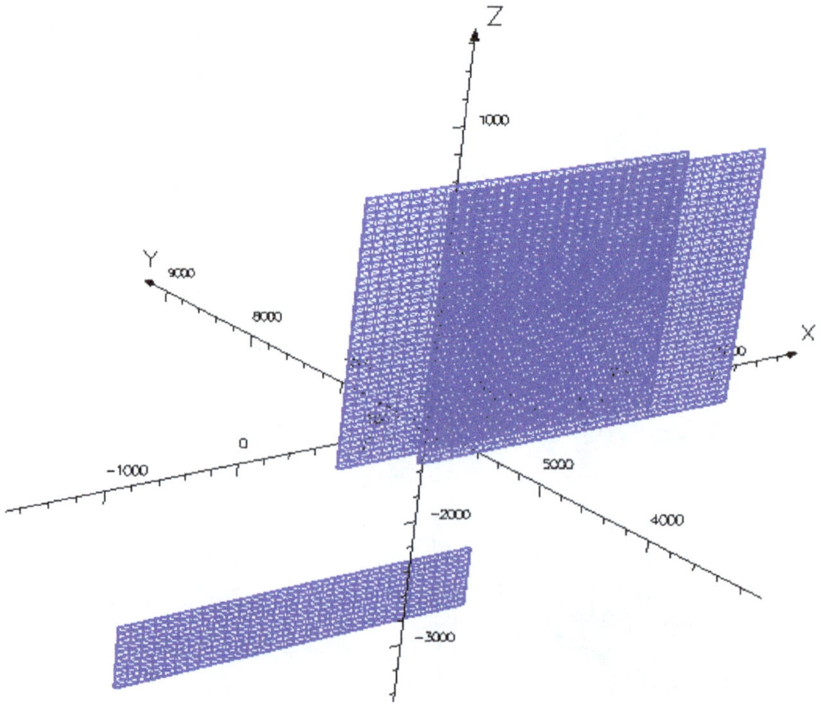

Figure 15. Position of edge against each other (non-folded edges).

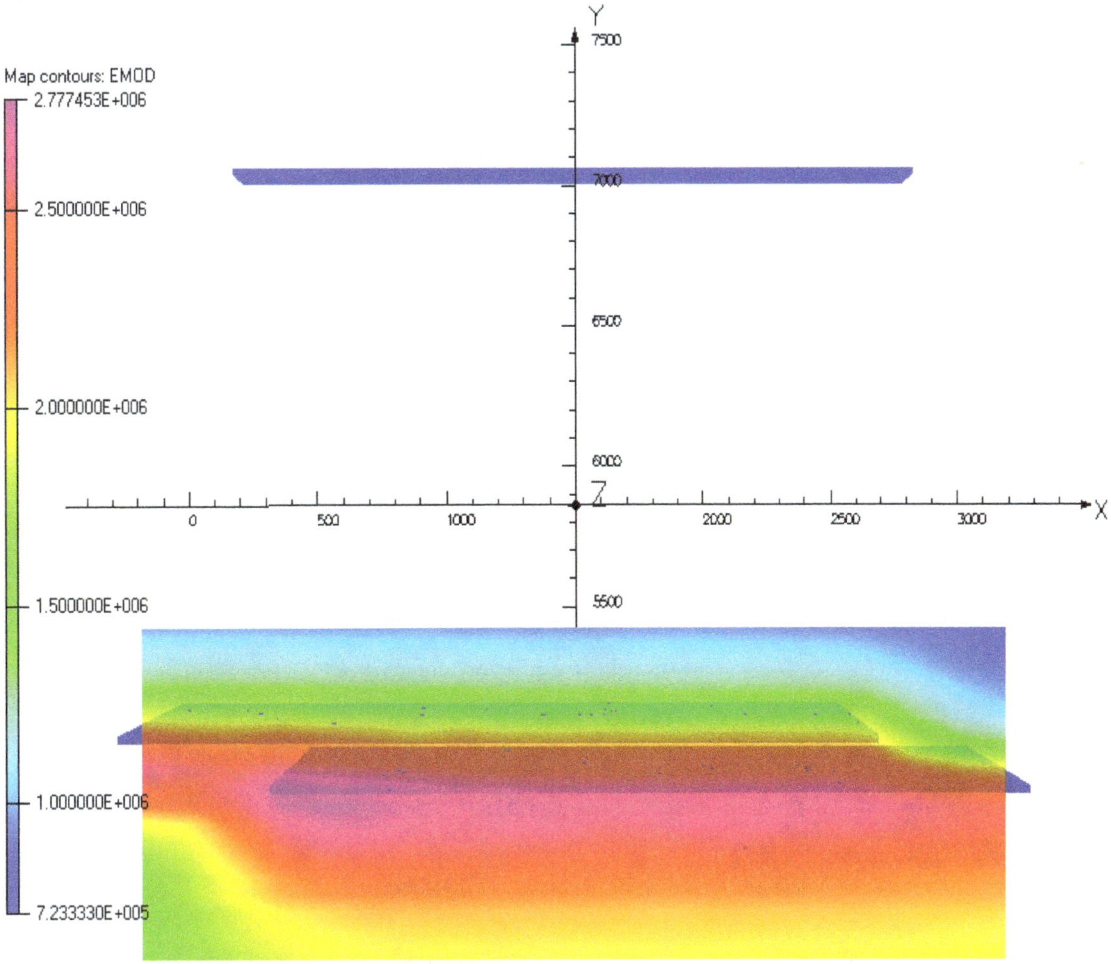

Figure 16. Electric field contour (non-folded edges).

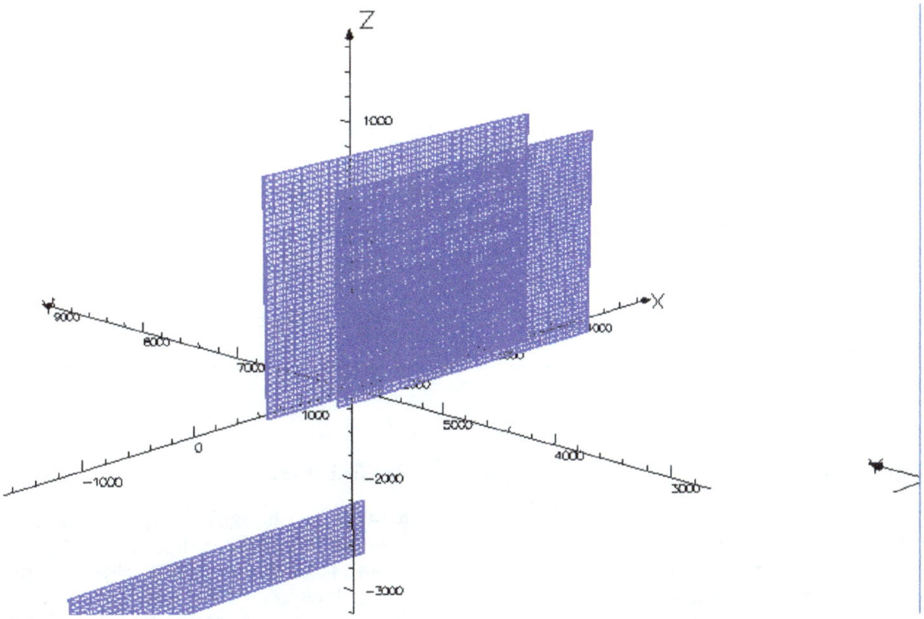

Figure 17. Relative positions of the non-folded and displaced edges.

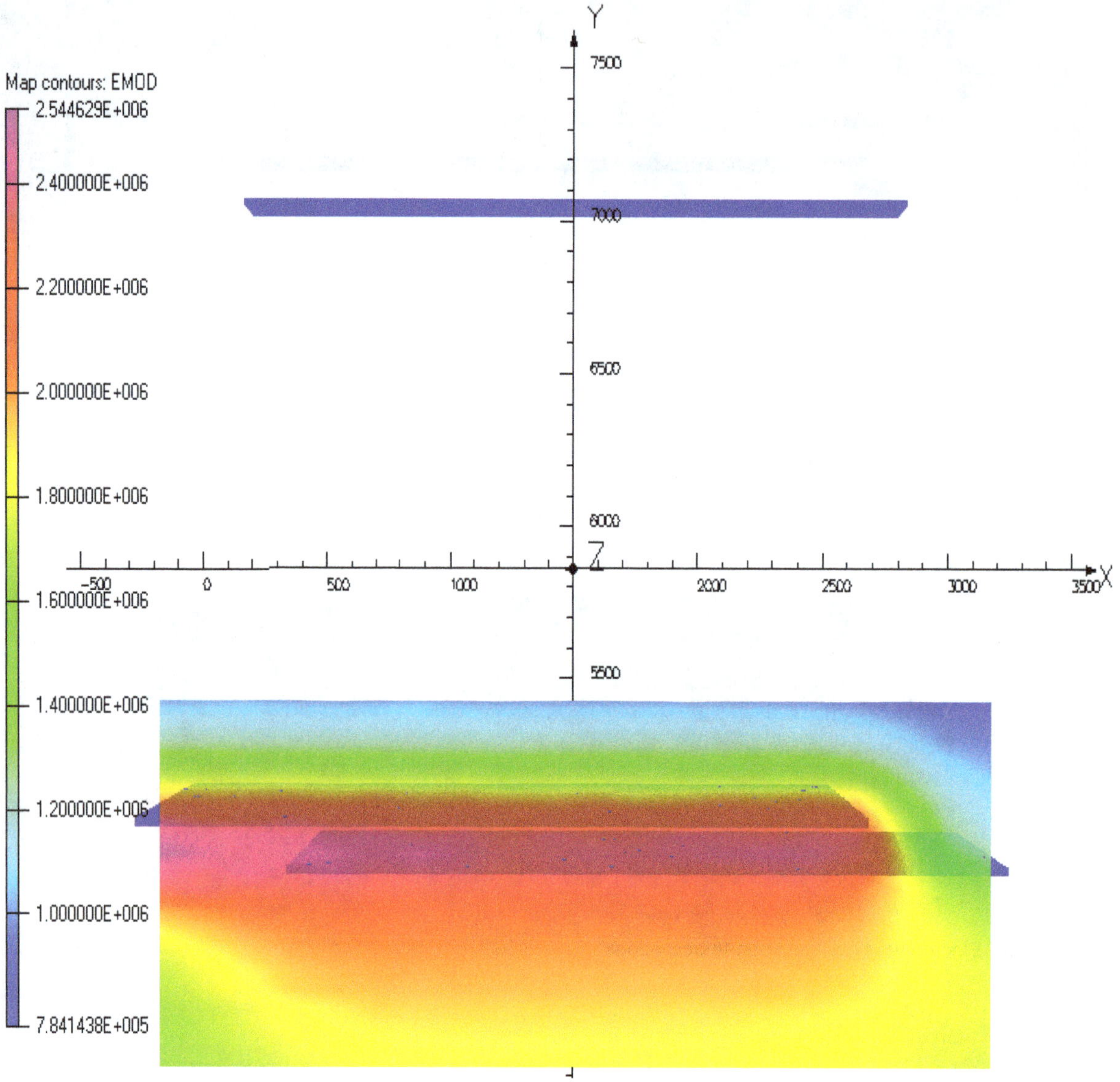

Figure 18. Electric field contour of non-folded and displaced edges.

Table 5. Maximum electric field values.

	E_{max}	E_{zmax}	E_{rmax}
Folded	2.85	-2.43	2.48
Non-folded	2.77	-2.36	2.13
Non- folded and displaced	-2.4	-2.17	2

lower than the outer edge better distributes the field around those edges. Contrarily, placing the inner edge at a position higher than the outer edge makes for the worst field distribution around the edges. Also, the folding of a foil edge affects the field intensity. Non-folded foil edge and the placing of the inner edge at a position lower than the outer edge greatly affect optimization of field intensity.

REFERENCES

ABB Company (2011). Resin impregnated paper bushings, product information. ABB manual, pp. 8-12.

Arora R, Mosch W (2006). High voltage insulation engineering. New Age International Pvt. Ltd., New Delhi, pp. 1-5.

Ellis KP (2004). Bushings for Power Transformers. ABB Company, pp. 9-15.

Faiz J, Ojaghi M (2002). Instructive Review of Computation of Electric

Fields using Different Numerical Techniques. Int. J. Eng. Edu., 18(3): 344-356.

Ghourab M, El-Makkawy S (1994). Analysis of electric field distribution in cavities within solid dielectric materials. IEEE Conf. Electrical Insulation and Dielectric Phenomena. Arlington, TX, USA, pp. 155-160.

Gremmel H, Kopatsch G (2001). ABB switchgear manual. ABB Calor Emag Schaltanlagen AG Mannheim and ABB Calor Emag Mittelspannung GmbH Ratingen, 10th edition, pp. 627-635.

Guoqiang Z, Yuanlu Z, Xiang C (1999). Optimal design of high voltage bushing electrode in transformer with evolution strategy. IEEE Trans. Magnetics. 35(3): 1690-1693.

Hsu Ya-Chen, P Lu W, Chun-Seng L (2011). Capability measures for Weibull processes with mean shift based on Erto's-Weibull control chart. Int. J. Phys.Sci., 6(19): 4533-4547.

Jumaat MZ, Ashraful-Alam MD (2011). Experimental and numerical analysis of end anchored steel plate and CFRP laminate flexurally strengthened reinforced concrete (r. c.) beams. Int. J. Phys Sci., 5(2): 132-144.

Kumar SP, Sharma S (2011). Substrate noise coupling in NMOS transistor for RF/analog circuits. Int. J.Phys. Sci., 6(9): 2285-2293.

Lesniewska E (2002). The use of 3-D electric field analysis and the analytical approach for improvement of a combined instrument transformer insulation system. IEEE Trans. Magnetics. 38(2): 1233-1236.

Meeker D (2006). Finite element method magnetics. Users Manual. 4: 155.

Monfared V (2011). A new analytical formulation for contact stress and prediction of crack propagation path in rolling bodies and comparing with finite element model (FEM) results statically. Int. J. Phys. Sci.,6(15): 3613-3618.

Monga S, Gorur RS, Hansen P, Massey W (2006). Design optimization of high voltage bushing using electric field computations. IEEE Trans. Dielectrics and Electrical Insulation. 13(6): 1217-1224.

Nirou Company (2011). Condenser bushings. Available online: http://www.niroutrans.com/site/335

Opera Version 14.0 User Guide. Vector Fields. 2011. http://www.cobham.com

Pompili M, Mazzetti C (2002). Partial discharge behavior in switching-surge-aged oil-paper capacitor bushing insulation. IEEE Trans. Dielectrics and Electrical Insulation. 9(1): 104-111.

Seghir T, Mahi D, Lebey T, Malec D (2006). Analysis of the electric field and the potential distribution in cavities inside solid insulating electrical materials. Proceeding of COMSOL. New York, USA, pp. 158-164.

Smith DG, McMeekin SG, Stewart BG, Wallace PA (2010). Transformer bushings-Modeling of electric field and potential distributions within oil impregnated paper with single and multiple spherical cavities. Universities Power Engineering Conf. (UPEC). Cardiff, Wales, UK, pp. 1-6.

Ueda M, Honda M, Hosokawa M, Takahashi K, Naito K (1985). Performance of contaminated bushing of UHV transmission systems. IEEE Trans. Power Apparatus and Systems. 104(4): 890-899.

Minimum time to use the fluid model in dc electrical discharges

B. Ardjani and B. Liani

Laboratory of Theoretical Physics, Science Faculty, Tlemcen University, P. O. Box 119, 13000 Tlemcen, Algeria.

In the study of thermal equilibrium of the electrons, it is very necessary to use the fluid model method. In this work, effect of the gas density, the electrons initial energy and the electric field on the electron thermal equilibrium is studied. It is found that the steady state can be achieved by increasing the gas density or the inter-electrode distance. If the gas density or the inter-electrode distance are not sufficient, steady state can be achieved by increasing the electric field. The initial energy of the electrons has great effect on the steady state if the electric field is weak and the difference between the initial energy and the electrons mean energy is great, however this problem can be solved by extending the time. It is necessary to make a best combination between the gas density, inter-electrode distance and the electric field to make a modeling by the fluid model method in the best cases of the steady state. A simple formula is given here to find the minimum time required to use the fluid model for E/N ≥ 500 Td.

Key words: Electrons thermal equilibrium, transport properties, fluid model.

INTRODUCTION

Fluid model is often used in the simulation of the dc electrical discharges, for example streamer discharges (Chao Li et al., 2007; Kulikovsky, 1995), dc glow discharges (Bogaerts, 1999; Baguer et al., 2003; Farouk et al., 2006).

The fluid model method cannot be used only if the steady state of the transport parameters is achieved. The equilibrium depends on different parameters such as the electric field strength, electron initial energy and the gas density, the effect of these parameters on the steady state of the transport parameters is well discussed in this work.

Also if the time is not sufficient the steady state is not achieved and the fluid model method cannot be used. A simple formula to find minimum time to use the fluid model method at high electric field is given here.

METHOD OF SIMULATION

The Monte Carlo collision method (MCC) described by Ardjani and Liani (2009) is used in this work. In the MCC method, firstly, the free flight time between two successive collisions is calculated by generating random numbers (between 0 and 1). The free flight time is used in the calculation of the equations of motion; therefore the electron energy changes under the acting of the electric field. The collision process (elastic, excitation, ionization, attachment) is selected by generating a random number between 0 and 1 and according to the electron energy, the electron energy decreases according to the collision nature. Finally several parameters are registered to be used in the calculation of the transport parameters as is well clarified by Ardjani and Liani (2009).

Table 1. Considered reactions of N_2 with his levels and energy range (Yousfi et al., 1988).

Reactions	Levels	Energy range (eV)
Momentum transfer		0 – 1000
Molecule rotation		0.03 – 3.6
Molecule vibration	v = 1	0.3 – 50
Molecule vibration	v = 1	1.65 – 3.6
Molecule vibration	v = 2	1.8 – 3.5
Molecule vibration	v = 3	2 – 3.3
Molecule vibration	v = 4	2.1 – 3.2
Molecule vibration	v = 5	2.2 – 3.3
Molecule vibration	v = 6	2.3 – 3.1
Molecule vibration	v = 7	2.4 – 3.4
Molecule vibration	v = 8	2.6 – 3.4
Molecule vibrational excitation		7 – 70
Molecule vibrational excitation		7.3 – 70
Molecule excitation		8 – 70
Molecule excitation		8 – 100
Molecule excitation		8.1 – 70
Molecule excitation		9 – 70
Molecule excitation		9 – 150
Molecule excitation		9 – 1000
Molecule excitation		9.1 – 50
Molecule excitation		11.5 – 100
Molecule excitation		12.92 – 50
Molecule excitation		13 – 1000
Molecule excitation		14 – 1000
Total ionization		16 – 1500

RESULTS AND DISCUSSION

The electrons are directed parallel to the *z* axis initially, but after the first collision the electrons are scattered randomly in 3-dimensional space. The gas used is the nitrogen, the reactions considered are: the momentum transfer, 1 molecule rotation, 9 molecule vibrations, 2 vibrational excitations, 11 excitations, total ionization. These are listed in the Table 1.

For comparison of the results with available literatures, Figure 1 shows the electrons energy and drift velocity calculated by this code with the calculations of Phelps and Pitchford (1985). Good agreement is shown specially for the drift velocity.

The gas density effect

Figure 2 shows the temporal mean energy of the electrons in the case of the nitrogen with different gas densities (T = 300°K and P = 0.1, 0.2, 0.4, 2 Torr) at E/N = 150 Td. At fixed time (100 ns) or rather at fixed inter-electrode distance the choice of the gas density has direct effect on the steady state and on the values of the transport parameters. Figure 2 shows that for 0.1 and 0.2 Torr the steady state is not achieved, but it begins approximately from 0.4 Torr, the best case of the steady state is for 2 Torr. The number of collisions increases proportionally to the increase of the gas density. The steady state is obtained when the number of collisions is sufficient to make good distribution of the energy between the electrons. As a result, at fixed inter-electrode distance the steady state is obtained by increasing the gas density, or at fixed gas density the steady state is obtained by increasing the inter-electrode distance.

The electrons initial energy effect

Figure 3 shows the temporal mean energy of the electrons in the case of the nitrogen at E/N = 20, 100, 150 Td with different initial energies of the electrons.

For E/N = 20 Td the steady state is achieved for $\varepsilon_0 = 1$ eV, but for $\varepsilon_0 = 4$ eV the mean energy is far to the steady state. The initial energy has great effect on the steady state when the difference between the initial energy and the mean energy is great. Generally the steady state is not influenced by the initial energy of the electrons for relatively high electric field (in this case approximately for E/N ≥ 100 Td), but if the initial energy increases the

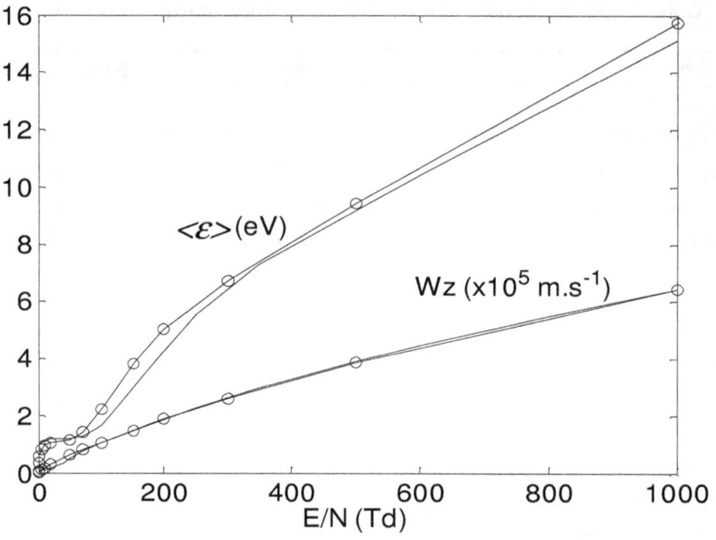

Figure 1. Mean energy and drift velocity as function of E/N(Td) for N_2 [solid line: this work; solid line with circles: Phelps and Pitchford (1985)].

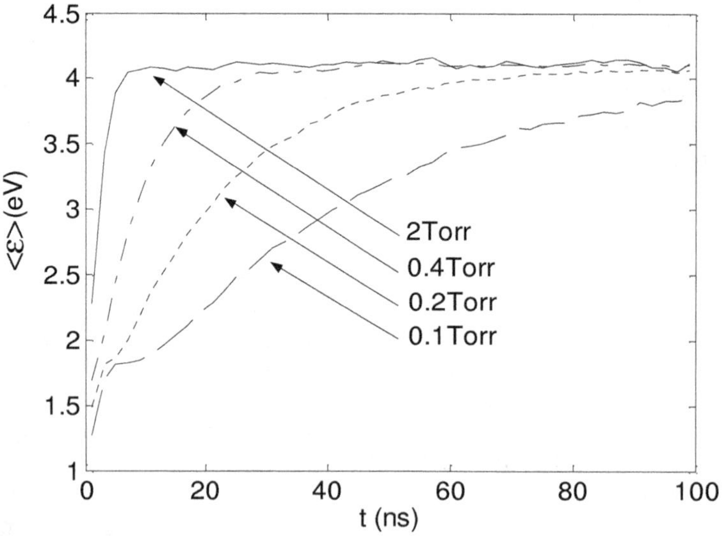

Figure 2. Temporal mean energy of the electrons in the case of the nitrogen with different gas densities (T = 300°K and P = 0.1, 0.2, 0.4, 2 Torr) at E/N = 150 Td.

steady state will be more delayed. Also the perfect steady state (which means that the stationary mean energy is from the beginning to the end) is obtained when the initial energy equals the mean energy and E/N ≤ 100 Td (for example, the cases of E/N = 20 Td, ε_0 = 1 eV and E/N = 100 Td, ε_0 = 2.5 eV). The perfect steady state allows the perfect use of the fluid model method; also the first region of non equilibrium must be greatly reduced to make the best use of the fluid model method. As a result it is necessary to best combine the gas density, the inter-electrode distance and the electric field to make a modeling of the electrical discharges by the fluid model method in the best cases of the steady state.

Figure 4 shows the temporal mean energy of electrons in the case of nitrogen at E/N = 20 Td with ε_0 = 4 eV, and with the time extended to 600 ns. Figure 4 shows that if the mean energy is not in the steady state since the initial energy is high the equilibrium can be achieved by extending the time or by increasing the gas density.

Figure 5 shows the electron energy distribution function

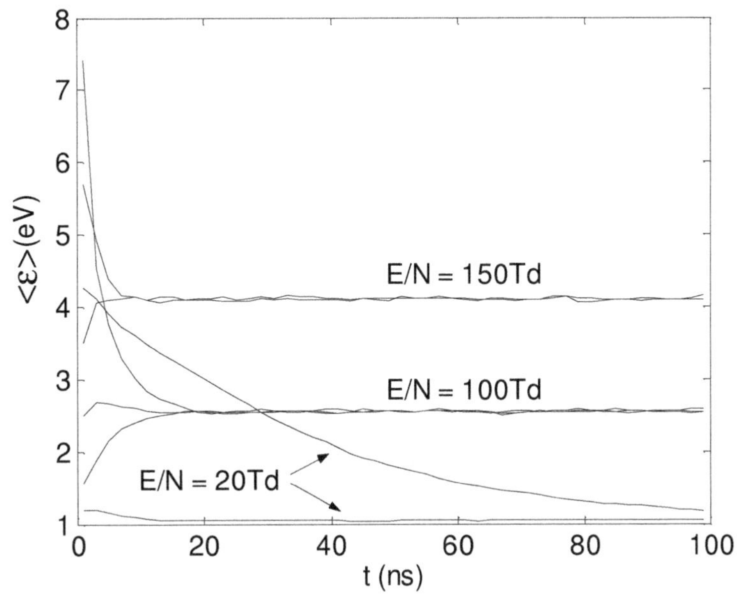

Figure 3. Temporal mean energy of electrons in the case of the nitrogen at E/N = 20 Td (ε_0 = 1, 4 eV), E/N = 100 Td (ε_0 = 1, 2.5, 7 eV), EN = 150 Td (ε_0 = 2.5, 4 eV) (P = 2 Torr).

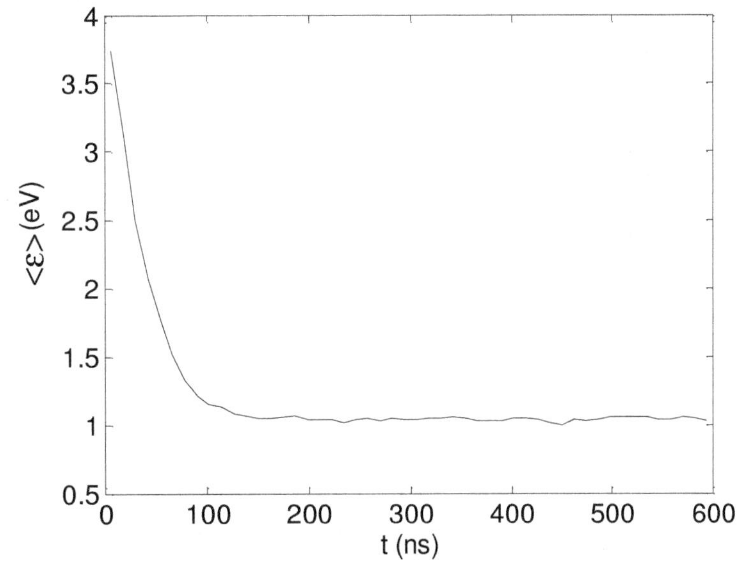

Figure 4. Temporal mean energy of electrons in the case of the nitrogen at E/N = 20 Td with ε_0 = 4 eV, and with the time extended to 600 ns (P = 2 Torr).

(EEDF) in the case of the nitrogen at E/N = 100 Td with different initial energies of the electrons. In the case of the nitrogen, the EEDF is Maxwellian for E/N ≥ 500 Td (Tas et al., 1997). Figure 5 shows that when the steady state is achieved and the electric field is relatively high (E/N ≥ 100 Td), and if the initial energy of the electrons changes the EEDF has a negligible change.

Figure 6 shows the electron energy distribution function (EEDF) in the case of nitrogen at E/N = 20 Td with different initial energies of the electrons. Figure 6 shows that when the steady state of the mean energy changes by the change of the initial energy, the EEDF changes also, but if the time is extended (to 600 ns in this case) the EEDF goes more and more to the steady state EEDF.

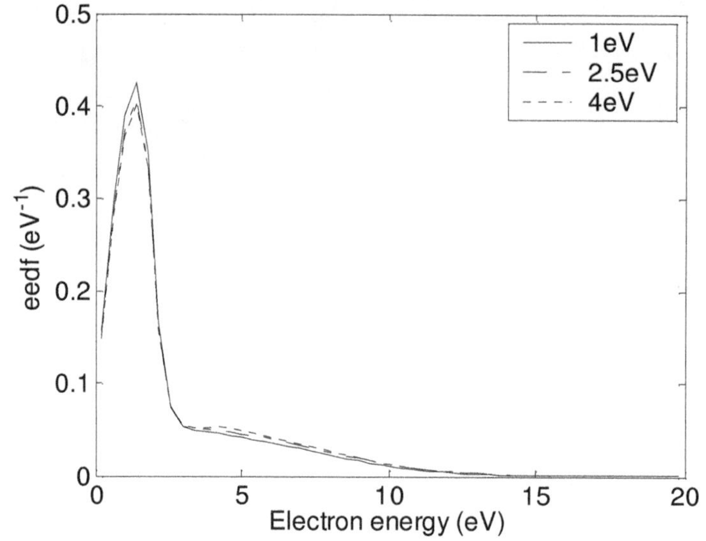

Figure 5. EEDF in the case of nitrogen at E/N = 100 Td with ε_0 = 1, 2.5, 4 eV.

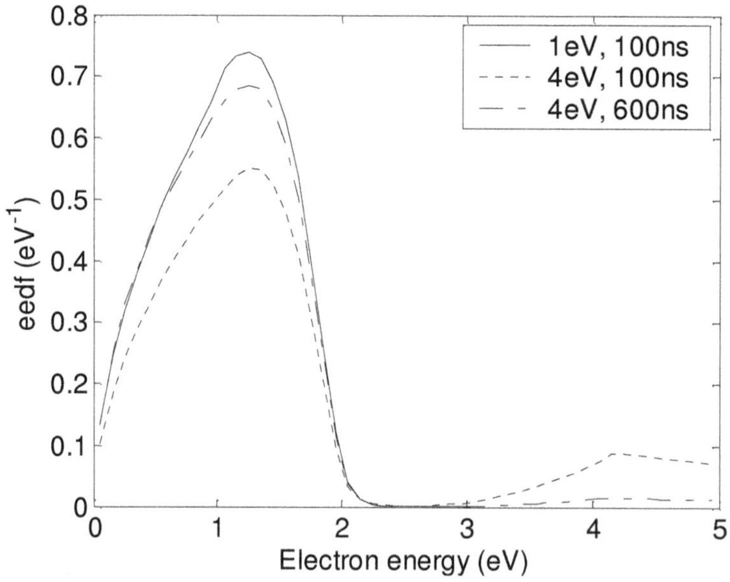

Figure 6. EEDF in the case of nitrogen at E/N = 20 Td with ε_0 = 1, 4 eV for t = 100 and 600 ns, respectively.

The electric field effect

Figure 7 shows the temporal mean energy of the electrons in the case of nitrogen and argon at different values of the electric field, Pg = 2 Torr.

Figure 7 show that the electric field has a role to achieve the steady state. In the case of N_2 when E/N = 150 Td the mean energy is not in the equilibrium, but for E/N = 300 Td the equilibrium is achieved and the best case of the steady state begins from E/N = 500 Td. Therefore if the gas density or inter-electrodes distance are not sufficient the equilibrium can be achieved by increasing the electric field.

Required time to achieve the electron thermal equilibrium

From Figure 7, for N_2, approximately the same behavior of the mean energy is obtained if E/N ≥ 500 Td.

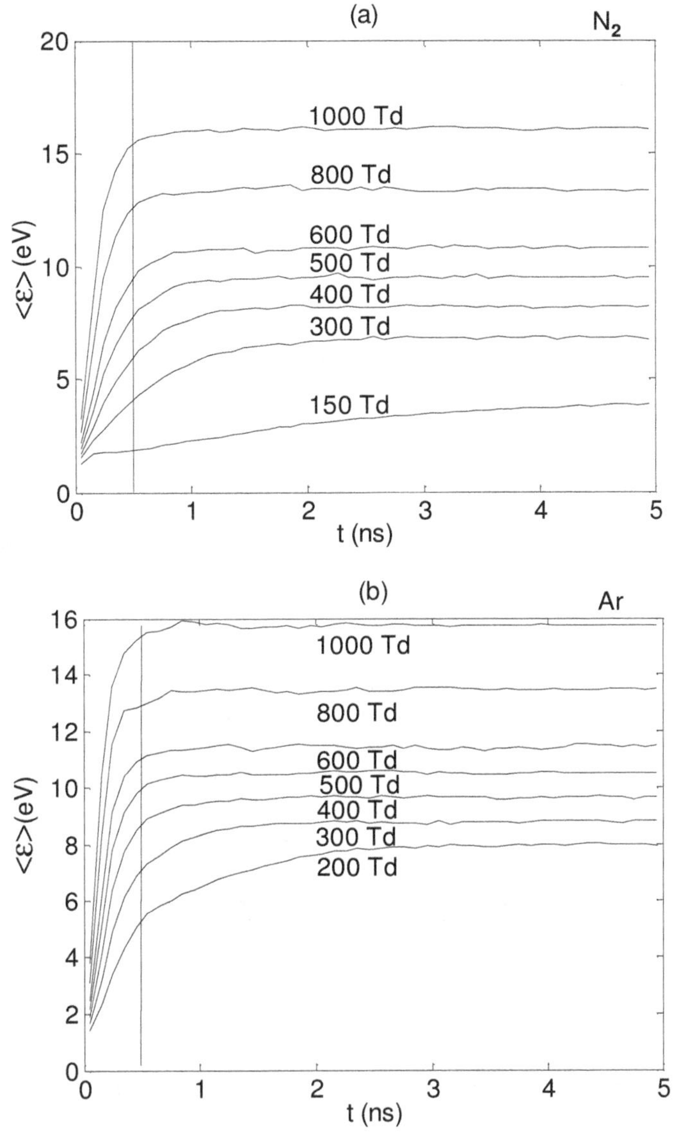

Figure 7. Temporal mean energy of the electrons at different values of E/N, P = 2 Torr: (a) nitrogen, (b) argon.

Therefore one can say that the steady state is independent of the electric field if E/N ≥ 500 Td, as it depends only on the gas density. For the argon gas the same behavior of the mean energy is obtained if E/N ≥ 400 Td. Also the time to achieve the steady state is inversely proportional to the gas density; therefore the time to achieve the steady state when E/N ≥ 500 Td can be deduced using the following relation:

$$t_{eq} = C_{et} / N_g \quad (1)$$

Where t_{eq} is the time required to achieve the steady state, $C_{et} = 3.14 \times 10^{14}$ is the equilibrium time coefficient and N_g is the gas density.

Also the equilibrium time can be expressed as a function of the gas pressure at ambient temperature (300°K) as follows:

$$t_{eq} = C_{etp} / P_g \quad (2)$$

Where $C_{etp} = 1 \times 10^{-8}$ is the pressure coefficient of the equilibrium time.

The values of C_{et}, C_{etp} given here are available for the nitrogen and the argon gas, but they can be changed slightly for other gases. The Equations (1) and (2) are available for E/N ≥ 500 Td in the case of nitrogen and E/N ≥ 400 Td in the case of argon; though the minimum

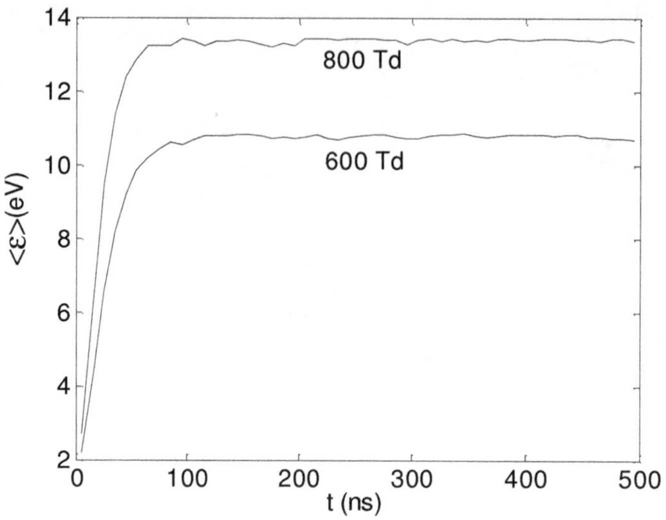

Figure 8. Temporal mean energy of the electrons in the case of nitrogen at E/N = 600 and 800 Td for Pg = 0.02 Torr.

E/N can be changed for other gases, E/N ≥ 500 Td is sufficient.

The equilibrium time given in Equations (1) and (2) is chosen to give best cases of the steady state, for modelling by the fluid model. This time is the minimum time to be used in this type of simulation.

For E/N ≤ 500 Td the time of simulation depends on the electric field; when E/N decreases the time of simulation must be increased. This makes it difficult to find a relation that gives the equilibrium time because the equilibrium time is not inversely proportional to the electric field.

Figure 8 shows the temporal mean energy of the electrons in the case of nitrogen at E/N = 600 and 800 Td, for Pg = 0.02 Torr. The relations (1) and (2) are tested by changing the gas pressure, for P_g = 0.02 Torr; the equilibrium time required is 500 ns. In the case of nitrogen at E/N = 600, 800 Td and P_g = 0.02 Torr the same behavior of the mean energy is obtained compared to the case of P_g = 2 Torr; it changes only the time of simulation.

Conclusion

In this work, effect of gas density, electric field and initial energy of the electrons on the electron thermal equilibrium is studied. It thus reveals that:

1) The steady state can be achieved by increasing the gas density or inter-electrode distance.
2) If the gas density or inter-electrode distance is not sufficient, the steady state can be achieved by increasing the electric field.
3) The initial energy of the electrons has great effect on the steady state if the electric field is low and the difference between the initial energy and the electrons mean energy is great, but this problem can be solved by extending the time.
4) It is necessary to best combine the gas density, inter-electrode distance and electric field to make a modeling of the electrical discharges by the fluid model method in the best cases of the steady state.
5) Simple formulas of the minimum time needed to achieve the electron thermal equilibrium are provided.

REFERENCES

Ardjani B, Liani B (2009). The validity of using the transport parameters of the Townsend discharge in the simulation of the rf and DBD discharges by the fluid model. FIZIKA A (Zagreb) 18(3):107-120.
Baguer N, Bogaerts A, Gijbels R (2003). Hollow cathode glow discharge in He: Monte Carlo–Fluid model combined with a transport model for the metastable atoms. J. Appl. Phys. 93(1).
Bogaerts A (1999). Comprehensive modelling network for dc glow discharges in argon. Plasma Sources Sci. Technol. 8:210-229.
Chao Li, Brok WJM, Ebert U, van der Mullen JJAM (2007). Deviations from the local field approximation in negative streamer heads. J. Appl. Phys. 101:123305.
Farouk T, Farouk B, Staack D, Gutsol A, Fridman A (2006). Simulation of dc atmospheric pressure argon micro glow-discharge. Plasma Sour. Sci. Technol. 15:676-688.
Kulikovsky AA (1995). Two-dimensional simulation of the positive streamer in N2 between I parallel-plate electrodes. J. Phys. D. Appl. Phys. 28:2483-2493.
Phelps AV, Pitchford LC (1985). Anisotropic scattering of electrons by Nz and its effect on electron transport. Phys. Rev. A 31(5):2932.
Tas MA, van Veldhuizen EM, Rutgers WR (1997). Plasma excitation processes in flue gas simulated with Monte Carlo electron dynamics. J. Phys. D: Appl. Phys. 30:1636-1645.
Yousfi M, Azzi N, Gallimberti I Stangherlin S (1988). Electron–molecule collision cross sections and electron swarm parameters in some atmospheric gases. Collection data n°1, Internal report. Toulouse Padova.

Convolutive cyclic voltammetry studies of PVA/CuI polymer composites at gold electrode

El-Hallag I. S.[1] , El-Mossalamy E.H.[2], El-Mansy M. K.[3] and Al-Harbi L. M.[4]

[1]Department of Chemistry, Faculty of Science, Tanta University, Tanta 31527, Egypt.
[2]Department of Chemistry, Faculty of Science, Benha University, Benha, Egypt.
[3]Department of Physics, Faculty of Science, Benha University, Benha, Egypt.
[4]Department of Chemistry, Faculty of Science King Abdulaziz University, P. O. Box 42805, Jeddah 21551, Saudi Arabia.

Cyclic voltammetry of PVA/CuI polymer composites was studied in acetonitrile as a solvent containing Tetrabutylammonium tetrafluoroborate (TBAFB). Cyclic voltammogram of PVA/CuI polymer composites exhibited unidirectional oxidative peak in the forward scan and irreversible reductive peak in the negative scan. The voltammetric behaviour explored that the electrode reaction of the investigated composite proceeds as EC mechanism. The highest occupied molecular orbital (HOMO) and the lowest unoccupied molecular orbital (LUMO) energy levels of the PVA/CuI nanocomposites were estimated from cyclic voltammetry. The chemical and electrochemical parameters of the investigated composite were determined and discussed using cyclic voltammetry, convolutive voltammetry and digital simulation methods.

Key words: Polymer composite, polyvinyl alcohol, cyclic voltammetry, convolutive voltammetry.

INTRODUCTION

Solid polymer electrolytes (SPEs) have been extensively studied for application in many electrochemical devices, such as cellular phones, smart credit cards, and laptop computers (Gray, 1991). The unique physical and chemical properties of nanoparticles, which are different from their bulk counterparts, offer excellent opportunities for chemical and biological sensing (Carusu, 2001; Storhoff and Mirkin, 1999). The facilitation of the electron transfer can significantly improve the sensitivity via introducing the nano-particles into the sensing interface (Kang et al., 2008; Park et al., 2005). The size controllability, chemical stability, high catalysis activity and surface tenability make them very advantageous for various electroanalytical and electrocatalytic applications (Park et al., 2005; Xiao et al., 2008).

Solar energy conversion devices based on organic semi-conductors has attracted great attention due to the advantages of light weight, flexibility and low cost of production with the possibility of fabricating large area devices based on solution processing. New materials and fabrication procedures leading to the substantial reduced cost of photovoltaic electricity could help drive a rapid expansion in the implementation of photovoltaic technology. The need to improve the light-to-electricity conversion efficiency requires the implementation of such materials and the exploration of new device architectures (Nielsen et al., 2010; Genes et al., 2007; Maggini et al., 2010; Krebs et al., 2010). It is well known that the electrical and optical properties of polymers can be improved to a desired limit through suitable doping (Blom et al., 1998). Further, polymers, on doping with metal nanoparticles, show novel and distinctive properties obtained from the unique combination of the inherent characteristics of polymers and properties of the metal nanoparticles (Gautam and Ram, 2010). Over the years, polyvinyl alcohol (PVA) polymers have attracted attention due to their variety of applications. PVA is a potential material having high dielectric strength, good charge storage capacity and dopant-dependent electrical and optical properties. It has carbon chain backbone with hydroxyl groups attached to methane carbons; these OH groups can be a source of hydrogen bonding and hence assist the formation of polymer composite (Yang et al., 2006).

Cuprous iodide (CuI) is a versatile candidate in band gap materials (CuI, CuSCN, and $CuAlO_2$) that were identified in the preparation of optical properties of thin film. CuI belongs to the I–VII semiconductors with zinc blend structure. Conducting materials and optically transparent films aroused much interest in the capability of application in electronic devices such as liquid crystal displays, photovoltaic devices, photothermal collectors and so on. The most interesting nature of this compound is that an inorganic semiconductor and its coordination chemistry let it readily couple with many inorganic and organic ligands as well (Yang et al., 2006; Tennakone et al., 1998; Kumara et al., 2001, Tonooka et al., 2002; Dherea et al., 2010; Kamat, 1994).

The crucial factors for the use of a semiconducting material as an active layer in the solar cells are the band gap and optical absorption of the material. The relatively large band gap of polymer such as PVA limits the absorption of near-infrared light and thus lowers the light harvesting and therefore cannot be used as an active layer in organic solar cells. The control over the band gap is necessary while designing new materials for organic solar cells. The band gap engineering allows one to design and synthesize new materials with maximum overlap of absorption spectrum with the solar emission spectrum. It is often found that the synthesis of low band gap polymer is not only the solution to address this problem but also the position of highest occupied molecular orbital (HOMO) and lowest unoccupied molecular orbital (LUMO) limits the open-circuit voltage (Voc) of the photovoltaic cell. These two properties of organic materials can be controlled by introducing nanoinorganic salts or introducing alternative electron rich and electron-deficient units in the polymer backbone (Kanimozhi et al., 2010; Peiova and Grozdanov, 2005).

In the present work, the PVA polymer was prepared with and without the inorganic salt CuI nanoparticles to produce p-type PVA/CuI nanocomposite with suitable energy band gap which matches with the solar energy spectra. Cyclic voltammetry technique was used to characterize the PVA/CuI polymer composite. The chemical and electrochemical parameters were calculated and discussed as unprecedented calculation. Digital simulation and convolutive voltammetry were used to support, verify and confirm the calculated parameters and the nature of electrode reaction.

EXPERIMENTAL PROCEDURE

Chemicals

Tetrabutylammonium tetrafluoroborate (TBAFB) (Sigma) were used without further purification. A polyvinyl alcohol polymer (PVA 99.9%), acetonitrile (99.9%), dimethyl sulfoxide (DMSO 99.9%) and CuI (99.5%) used in the present study were purchased from Sigma–Aldrich. Solution of polyvinyl alcohol (PVA) was prepared by adding deionized distilled water to solid PVA $(C_2H_4O)_x$ (where x = 30,000–70,000, average molecular weight) and then stirred by a magnetic stirrer at room temperature for 2 h. The composite film was formed by drop casting of composite solution in dimethyl sulfoxide (DMSO) on the working electrode, Indium Tin Oxide, (ITO). In order to remove the residual Dimethyl sulfoxide (DMSO) from the composite film, we heated the composite film at 50°C over-night.

The structure of the polymer electrolyte layers was investigated and confirmed in previous work (Sheha et al., 2012) using X-ray diffraction via ShIMADZU diffractometer type XRD 6000, wave length 1.5418 °A. In addition, the surface morphology of these nanocomposites was also examined using scanning electron microscope, SEM (JOEL-JSM Model 5600) (Sheha et al., 2012).

Instrumentation

The electrochemical cyclic voltammetry was conducted on the electrochemical instrument linked to an EG & G model 283 Potentiostat at 20°C in a 0.1 mol. Acetonitrile solution of tetrabutylammonium tetrafluoroborate (TBAFB) as background electrolyte at potential scan rate of 100 mV/s using Ag/AgCl reference and platinum wire as counter electrodes, respectively. Indium tin oxide was used as working electrode.

Convolutive voltammetry and digital simulation of the data for cyclic voltammetric experiments was performed on PC computer using EG & G Condesim software package. The simulation procedure was carried out using finite differences techniques. Algorithms for the simulation program were coded and implemented into the condesim software package supplied by EG & G. A direct method of the determined electrochemical parameters was performed by generating the simulated cyclic voltammogram of PVA/CuI using the average values of electrochemical parameters extracted experimentally and comparing it with the voltammogram recorded experimentally.

RESULTS AND DISCUSSION

Cyclic voltammetry study

Cyclic voltammetry was performed on the PVA/CuI in order to assess its electroactivity. Cyclic voltammetry revealed the presence of one oxidation and one reduction waves as illustrated in the voltammogram shown in Figure 1. The range of the scan was between +2 V to -2 V with a scan rate of 0.1 $V.s^{-1}$. Cyclic voltammetry (CV) confirmed the presence of a charge capacity (area under CV curve) for the PVA/CuI. The curve also indicated the presence of an intrinsic redox reaction between the electrode and the PVA/CuI, at the potential applied between +2 V and -2 V, resulting in an ion exchange between the electrodes and the electrolytes in the solvents carrying the mobile charged carriers to and from the PVA/CuI.

Cyclic voltammogram of CuI/PVA polymer composite recorded in 0.1 mol L^{-1} TBAFB/ acetonitrile at indium tin oxide (ITO) electrode and scan rate of 0.1 V s^{-1} exhibited a single anodic peak (E_{pa} = 1.370 V) in the positive scan

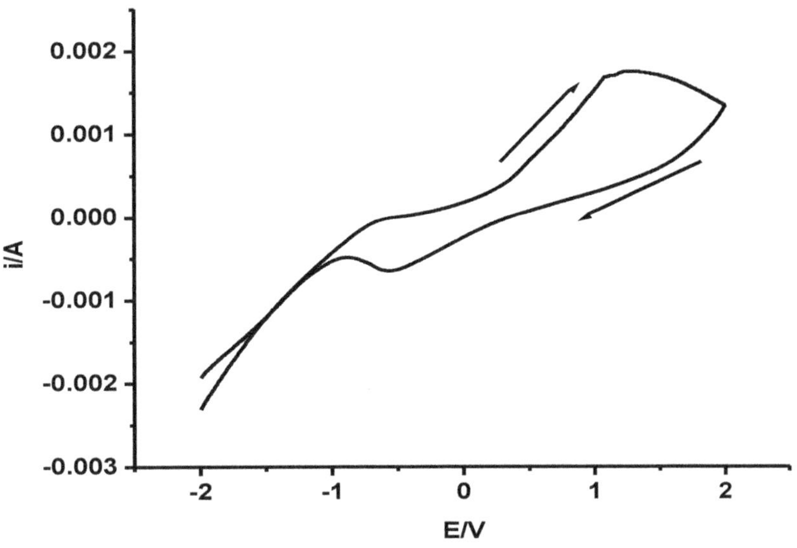

Figure 1. Cyclic voltammogram of PVA/CuI nanocomposite in CH3CN/0.1 M Tetrabutylammonium tetrafluoroborate **(**TBAFB**)** at sweep rate of 0.1 V.s^{-1}.

Table 1. Values of the electrochemical parameters of the investigated PVA/CuI polymer composite at ITO electrode.

Electrochemical parameter	Method							
	CV		Convolution		Deconvolution		Simulation	
	Ox	Red	Ox	Red	Ox	Red	Ox	Red
k_s x 10^6 (cm.s^{-1})							3.53	2.65
D x10^6(cm^2.s^{-1})	2.32	3.23	2.75	3.26			2.13	3.45
E^0, V							0.71	-0.23
k_c , s^{-1}							0.05	0.04
Ep- Ep/2, V	0.550	0.252					0.554	0.257
w$^{1/2}$, V	0.397	0.320						
i_p,x10^3 A	1.70	1.21					1.73	1.25
E_p, V	1.35	0.50					1.36	0.52
I_{lim} x10^3, (A.s$^{-1/2}$)			4.1	2.5				

and a unidirectional cathodic peak (E_{pc} = - 0.510 V) in the negative scan at 20°C (Figure 1). Under normal circumstances, PVA exhibits one oxidation peak (Kanimozhi et al., 2010). The voltammogram depicted in Figure 1 exhibited two single redox waves. These redox waves may be due to more intricate reactions occurring between the PVA/CuI, the electrolytes and the electrode. The oxidative and reductive waves of the recorded cyclic voltammogram may be attributed to slow electron transfer followed by fast chemical reaction as indicated from the absence of counter peaks coupled with anodic and cathodic processes, that is, the electrode behaviour of the CuI/PVA polymer composites solution proceed as EC_{irr} mechanism for oxidation and reduction steps (El-Hallag and Hassanien, 1999).

The formal potential (E$^{0'}$), heterogeneous rate constant (k_s), the symmetry coefficient (α) and the homogeneous chemical rate constant (k_c) of oxidation and reduction steps were determined from simulated voltammograms and cited in Table 1. The generated theoretical cyclic voltammogram is indicated in Figure 2. In the selected scan rate (0.1 V.s^{-1}), the difference in the peak potential and half peak potential (E_p - $E_{p/2}$) of the anodic and the cathodic reactions were found to be 550 and 252 mV respectively. This behaviour may be due to the sluggish nature of the rate of electron transfer in addition to some uncompensated solution resistance of the CuI/PVA polymer composites solution (Oldham, 1983). Also the values of half peak width indicate that the oxidative process is slower than the reductive process.

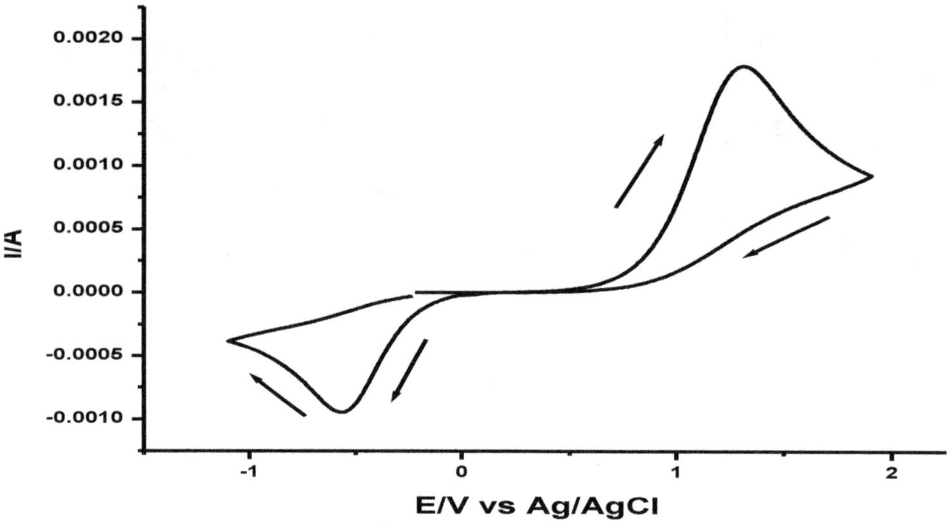

Figure 2. Simulated cyclic voltammogram of PVA/CuI polymer composite at sweep rate of 0.1 $V.s^{-1}$.

Effective charge transports from donor to acceptor component and charge collection at the electrodes are important parameters for designing and optimization of organic bulk heterojunction photovoltaic devices. In this regard, the electrochemical data gives valuable information and allow the estimation of relative position of HOMO/LUMO levels of the materials used for device fabrication. The knowledge of these levels is required for finding suitable donor-acceptor combination for the efficient bulk hetero-junction photo-voltaic device based on organic materials. The method of evaluation the HOMO and LUMO energy levels from the onset oxidation and reduction potentials have been proposed in literatures (Meng and Wudl, 2001). The position of HOMO and LUMO level has been determined from the analysis of redox potential behavior observed in cyclic voltammogram curve shown in Figure 1, using the expressions (1) and (2).

$$E_{HOMO} = -E_{OX}^{onset} - 4.75 \text{ eV} \quad (1)$$

$$E_{LUMO} = -E_{red}^{onset} - 4.75 \text{ eV} \quad (2)$$

Values of E_{OX}^{onset}, E_{red}^{onset}, E_{HOMO} and E_{LUMO} are 0.8, -0.23, -5.55 and -4.52 V respectively. Energy gap (E_g = E_{LUMO} - E_{HOMO}) in this material is found to be 1.05 eV. As indicated the cyclic voltammetry measurements have been performed to ascertain the position of both HOMO and LUMO levels which illustrated a movement of HOMO level towards vacuum level.

Convolution-deconvolution voltammetry

The convolution theorem finds use in the situation where it is required to perform the inverse transformation on a function which is the product of two functions of the Laplace variable each of which individually have known inverse transformations. In such a situation, the convolution theorem gives (Oldham, 1983; El-Hallag and Ghoneim, 1996)

$$L^{-1}[f_s(s).g_s(s)] = \int_t^0 G(u)\, F(t-u)\, du \quad (3)$$

in which f_s, g_s are the Laplace transform of the functions F and G, the variable u is a dummy variable which is lost when the definite integral is evaluated. For the following reaction, in which a given species undergoing only electron transfer and no subsequent processes other than 'linear'

$$A + ne \longleftrightarrow B \quad (4)$$

diffusion out in the solution from a planar electrode, that is, the Fick's Second Law is expressed as (El-Hallag and Ghoneim, 1996)

$$[\partial C_A/\partial t]_x = D_A [\partial^2 C_A/\partial x^2]_x \quad (5)$$

then the solution of the above via Laplace methods yields

$$(C^{bulk} - C^s) = I_1 / n\, FSD^{1/2}{}_A \quad (6)$$

and

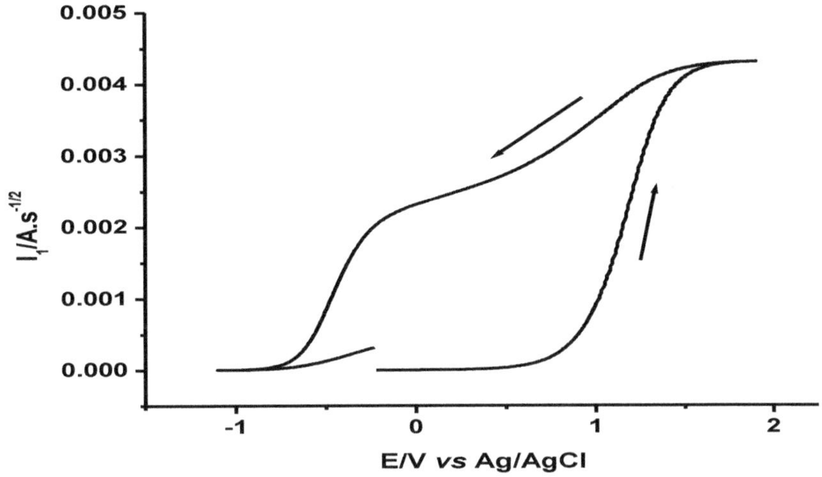

Figure 3. Convolution voltammetry (/1) of PVA/CuI polymer composite at sweep rate of 0.1 V.s^{-1}.

$$C^{bulk} = I_{lim} / n\ FSD^{1/2}{}_A \tag{7}$$

where C^{bulk} and C^s are the bulk and surface concentrations respectively and the convolution I_1 is given by $I_1 = i*(\pi t)^{-1/2}$ or more 'fully' as:

$$I_1(t) = \pi^{-1/2} \int_0^t i(u) / (t-u)^{1/2} du \tag{8}$$

where I_1 (t) is the convoluted current at the total elapsed time (t), i (u) is the experimental current at time u and I_{lim} is the limiting value of I_1 at 'extreme' potentials; that is, when the concentration at the electrode C^s is effectively reduced to zero by rapid redox conversion and the current is thus controlled solely by the maximum rates of diffusion to and from the electrode. In the case of electron transfer followed by homogeneous reaction, the species here are produced by electron transfer at a planar electrode and convert into new product in the bulk via a chemical reaction with rate constant k_c. The Fick's Second Law expression is now (El-Hallag, 2009; Blagg et al., 1985):

$$[\partial C_B/\partial t]_x = D_B[\partial^2 C_B/\partial x^2]_x - k_c C_B \text{ (at x)} \tag{9}$$

and solution via Laplace methods gives here, necessarily starting at zero concentration in the bulk, the following electrode concentration in the bulk:

$$C^s{}_B = I_2 / nFAD^{1/2}{}_B \tag{10}$$

where the 'kinetic' convolution I_2 is given by Blagg et al., (1985).

$$I_2(t) = \pi^{-1/2} \int_t^0 [\, i(u) \exp(-kc(t-u))] / (t-u)^{1/2} du \tag{11}$$

Thus in the I_2 convolution at time t, each segment of i(u) is scaled by dividing by the square root of the time which has elapsed from t to the time u to which the segment refers and likewise is scaled by the exponential factor $\exp(-k_c(t-u))$. Thus I_2 for example now goes to a plateau (at zero) on return of the sweep in cyclic voltammetry and this property allows determination of k_c in the case of the appearance of counter peaks coupled with anodic and/or cathodic peaks, otherwise the k_c is determined from digital simulation .

The diffusion coefficient of the investigated composite was determined, after applying background subtraction and correction for uncompensated resistance, from combination between cyclic voltammetry and convolution voltammetry via the following Equation 12 (Blagg et al., 1985; Doetsch, 1953; El-Hallag et al., 2000).

$$I_{lim} = i_p / 3.099(\alpha n_a v)^{1/2} \tag{12}$$

where I_{lim} is the limiting value achieved for I_1 when the potential is driven to a sufficiently extreme value past the peak, i_p is height of peak current of the cyclic voltammogram and the other terms has their usual meanings. The I_1 convolution of the investigated compound illustrated in Figure 3 shows a distinct separation between the forward and reverse sweep and clearly indicates the sluggishness of electron transfer of the electrode pathway. The reverse sweep of the I_1 convolution of both anodic and cathodic peaks does not return to zero due to the fast chemical reaction which

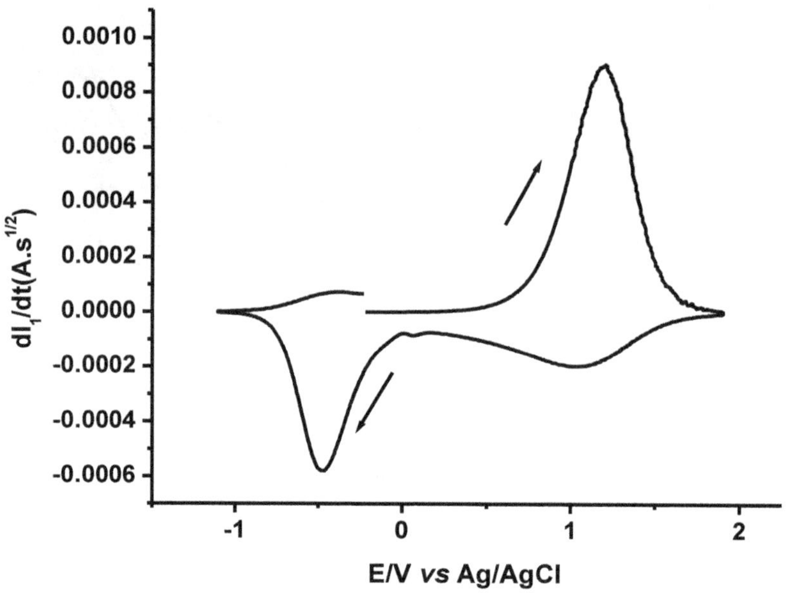

Figure 4. Deconvolution voltammetry (d/1/dt) of PVA/CuI polymer composite at sweep rate of 0.1 $V.s^{-1}$.

appears at time scales of the experiment. Value of the diffusion coefficient D evaluated via Equations 12 was found at 3.45 ± 0.5 $x10^{-6}$ cm^2s^{-1}. The homogeneous chemical rate constants (k^{ox}_c and k^{red}_c) of the chemical step that follows the charge transfer of both anodic and cathodic electrochemical reactions was calculated via digital simulation and listed in Table 1.

The deconvolution of current (dI_1/dt) can be expressed as the differential of the I_1 convolution. In more general terms, deconvolution is a kin to semi-differentiation in a similar manner to considering $t^{-1/2}$ convolution as semi-integration. The relationship between $t^{-1/2}$ convolutions and deconvolutions is indicated in the following scheme:

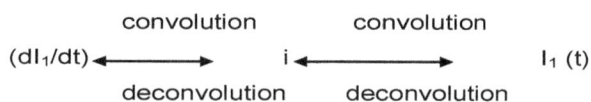

$$(dI_1/dt) \underset{\text{deconvolution}}{\overset{\text{convolution}}{\longleftrightarrow}} i \underset{\text{deconvolution}}{\overset{\text{convolution}}{\longleftrightarrow}} I_1\ (t)$$

Figure 4 indicates the deconvolution votammogram of the investigated polymer composite at 0.1 $V.s^{-1}$. It was found that the peak width ($w^{1/2}$) of andodic and cathodic process are equal 397 ± 4 mV and 320 ± 4 mV respectively. The measured values of $w^{1/2}$ are more than 90/n mV expected for one electron nernstain process confirming the slow nature of electron transfer (Saveant and Tessier, 1975).

Conclusion

In this article, CuI/PVA nanopolymer composites were prepared. The characterization of the composite using cyclic voltammetry and convolutive voltammetry was done as first time. The electrochemical band gap estimated from the cyclic voltammetry agrees well with the trend observed with the optical band gap established in literature. This can be easily understood in terms of the HOMO and LUMO energy levels of composites with respect to the work functions of both electrodes. From cyclic voltammetric measurements, it was concluded that the oxidation and reduction process proceeds as slow electron transfer followed by fast chemical process. The chemical and electrochemical parameters of polymer composite was determined and discussed.

REFERENCES

Blagg LA, Carr SW, Cooper GR, Dobson ID, Gill JB, Goodal DC, Shaw BL, Taylor N, Boddington T (1985). A mechanistic study on complexes of type mer-[$Cr(CO)_3(\eta^2$-L–L)(σ-L–L)] (where L–L = $Ph_2PCH_2PPh_2$, $Ph_2PNHPPh_2$, or $Ph_2PNMePPh_2$) using spectr-oscopic and convolutive electrochemical techniques. J. Chem. Soc. Dalton Trans. P. 1213.

Blom PWM, Schoo HFM, Matters M (1998). Electrical characterization of electroluminescent polymer/nanoparticle composite devices. Appl. Phys. Lett. 73:3914-3916.

Carusu F (2001). Nanoengineering of Particle Surfaces. Adv. Mater. 13:11- 22.

Dherea SL, Latthea SS, Kappenstein C, Mukherjee SK, Venkateswara Rao A (2010). Comparative studies on p-type CuI grown on glass and copper substrate by SILAR method. Appl. Surf. Sci. 256:3967–3971.

Doetsch G (1953). Laplace Transformation, Dover, New York.

El-Hallag IS (2009). Cyclic voltammetry, convolutive voltammetry, chronopotentiometry and digital simulation studies of [Pt(C≡C $tol)_2$ $(dppm)_2$ -Ir$(CO)_2]^+PF^{-6}$ complex. Chin. Sci. Bull. 541:3801-3807.

El-Hallag IS, Ghoneim MM (1996). Electrochemical investigation of N,N′-Propylene-*bis*-(salicylideneiminato) Mn(III) in phosphate buffer solutions. Monatsh fur Chemie. 127: 487 - 494.

El-Hallag IS, Ghoneim MM, Hammam E (2000). New method for the investigation of CE system via convolutive voltammetry combined with digital simulation. Anal. Chim. Acta. 414:173-180.

El-Hallag IS, Hassanien AM (1999). Electrochemical Studies of the Complex (OC-6-22)-W$(CO)_3$(dppm)$_2$ at a Glassy Carbon Electrode in CH_2Cl_2. Collect. Czech. Chem. Commun. 64:1953-1965.

Gautam A, Ram S (2010). Preparation and thermomechanical properties of Ag-PVA nanocomposite films. Mater. Chem. Phys. 119:266–271.

Genes S, Neugebauer H, Sariciftci NS (2007). Conjugated Polymer-Based Organic Solar Cells. Chem. Rev. 107:1324–1338.

Gray FM (1991). Solid Polymer Electrolytes - Fundamental and Technological Application, VCH, New York.

Kamat PV (1994). Interfacial charge transfer processes in colloidal semiconductor systems. Prog. React. Kinet. 19:277–316.

Kang Q, Yang L, Cai Q (2008). An electro-catalytic biosensor fabricated with Pt–Au Nanoparticle-decorated titania nanotube array. Bioelectrochem. 74:62-65.

Kanimozhi C, Balraju P, Sharma GD, Patil S (2010). Synthesis of Diketopyrrolopyrrole Containing Copolymers: A Study of Their Optical and Photovoltaic Properties. J. Phys. Chem. B. 114:3095–4003.

Krebs FC, Tromholt T, Jorgensen M (2010). Upscaling of polymer solar cell fabrication using full roll-to-roll processing. Nanoscale. 2:873–886.

Kumara GRRA, Konno A, Senadeera GKR, Jayaweere PVV, De Silva DBRA, Tennakone T (2001). Dye-sensitized solar cell with the hole collector p-CuSCN deposited from a solution in n-propyl sulphide. Sol. Energy Mater. Sol. Cells. 69:195–199.

Maggini R, Po M, Camaioni N (2010). Polymer Solar Cells: Recent Approaches and Achievements. J. Phys. Chem. C. 114:695–706.

Meng H, Wudl F (2001). A Robust Low Band Gap Processable n-Type Conducting Polymer Based on Poly (isothianaphthene). Macromolecules. 34:1810-1816.

Nielsen T.D, Fyenbo J, Wadstrøm M, Marie S, Pedersen M.S (2010). Manufacture, integration and demonstration of polymer solar cells in a lamp for the "Lighting Africa" initiative. Energy Environ. Sci. 3:512-525.

Oldham KB (1983). The extraction of kinetic parameters from chronoamperometric or chronocoulometric data. J. Electroanal. Chem. 145:9-20.

Park JE, Atobe M, Fuchigami T (2005). Sonochemical synthesis of conducting polymer–metal nanoparticles nanocomposite. Electrochim. Acta. 51:849-854.

Peiova B, Grozdanov I (2005). Three-dimensional confinement effects in semiconducting zinc selenide quantum dots deposited in thin-film form. Mater. Chem. Phys. 90:35–46.

Saveant JM, Tessier D (1975). Convolution potential sweep voltammetry V. Determination of charge transfer kinetics deviating from the Butler-Volmer behaviour. J. Electroanal. Chem. Interfacial Electrochem. 65:57-66.

Sheha E, Khoder H, Shanap TS, El- Shaarawy MG, El-Mansy M.K (2012). Structure, dielectric and optical properties of p-type (PVA/CuI) nanocomposite polymer electrolyte for photovoltaic cells. Optik – Int. J Light and Electron Optics123:1161-1166.

Storhoff J, Mirkin CA (1999). Programmed Materials Synthesis with DNA. Chem. Rev. 99:1849-1862.

Tennakone K, Kumara GRRA, Kottegoda IRM, Perera VPS, Aponsu GMLP, Wijayantha KGU (1998). Deposition of thin conducting films of CuI on glass. Energy Mater. Sol. Cells. 55:283-289.

Tonooka K, Shimokawa K, Nishimura O (2002). Properties of copper–aluminum oxide films prepared by solution methods. Thin Solid Films, 411:129–133.

Xiao F, Zhao F, Li J, Liu L, Zeng B (2008). Characterization of hydrophobic ionic liquid-carbon nanotubes–gold nanoparticles composite film coated electrode and the simultaneous voltammetric determination of guanine and adenine. Electrochim. Acta. 53:7781–7788.

Yang Y, Zi Wang Z, Yan Gao Y, Liu T, Hu C, Dong Q (2006). Synthesis of fluorinated diblock copolymer by ATRP and its application of PAVAc polymerization in ScCO2. J. Appl. Polym. Sci. 2:1146–1151.

19

Optimal lead-lag controller designing for reduction of load current total harmonic disturtion and hanmonic with voltage control using honey bee mating optimization (HBMO)

Hamdi Abdi[1] and Ramtin Rasoulinezhad[2]

[1]Electrical Engineering Department, Faculty of Engineering, Razi University, Kermanshah, Iran.
[2]Department of Electrical Engineering, Science and Research Branch, Islamic Azad University, Kermanshah, Iran.

Increase in world demand load has resulted in new distributed generation (DG) that has entered into the power system. One of the most renewable energies is fuel cell, which is connected to power system using a power electronics interface in microgrid or standalone condition. The highest problem of this switching interface based DGs is the power quality (PQ) and harmonics of currents. Also, the voltage of the DGs should be controlled in islanding condition. In this paper, we present a Lead-Lag optimal controller for controlling one of the most important types of fuel cell, namely proton exchange membrane fuel cell (PEMFC) in islanding mode operation for reducing PQ problems and voltage control. At first, the introduction and implementation of the PEMFC is present and next, during system load variations the proposed controller is designed. The controller should be designed against the demand load variations of fuel cell. Here, the lead-lag controller is used when its coefficients are optimized based on honey bee mating optimization (HBMO). In order to use this algorithm, at first, the problem is written as an optimization problem which includes the objective function and constraints, and then to achieve the most desirable controller, HBMO algorithm is applied to solve the problem. Simulation results are done for various loads in time domain, and the results show the efficiency of the proposed controller in contrast to the previous controllers.

Key words: Power system, fuel cell.

INTRODUCTION

Rising of fossil fuel cost and their probable depletion, air pollution, global warming phenomenon and severe environmental problems that caused distributed energy sources have gained the attention of many nations in producing electricity. High efficiency and very low emissions can be satisfied in fuel cell-based power generation systems. Moreover, fuel cells have a superior dynamic response, good stability and low noise. Proton exchange membrane fuel cell (PEMFC) can be a great alternative for power generating sources in the coming years, especially in the automotive, distributed power generation, and portable electronic applications (Alireza and Alireza, 2011).

PEMFC is composed of cathode, anode and electrolyte between the anode and cathode. Hydrogen gas (H_2), which is obtained from methanol (CH_3OH), is inserted into the end of the anode blade and oxygen or air at the end of the positive electrode of the cell (cathode) (Mo

Zhigun et al., 2005).

In the previous literature, various models have been developed for the PEMFC system dynamic modeling, analysis, control and operation. A type of fuzzy controller that controls the fuel cell output voltage is considered by Mo Zhigun et al. (2005). BP and RBF networks control strategy for voltage and current control of the fuel cell is used by Yanjun et al. (2006). The development of a computer model for simulating the transient operation of a tubular solid oxide fuel cell (SOFC) is described by David et al. (1999). The power quality in an FC-based on power system is affected by the harmonic contents of the current waveform injected to the load / grid by the inverter and also by the harmonic currents produced by the non linear loads connected to the system (Tanrioven and Alam, 2006). In addition, the harmonics injected by the inverter would increase in the FC connected to a distribution generation DC bus with devices such as photovoltaic and wind turbines. The electrochemical and thermal parts of the model were developed and verified separately before they were combined to form the transient model. The model includes the electrochemical, thermal, and mass flow elements that affect SOFC electrical output. A nonlinear lumped-parameter mathematical model of direct reforming carbonate fuel cell stack is considered by Michael et al. (2001). Analytical detailed active and reactive power output of a stand-alone PEM fuel cell power plant (FCPP) is controlled (El-Sharkh et al., 2004). The validity of the analysis in this paper is verified when the model is used to predict the response step changes in the active load and reactive power demand and actual active and reactive load profile.

The ripple current propagation path is analyzed by Changrong and Jih-Sheng (2007), who derived its linear AC model. Equivalent circuit model and ripple current reduction with passive energy storage and advanced active control technique are then proposed to incorporate a current control loop in the DC–DC converter, which is used for this goal. A fully integrated modeling approach that lends itself to parallelism is introduced by Abdelkrim et al. (2010). Simulation time reduction with parallel computing is achieved with this modeling. Gemmen (2003) suggested that the ripple current be limited to less than 10%. Passive energy storage compensation method was suggested and tested extensively by Schenck et al. (2005). Active compensation with external bidirectional DC–DC converter method was suggested by Novaes and Barbi (2003) and Monti et al. (2002). These methods require externally added components or circuits and are not preferred.

To produce electrical energy from the fuel cell, it is essential that the output voltage of cell be kept constant for different loads to supply high quality power to the loads. Also the Power Quality (PQ) problems should be solved in islanding mode operation of fuel cell. . In this paper, a simple Lead-Lag controller is proposed for fuel cell voltage control, reducing the current Total Harmonic Distortion (THD) and Current Harmonic (CH) reduction. The proposed controller is design based on HBMO algorithm. In order to achieve the optimal Lead-Lag controller at first, the problem is converted to optimization problem and then is solved by using HBMO algorithm. The main goal of this optimization problem is to regulate the voltage of PEM. The advantages of the proposed control are as follows: 1- controllers are simple, 2- its robustness against load changes, 3- has the desired control features, 4- has fast transient response and 5- has zero steady error.

Dynamic model of the fuel cell

Firstly, to study the dynamic behavior of the fuel cell, the scheme, structure and modeling of the fuel cell should be done. Figure 1 shows the schematic model of the fuel cell system proposed as voltage controller that is applied in this paper. The mass of the anode and cathode in the figure is considered as a sole compression of anode and cathode (Noradin et al., 2012). The dynamic model of the PEMFC system is based on Akbar et al. (2012). The equal output voltage of the PEMFC system is extracted by deducing the voltage drops from the regressive voltage. Equation (1) expresses how to calculate the fuel cell output voltage (Noradin et al., 2012).

$$V_s = n(E_{reversable} - V_{act} - V_{ohmic} - V_{con}) \tag{1}$$

Where, V_s is the accumulated fuel cell outandut voltage in volts, n is the existing cells in the accumulated fuel cell, V_{act} is the voltage drop resulting from anode and cathode activity in volts, V_{ohmic} is the ohmic voltage drop in volts, which is a certain amount of resistance in the transfer of electrons and protons in the electrolyte between the anode and cathode. V_{con} results from the mass transfer of oxygen and hydrogen. $E_{reversable}$ in equation (1) is calculated through the following Equations (1) and (10).

$$E_{reversable} = 1.229 - 0.85\times10^{-3}(T - 298.15) + 4.3085\times T\times[\ln(PH_2 + 0.5\ln(PO_2)] \tag{2}$$

Where, T is the cell temperature in Kelvins, PH_2, PO_2 are effective partial pressure (atm) of hydrogen and oxygen gases respectively that can be calculated by the following equation,

$$PO_2 = P_c - P_{H_2o}^{sat} - P_{N_2}^{channel}\exp\left(\frac{0.291\left(\frac{i}{A}\right)}{T^{0.932}}\right) \tag{3}$$

$$PH_2 = 0.5P_{H_2o}^{sat}\left(\frac{1}{\exp\left(\frac{1.635\times\left(\frac{i}{A}\right)}{T^{1.334}}\right)\cdot\left(\frac{P_{H_2o}^{sat}}{P_a}\right)} - 1\right) \tag{4}$$

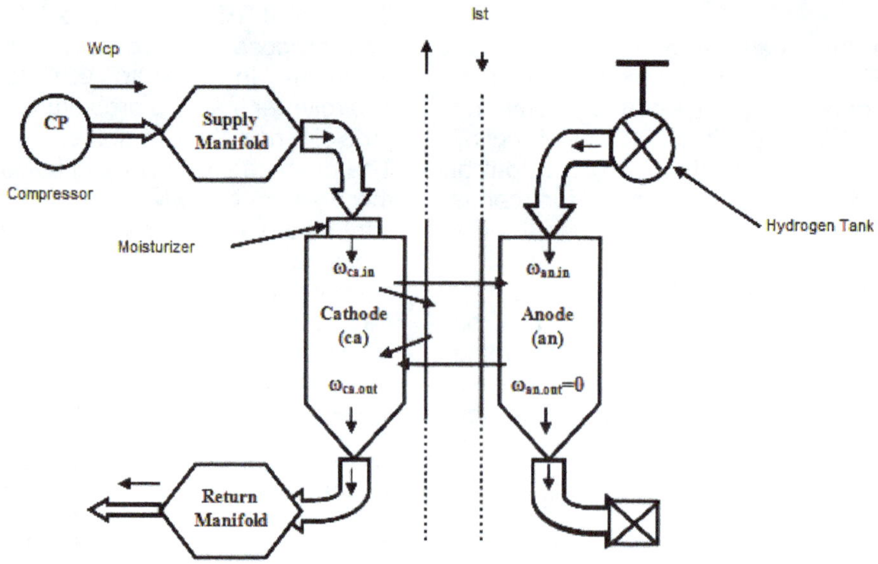

Figure 1. Schematic diagram of fuel cell.

Where, P_a and P_c are the anode and cathode inlet pressure in atmospheres, A is the effective electrode area in Cm^2, i is the current of each cell in amperes, $P_{H_2O}^{sat}$ is the amount of saturated steam pressure whose value depends on the fuel cell. $P_{N_2}^{channel}$ is the partial pressure of N_2 in the cathode gas flow channels in atmospheres which can be calculated by the following equation,

$$P_{N_2}^{channel} = \frac{0.79}{0.21} PO_2 \quad (5)$$

All the amounts used in this article are the same with data available in the work of Noradin Ghadimi, (2012).

Honey bee mating optimization

Honey bee is a social insect that can survive only as a member of a community or a colony. The colony inhabits an enclosed cavity. A colony of honey bees consist of a queen, several hundred drones, 30,000 to 80,000 workers and broods during the active season. A colony of bees is a large family of bees living in one bee-hive. The queen is the most important member of the hive because she is the one that keeps the hive going by producing new queen and worker bees (Taher, 2011). Drones' role is to mate with the queen. Tasks of worker bees are several such as: rearing brood, tending the queen and drones, cleaning, regulating temperature, gathering nectar, pollen, water, etc. Broods arise either from fertilized (represent queen or worker) or unfertilized (represent drones) eggs. The HBMO Algorithm is the combination of several different methods corresponded to a different phase of the mating process of the queen. In the marriage process, the queens mate during their mating flights far from the nest. A mating flight starts with a dance performed by the queen who then starts a mating flight during which the drones follow the queen and mate with her in the air. In each mating, sperm reaches the spermatheca and accumulates there to form the genetic pool of the colony. The queen's size of spermatheca number equals to the maximum number of matings of the queen in a single mating flight is determined. When the queen mates successfully, the genotype of the drone is stored. At the start of the flight, the queen is initialized with some energy content and returns to her nest when her energy is within some threshold of zero or when her spermatheca is full. In developing the algorithm, the functionality of workers is restricted to brood care, and therefore, each worker may be represented as a heuristic which acts to improve and/or take care of a set of broods. A drone mates with a queen probabilistically using an annealing function (Yannis et al., 2011):

$$P_{rob}(Q,D) = e^{-\frac{\Delta(f)}{s(t)}} \quad (6)$$

Where Prob (Q, D) is the probability of adding the sperm of drone D to the spermatheca of queen Q (that is, the probability of a successful mating); Δ (f) is the absolute difference between the fitness of D (that is, f (D)) and the fitness of Q (that is, f (Q)); and S (t) is the speed of the queen at time t. It is apparent that this function acts as an annealing function, where the probability of mating is high when both the queen is still in the start of her mating–flight and therefore her speed is high, or when the fitness of the drone is as well as the queen's. After each transition in space, the queen's speed, S(t), and energy E(t) decay using the following equations:

$$S(t+1) = \alpha \times s(t)(2) \quad (7)$$

$$E(t+1) = E(t) - \gamma \quad (8)$$

Where α is a factor and γ is the amount of energy reduction after each transition. Also, Algorithm and computational flowchart of HBMO method to optimize the PEM controller parameters is presented in Figures 2 and 4. Thus, HBMO algorithm may be constructed in the following five main stages:

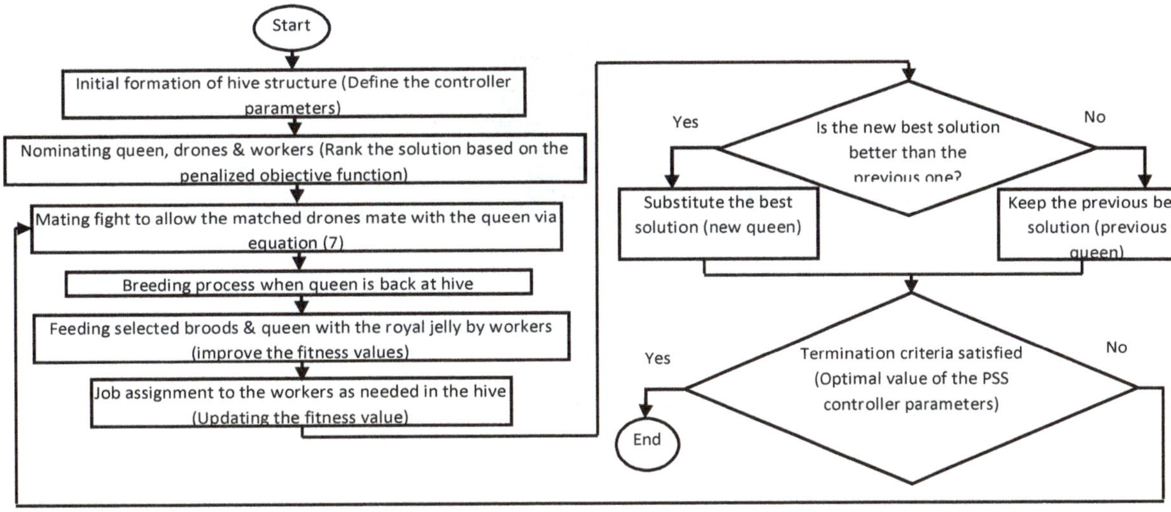

Figure 2. Algorithm and computational flowchart of HBMO.

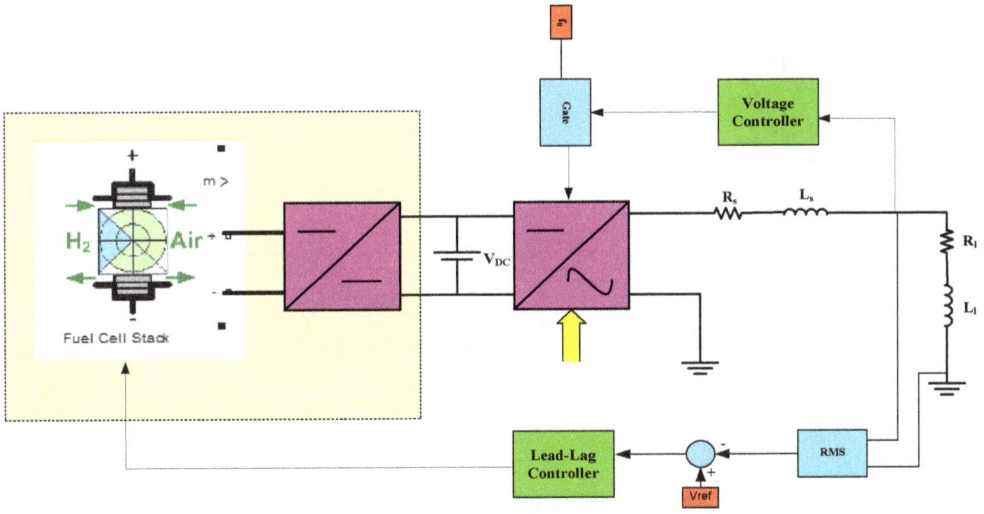

Figure 3. Single line diagram of DG, local load, controller and grid.

(i) The algorithm starts with the mating–flight, where a queen (best solution) selects drones probabilistically to form the spermatheca (list of drones). A drone is then selected from the list at random for the creation of broods.
(ii) Creation of new broods by crossover ring the drones' genotypes with the queen's.
(iii) Use of workers (heuristics) to conduct local searches on broods (trial solutions).
(iv) An adaptation of workers' fitness based on the amount of improvement achieved on broods.
(v) Replacement of weaker queens by fitter broods.

Study system description

Schematic diagram of an electronically coupled PEMFC, DG unit is depicted in Figure 3. In this figure, DG unit is connected to load via a low pass filter R_t, L_t and VSC. Parameters of the system are listed in Table 1.

Table 1. Parameters of DG, local load and grid.

Parameter	Value
R	3.8 Ω
Rt	0.075 mΩ
Lt	15 μH
VSC rated power	5 kW
PWM carrier frequency	1,980 HZ
f0	50 HZ
Nominal Load voltage	110 V (rms)

This system should be working in islanding mode. Depending on the load value, voltage of local load may be increased or decreased and even it may approach instability. Although, in the case of rated load, the voltage and the frequency would remain in nominal value.

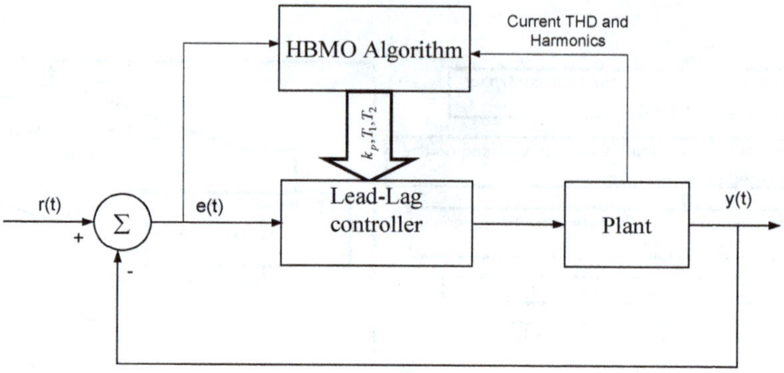

Figure 4. Schematic of the proposed controller designing.

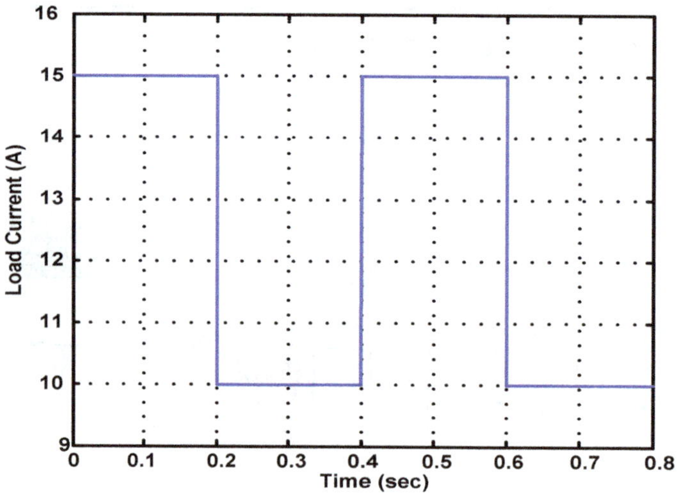

Figure 5. Worst case of load variation of PEMFC.

Also, based on load current, the harmonic current can be produced. Thus, it is necessary for system, different controller appropriate for improving system performance.

Using HBMO to adjust controller parameters

Due to the development of system controllers, the conventional controllers are used widely in power system applications. Applying conventional controllers is simple against the new controllers of power systems (Akbar et al., 2012). The Lead-Lag controllers are widely used in most cases of power system controllers which compensate very well. One of the highest benefits of these controllers is the easily implemention in analog and digital systems. If these controllers are designed optimally, indubitable they become one of the most implemented controllers in modern systems. This paper introduces a new optimal Lead-Lag controller, which is used by HBMO algorithm for designing the controller of proton exchange membrane fuel cell in order to control OD voltage and improve the PQ. The overall controller schematic is shown in Figure 4.

Lead-Lag general controller is expressed in Equation (9) in which the controller k_p, T_1, T_2 parameters should be optimized using the proposed algorithm. In the load variations, it is obvious that the transient mode of the PEMFC system depends on the controller parameters. The conventional controller designing method is not viable to be implemented because this system is an absolute nonlinear. So these methods would have no efficient performance in the system.

$$G_c(s) = k_p \frac{1 + sT_1}{1 + sT_2} \tag{9}$$

In order to design an optimal controller using HBMO for the fuel cell from the load current curve, we consider the worst condition for load design controllers for these conditions. Figure 5 depicts the worst condition for a load current in the system for voltage equal to 200V. At first, the problem should be written as an optimization problem and then by applying the proposed optimization method, the best Lead-Lag controller is achieved. Selecting objective function is the most important part of this optimization problem. This is because choosing different objective functions may completely change the particle's variation state. In optimization problem, we considered the voltage error signal, total harmonic distortion of current and 3rd harmonic of current in order to achieve the best controller.

$$J = \int_{tstart}^{tsim} \left[\left| V_{ref} - V_{out} \right| + \left| THD_i \right| + \left| I_h^{3th} \right| \right] dt \tag{10}$$

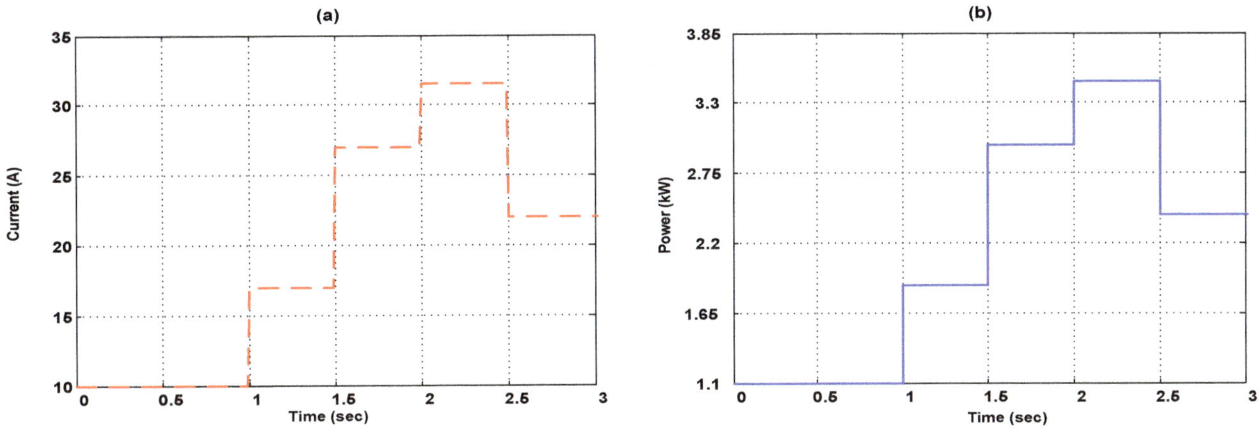

Figure 6. Load variation considering constant voltage for the fuel cell a) current b) demand power.

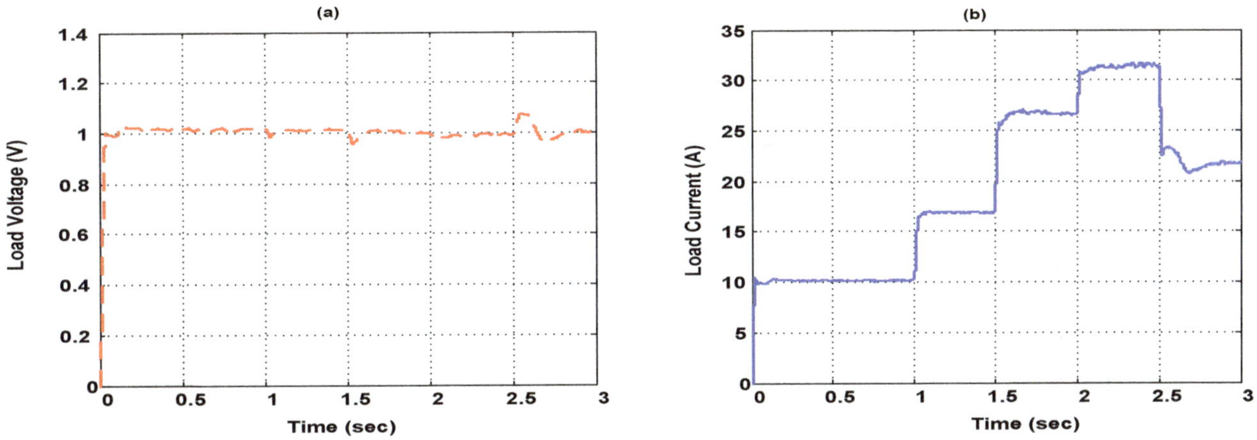

Figure 7. the results of proposed controller for load changing condition a) output voltage and reference voltage b) anode and cathode gas pressure with load current.

Where, *tsim* is the simulation time in which objective function is calculated, V_{out} is the real output voltage and V_{ref} is the reference voltage, THD$_i$ is the current total harmonic distortion and $\left|I_h^{3th}\right|$ is the 3rd current harmonics. We are reminded that when the objective function is a small amount, in this case the answer will be more optimized. Each optimizing problem is optimized under a number of constraints. Problem constraints should be expressed as:

$$\begin{aligned} &Minimize J \ subject \ to \\ &k_p^{\min} \le k_p \le k_p^{\max} \\ &T_1^{\min} \le T_1 \le T_1^{\max} \\ &T_2^{\min} \le T_2 \le T_2^{\max} \end{aligned} \tag{11}$$

Where, T_1, T_2 are in the interval [0.01 2] and k_p in the interval [1 500]. In this problem, the number of particles, the dimension of the particles, and the number of repetitions are selected as 40, 3, 80, respectively. After optimization, results are determined as:

$$k_p = 134.1 \ , T_1 = 0.077 \, , T_2 = 1.7240 \tag{12}$$

Simulation results

The load curve variation for fuel cells is considered in order to show good performance of the proposed algorithm. Desired load current is plotted in Figure 6(a) and in Figure 6(b); the amount of fuel cell power demand or load power variation is displayed. Desired load is considered under the constant output voltage, while the current is changing between the range of 10 to 32 amperes and its variations are considered to show the performance of the proposed controller in transient times. Simulation output results obtained from the proposed algorithm which is expressed in Equation (12) are shown in Figures 7, 8 and 9. Figure 7(a) depicts PEMFC's

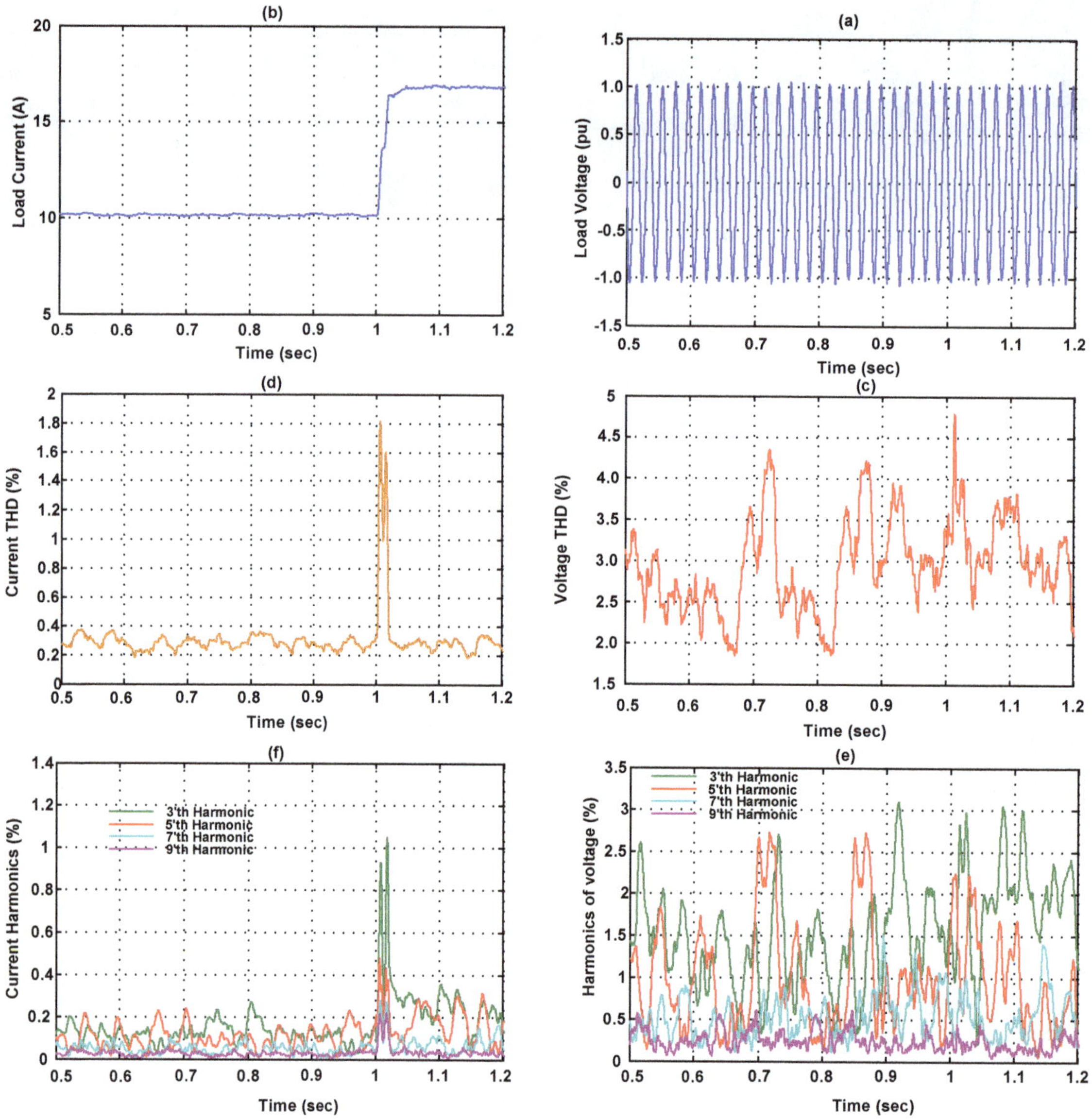

Figure 8. Transient response in load changing using proposed controller a) reducing of demand power b) increasing of demand power.

output voltage which is about 1 pu. Figure 7(b) shows the variation of load current of demand load that reaches the reference current signal. From this Figure, it can be seen that by changing load current, gas pressure in the anode and cathode change quickly to keep stable the output voltage of the fuel cell at 1 pu and this shows good performance of the proposed controller albeit simplicity. Also, according to an output voltage of load and reference voltage, it is obvious that controller response is appropriate and it could follow the reference voltage properly.

In Figure 8(a) and (b), the load voltage and current are plotted in increasing load variation condition, respectively. From the figure, it is obvious that the proposed controller can control the load voltage. Figure 8(c) and (d) show the THD value of voltage and current, respectively. It can be seen that, the THD of the voltage is lesser than 5% and THD of current is lesser than 2%. The harmonics of voltage and currents are plotted in Figure 8(e) and (f), respectively. These results have shown the high efficiency of the proposed algorithm. At t=2.5 s, reducing step in load switching occurrs. Figure 9 depicts the results of the proposed algorithm for this condition. Figure 9(a) and (b) show the load voltage and current in reducing load variation condition, respectively. From the figure, it is obvious that the proposed controller can control the load voltage. The THD value of voltage and current is plotted in Figure 9(c) and (d), respectively. It

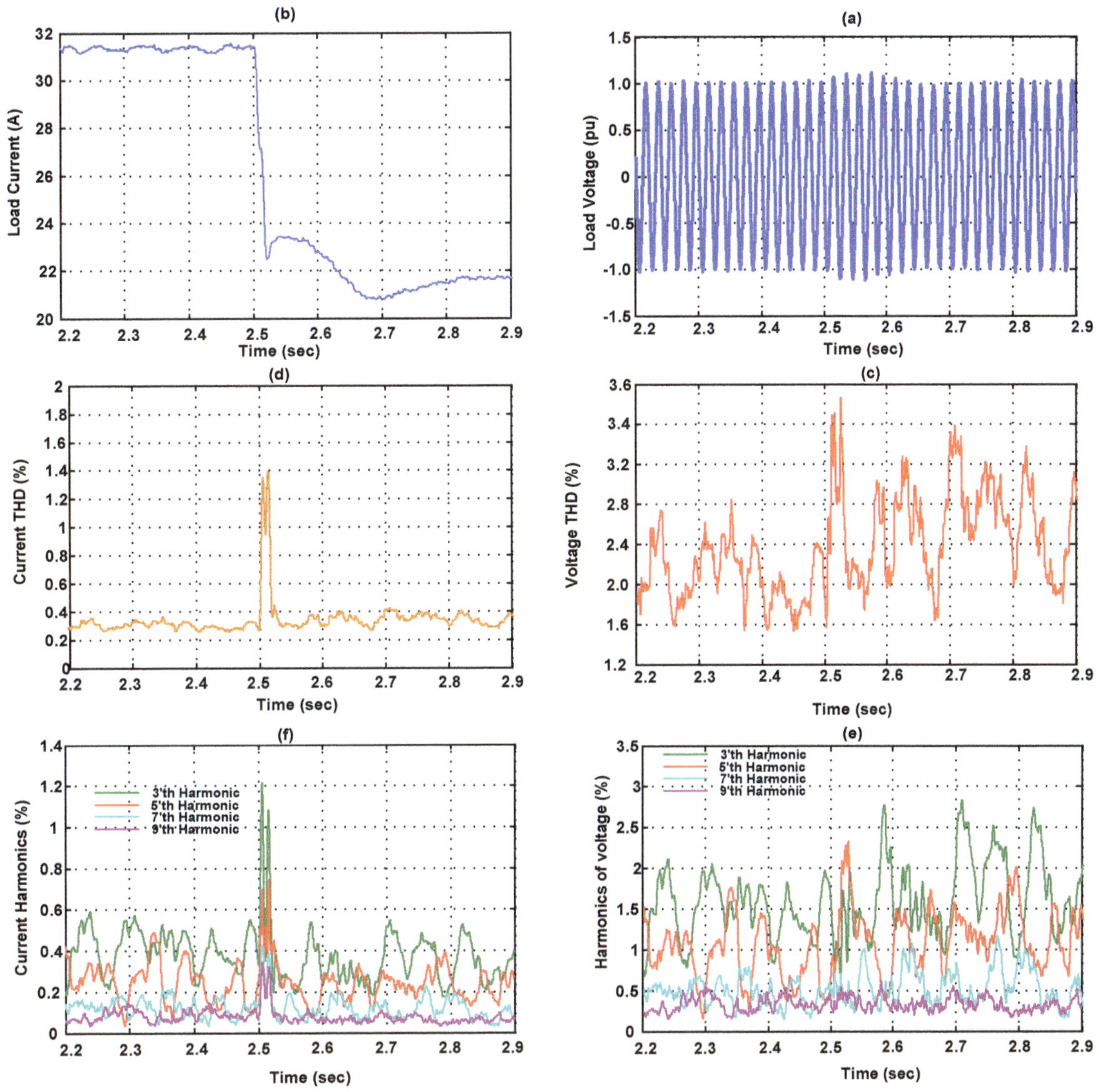

Figure 9. Transient response in applying the disturbance in anode and cathode gas.

can be seen that, the THD of the voltage is lesser than 3.6% and THD of current is lesser than 1.4%. The harmonics of voltage and currents are plotted in Figure 9(e) and (f), respectively. From these results, one can say the proposed controller is an optimal controller for PEMFC system.

CONCLUSION

Optimal controller designing for voltage control and power quality improvement using HBMO based Lead-Lag controller was proposed in this paper. For simplicity, easy implementation, high efficiency features of the Lead-Lag controller, this controller is chosen in this paper. To obviate the problem of the previous controller HBMO algorithm was utilized to design the Lead-Lag controller to have the most optimized state. In solving this problem, the problem was first written in the form of the optimization problem in which its objective function was defined and written in time domain; and then the problem was solved using HBMO algorithm. The objective function contains three parts, namely: voltage error signal, total harmonic distortion of current signal and the 3rd harmonic signal of current. The most optimal mode for gain coefficient and controller zero and pole was determined using the algorithm.

REFERENCES

Abdelkrim S, Jaafar G, Rachid O, Olivier S, El-Sayed H (2010). Modeling and Simulation of PEM Fuel Cell Thermal Behavior on Parallel Computers. IEEE Trans. Energy Conver. 25(3):768-777.

Akbar H, Mehdi M, Ramtin RN, Noradin G (2012). Designing PID Controller for Fuel Cell Voltage Using Evolutionary Programming Algorithms, J. Basic. Appl. Sci. Res. 2(2):1981-1987.

Alireza A, Alireza R (2011). Artificial immune system-based parameter extraction of proton exchange membrane fuel cell. Electrical Power Ener. Syst.33:933–9384.

Changrong L, Jih-Sheng L (2007). Low Frequency Current Ripple Reduction Technique With Active Control in a Fuel Cell Power System With Inverter Load. IEEE Trans. Power Electronics, 22(4):1429.

David J, Hall R, Gerald C (1999). Transient Modeling and Simulation of a Tubular Solid Oxide Fuel Cell, IEEE Trans. Energy Conversion, 14(3):749-753.

El-Sharkh MY, Rahman A, Alam MS, Sakla AA, Byrne PC, Thomas T (2004). Analysis of Active and Reactive Power Control of a Stand-Alone PEM Fuel Cell Power Plant,, IEEE Trans. Power Syst. 19(4):2022-2028.

Gemmen RS (2003). "Analysis for the effect of inverter ripple current on fuel cell operating condition," J. Fluids Eng. 125(3):576–585.

Michael DL, Kwang YL, Hossein G (2001). An Explicit Dynamic Model for Direct Reforming Carbonate Fuel Cell Stack. IEEE Trans. Energy Conver.16(3):289-295.

Monti A, Santi E, Ponci F, Franzoni D, Patterson D, Barry N (2002). "Fuel cell based domestic power supply-a student project," in Proc. IEEE Power Electronics Specialists Conf., Cairns, Australia, Jun. 2002, pp. 315–320.

Noradin G (2012). Genetically tuning of lead-lag controller in order to control of fuel cell voltage, Sci. Res. Essays 7(43):3695-3701.

Noradin G, Mohammad M, Rasoul G (2012). Adjusting Parameters of Lead Lag Controller Using Simulated Annealing to Control Fuel Cell Voltage, Res. J. Infor.Technol. 4(1):23-26.

Novaes YR, Barbi I (2003). "Low frequency ripple current elimination in fuel cell systems," in Proc. Fuel Seminar Special Session on Fuel Cell Power Conditioning, Miami, FL, 2003, pp. 21–27.

Schenck M, Stanton K, Lai JS (2005). "Fuel cell and power conditioning system interactions," in Proc. IEEE Applied Power Electronics Conf., Austin, TX, Mar. 2005, pp. 114–120.

Taher N (2011). "An efficient multi-objective HBMO algorithm for distribution feeder reconfiguration," Expert Syst. Applications, 38(3):2878–2887.

Tanrioven M, Alam MS (2006). "Modeling, Control, and Power Quality Evaluation of a PEM Fuel Cell –Based Power Supply System for Residential Use." IEEE Trans. Industry Applications, 42:6. November/December 2006.

Yanjun L, Houjun W, Zhijian D (2006). "Using Artificial Neural Network to Control the Temperature of Fuel Cell"IEEE Conference 2006, pp. 2159-2162.

Yannis M, Magdalene M, Georgios D (2011). "Honey bees mating optimization algorithm for the Euclidean traveling salesman problem," Infor. Sci. 181(20):4684–4698.

Zhigun M, Xinjian Z, Guangyi C (2005). "Design and Simulation of Fuzzy Controller for PEMFCs" IEEE Conference. 2005, pp. 220-224.

Performance analysis of a Ĉuk regulator applying variable switching frequency

Md. Nazmul Hasan[1] , Md. Shamimul Haque Choudhury[1,2], M. Shafiul Alam[1,2] and Muhammad Athar Uddin[1,2]

[1]International Islamic University Chittagong (Dhaka Campus), 147, Green road, Dhaka-1205, Bangladesh.
[2]Bangladesh University of Engineering and Technology, Bangladesh.

This paper examines the performance of Ĉuk regulator to convert three phase alternating current (ac) to direct current (dc). Firstly, three phase ac is rectified with bridge diode. The rectified dc voltage is applied to input of a Ĉuk regulator. Three phase harmonics filter is introduced at the input of the diode rectifier to improve the shape of the input current. The main technique applied here to improve the performance of the regulator is a variable switching frequency which is applied to the gate of the IGBT; that is, at different duty cycle of the regulator, the switching frequency is different. The Total Harmonics Distortion (THD) of the input current and the efficiency of the regulator are observed. The results shown here are simulated with Pspise.

Key words: Ĉuk regulator, Total Harmonics Distortion (THD), duty cycle.

INTRODUCTION

The demand of high quality power is increasing day by day in both commercial and industrial purpose. A great amount of work has already been done concerning the three-phase pulse width modulation (PWM) boost rectifier (Ooi et al., 1985; Wu et al., 1988; Wu et al., 1991; Habetler, 1993; Blasko and Kaura, 1997) and the buck rectifier (Kataoka et al., 1979; Busse and Holtz, 1982; Wiechmann et al., 1984; Ziogas et al., 1985). The disadvantage of these buck rectifiers is not being capable of providing more voltage than the input voltage. Another technique is boost type, which have the limitation of not being capable of providing output voltage lower than the input voltage. To overcome this limitation, a three phase PWM buck-boost rectifiers has been proposed in (Kikuch and Thomas, 2002). High order harmonic control analysis has been done. Combination of rectification and inversion function increases the complexity of operation and control of the scheme. Pulse generation is difficult because it is divided into three classes which increase difficulty of implementation. But, simplicity is a high consideration to design a regulator. Moreover, input current was found very high and no analysis has been reported to reduce the input current. A Ĉuk regulator was also proposed (Ruma, 2008) to improve the quality of the input current. This allows both steps up and step down control of rectifier output voltage. However, the study did not provide solution to input currents worsening shape with voltage control by duty cycle change. A switch mode regulator based on Ĉuk principle has been presented in (Alomgir, 2005) to regulate ac voltage to a desired value irrespective of the input voltage and load. But the efficiency was poor and no analysis has been done to improve input current. An improved voltage regulator and three phase rectifier based on boost topology are proposed in (Ahmed, 2006; Abedin et al., 2006) respectively. Ahmed (2006) has the problem of low

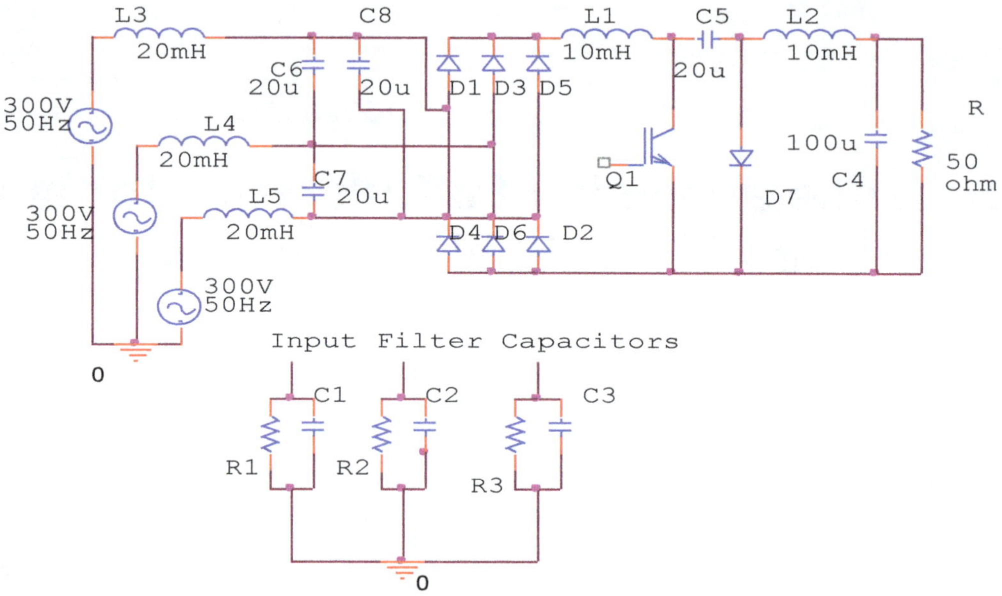

Figure 1. Circuit diagram of three phase Ĉuk rectifier with passive input filter and variable carrier frequency.

efficiency and Abedin et al. (2006) is not practically implementable due to large voltage drop across filters. The main objectives of this research work are (a) to propose a new control strategy to improve the performance of a Ĉuk regulator, (b) to simulate and study the new scheme under proposed control strategy for input current improvement of three phase diode rectifier, (c) total design of the input filter to reduce total harmonics distortion (THD) and hence reduce the input current of the converter, and (d) to increase the overall efficiency of the converter.

PROPOSED ĈUK REGULATOR

A Ĉuk regulator was investigated here to improve the overall performance to overcome the limitation of the fixed output voltage. A new strategy was introduced into the work. To avoid wide range of harmonics at the input current, the variable frequency control scheme was applied here by which the harmonics at the input current was limited within a certain range for all duty cycle (Figure 1). It is expected that the study will yield a three phase rectifier with improved power quality which is practically implementable for medium power application.

RESULTS AND DISCUSSION

The shape of the input current indicates that the performance of the regulator was improved. The simulated input current with 60% duty cycle is shown in Figure 8. Distortion free sinusoidal input current is a major consideration in a rectifier design. Many techniques have been developed by researchers in previous works. But large input filter and complex control strategy was the limitation for practically implementing of these regulators. Another important thing is, regulated output voltage both below and above the input voltages are required in many cases. Only Ĉuk and Buck-Boost regulators are able to supply regulated voltage below and above the input voltage. In this paper, a Ĉuk regulator is proposed for improvement of input current and efficiency of a three phase rectifier. At first a three phase full wave diode rectifier has been studied. The input current was found non sinusoidal pulsating and THD was found 25%. A passive filter was introduced to improve the performance of the rectifier. THD was improved to 2% and the efficiency was improved to 94%. But input for this performance 100uf input capacitor is required which is very large and also draw high input current. As a result, the VA rating of the rectifier increases and weight becomes large. The output voltage was not controllable.

To overcome the problems, a Ĉuk regulated three phase rectifier has been studied without input filter. It was observed that the input current was highly distorted with large THD, though the efficiency was good. The output voltage is controllable. To improve the shape of input current of Ĉuk regulated three phase rectifier with passive input filter was studied. The switching frequency was kept constant. It was found from the analysis that the THD has improved for many of the duty cycles, but the overall efficiency of the regulator was not acceptable at all duty cycles. It was also observed that efficiency and THD cannot be kept at the desired level simultaneously with change in duty cycle.

To improve the overall efficiency and to maintain the

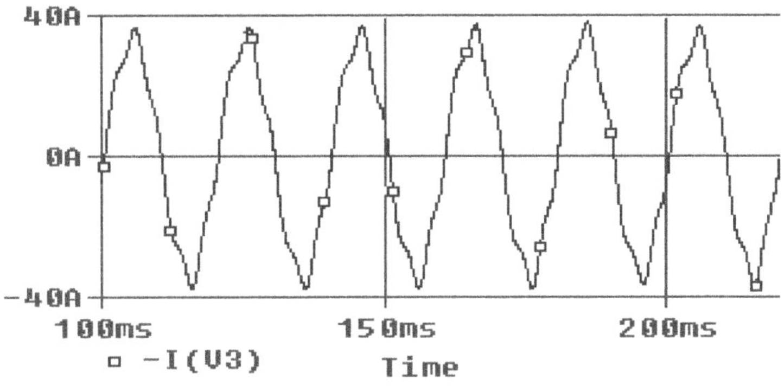

Figure 2. Input current of Ĉuk regulated three phase rectifier with passive input filter (duty cycle=60% and switching frequency= 2kz).

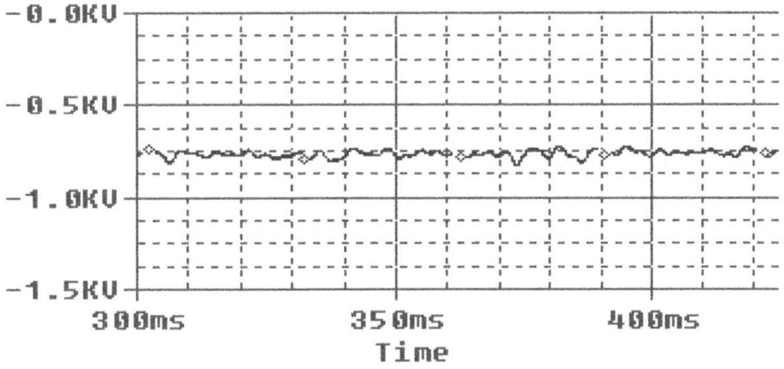

Figure 3. Output voltage of Ĉuk regulated three phase rectifier with passive input filter.

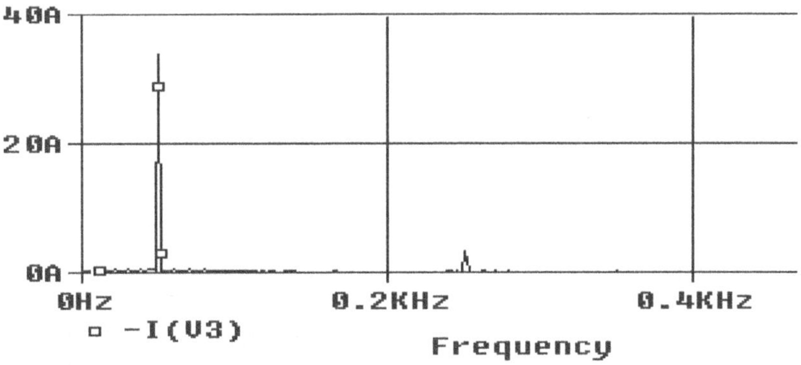

Figure 4. Typical FFT of input current of a Ĉuk regulated three phase rectifier.

THD of the input current at acceptable limit, a new topology was proposed and studied. A mixed passive filter was introduced at the input side of a Ĉuk regulator. At the same time, the switching frequency was varied from low to high frequency together with the variation duty cycles. It was found that highest THD was 7.533461% for 60% duty cycle which is below the tolerance level. The output voltage was varied from 187 volts to 770 volts with equal or more than 80% efficiency at all duty cycle. The value of input current was also acceptably low.

Input current, output voltage, FFT of input current, efficiency and power factor of the proposed Ĉuk regulator are shown in Figures 2 to 6 respectively. The overall performance of the proposed Ĉuk regulator can be understood in Table 1. The duty cycle versus power

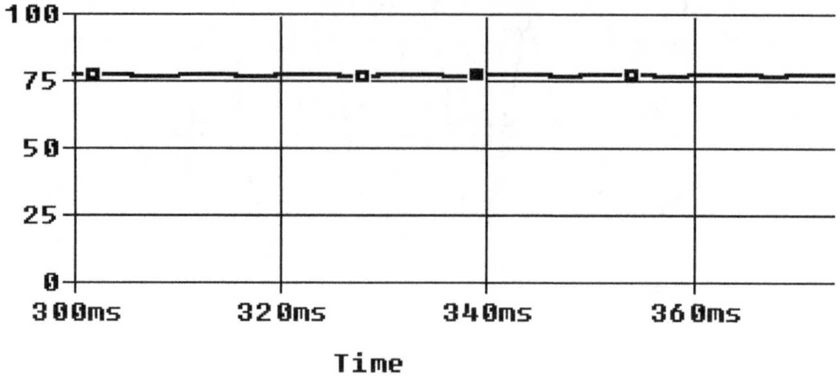

Figure 5. Efficiency of a Ĉuk regulated three phase rectifier.

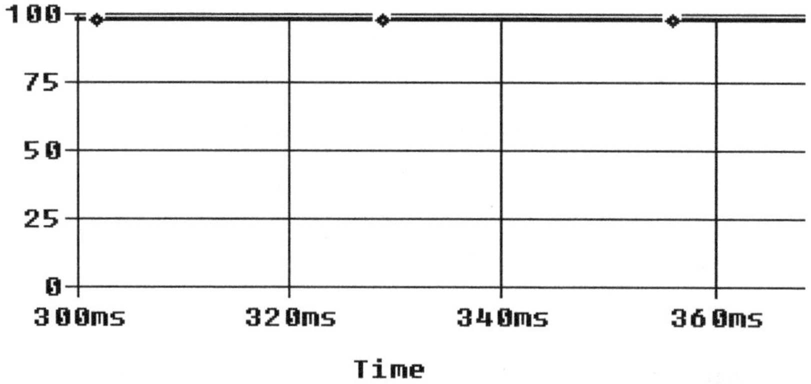

Figure 6. Power factor of Ĉuk regulated three phase rectifier with passive input filter.

Table 1. Performance parameter of Ĉuk regulator.

Duty cycle(%)	Frequency (KHz)	Input voltage (Volt)	THD (%)	PF (%)	Efficiency (η)%	Output voltage (Volt)
10	.75	300	3.144534E+00	21	82	187
20	1	300	5.868050E+00	47	94	310
30	1.25	300	5.265780E+00	65	99	400
40	1.5	300	5.247629E+00	76	92	460
50	1.75	300	7.149165E+00	91	91	600
60	2.00	300	7.533461E+00	98	80	770

factor, efficiency, output voltage, THD and switching frequency are shown in Figures 7 to 11.

Control strategy

A microcontroller based control system was introduced here. PWM modulation technique is applied to vary the duty cycle. Different duty cycles at different frequency have been generated by the microcontroller based circuit system. Practically generated pulse and flow chat of the program for generating pulse is given in Figures 12 and 13 respectively.

Conclusion

Three-phase PWM Ĉuk rectifiers have been investigated

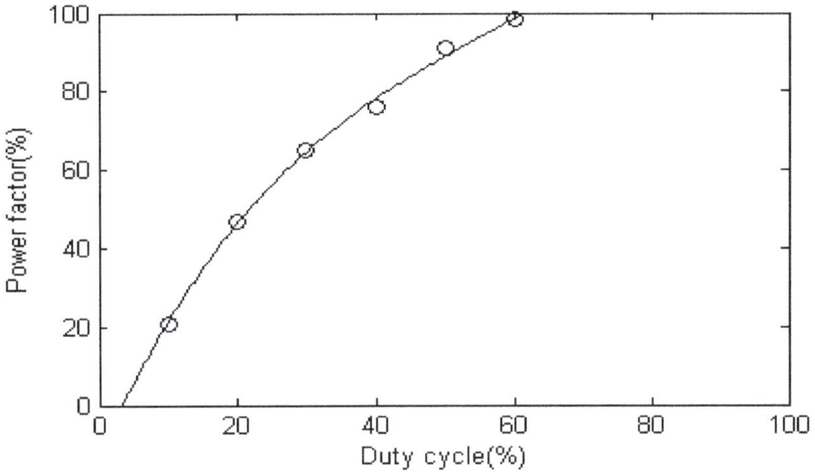

Figure 7. Duty cycle versus power factor.

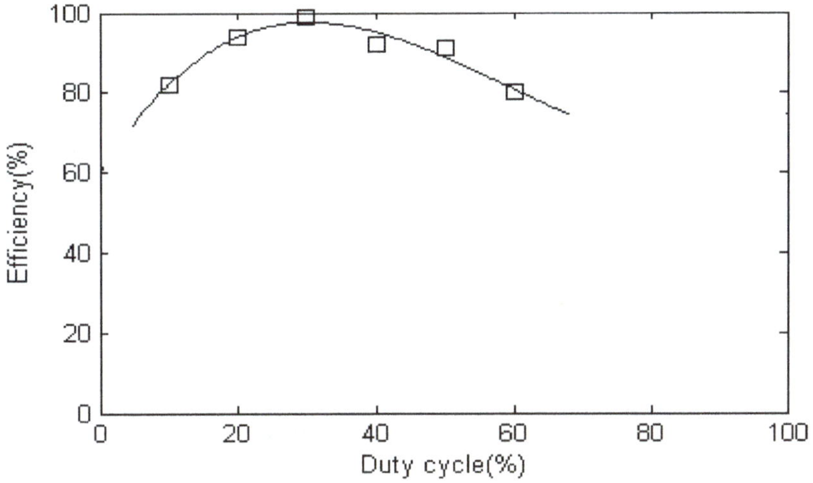

Figure 8. Duty cycle versus efficiency.

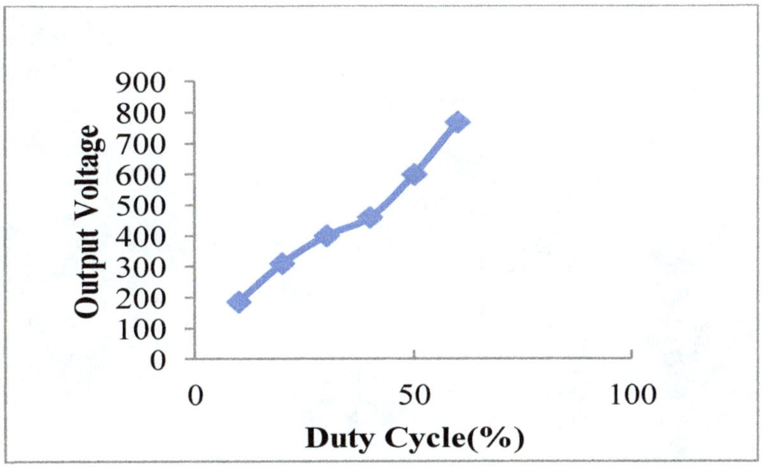

Figure 9. Duty cycle versus output voltage.

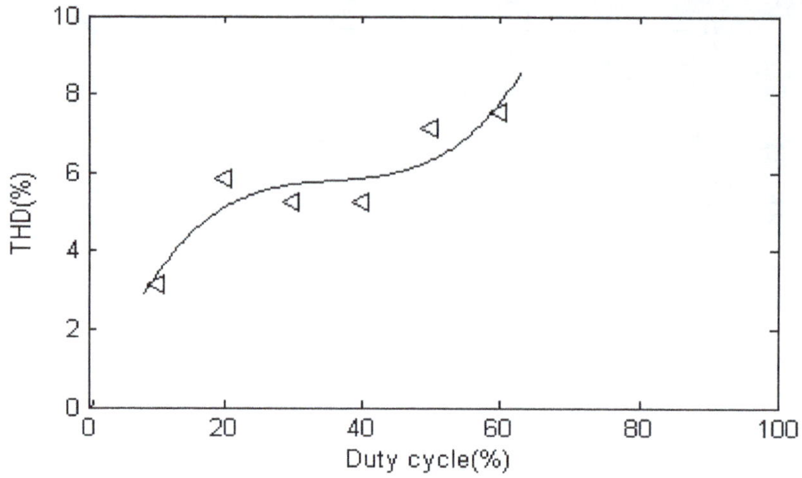

Figure 10. Duty cycle versus THD (%).

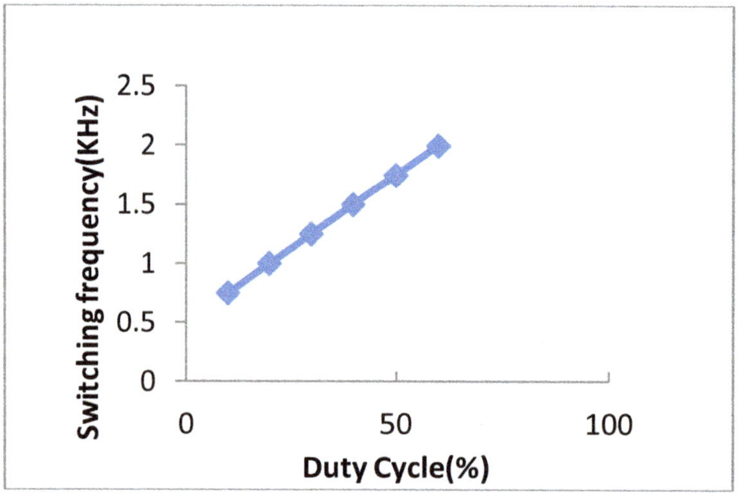

Figure 11. Duty cycle versus switching frequency.

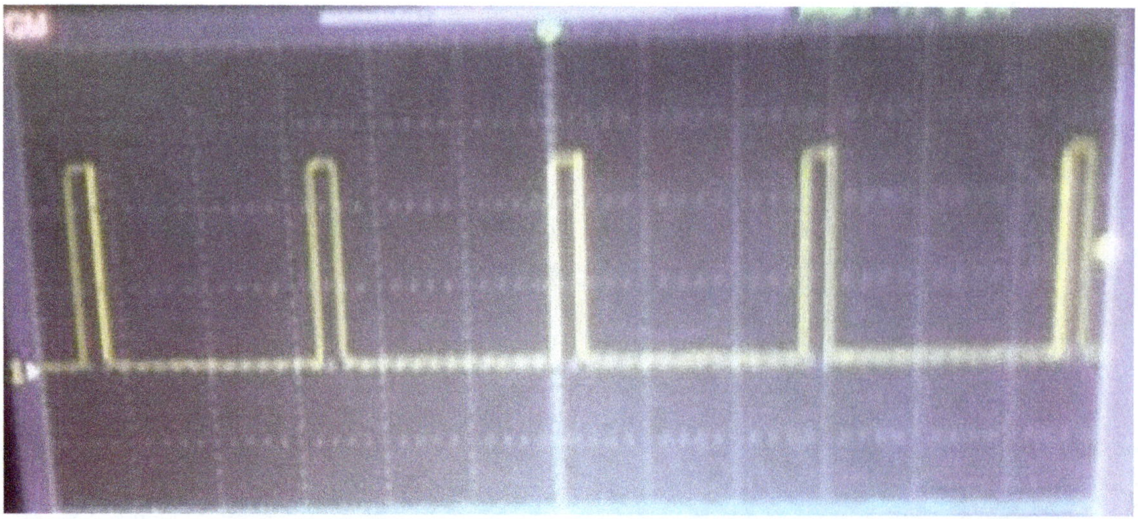

Figure 12. Microcontroller generated pulse.

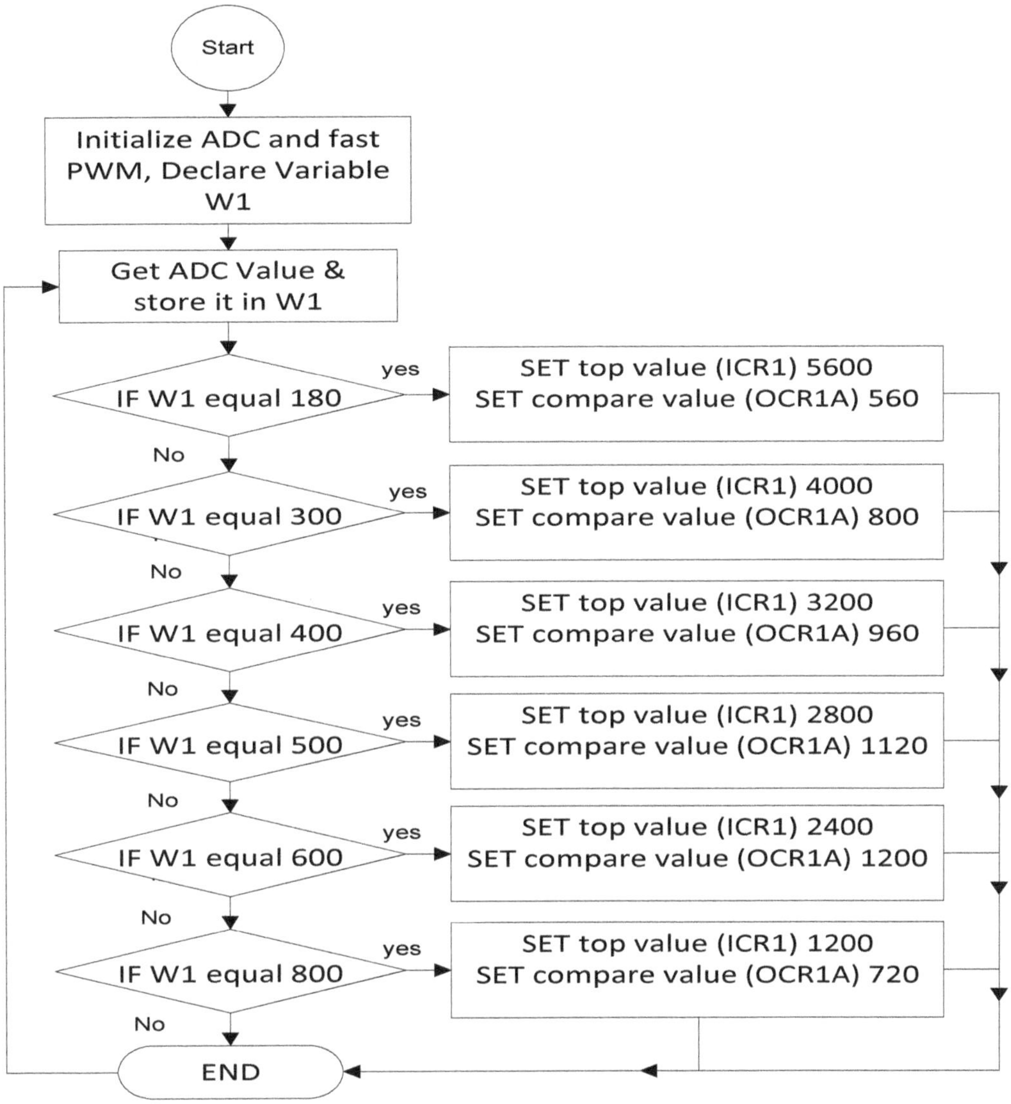

Figure 13. Flow chat of the program for generating pulse.

in this paper. The converters of interest have the properties of:

(1) Capabling both voltage step-up and step-down.
(2) Efficiency over 80% at all duty cycle.
(3) Almost sinusoidal input current. The simulated results have been presented here.

ACKNOWLEDGEMENTS

The authors acknowledge gratefully the support and facilities extended by the International Islamic University Chittagong (Dhaka Campus) and Department of Electrical and Electronic Engineering, Bangladesh University of Engineering and Technology, Dhaka, Bangladesh.

REFERENCES

Abedin AH, Ahmed MR, Alam MJ (2006). Improvement of input side current of three phase rectifier combining active and passive filters. J. Electric. Eng. IEB, EE 33(I- II):87-90, December 2006.

Ahmed MR (2006). Design of a switch mode ac Voltage regulator with improved power factor. M. Sc. Engineering Thesis, BUET, Department of EEE.

Alomgir H (2005). AC voltage regulation by Ĉuk switch mode power supply. M.Sc. Engineering Thesis, Department of EEE, BUET.

Blasko V, Kaura V (1997). A new mathematical model and control of athree-phase ac–dc voltage source converter. IEEE Trans. Power Electron. 12:116–123.

Busse A, Holtz J (1982). Multiloop control of a unity power factor switching ac to dc converter. Proc. IEEE PESC. 82:171–179.

Habetler TG (1993). A space vector-based rectifier regulator for ac/dc/ac converters. IEEE Trans. Power Electron. 8:30–36.

Kataoka T, Mizumachi K, Miyairi S (1979). A pulse width controlled ac-to-dc converter to improve power factor and waveform of ac line current. IEEE Trans. Ind. Appl. IA-15:670–675.

Kikuch L, Thomas LA (2002). Three-phase PWM Buck-Boost rectifiers with power regenerative capability. IEEE Trans. Ind. Appl. 1(5):

1361-1369.
Ooi BT, Salmon JC, Dixon JW, Kulkarni AB(1985). A 3-phase controlled current PWM converter with leading power factor. Int Conf. Rec. IEEE-IAS Annu. Meet. pp. 1008–1014.
Ruma B (2008). Input Current Improvement of a Three Phase Rectifier by Ĉuk regulator. M.S.c Engg thesis, Department of EEE, BUET, Dhaka, Bangladesh.
Wiechmann EP, Ziogas PD, Stefanovic VR (1984). A novel bilateral power conversion scheme for variable frequency static power supplies. Proc. IEEE PESC 84:388–396.
Wu R, Dewan SB, Slemon GR (1991). Analysis of an ac-to-dc voltage source converter using PWM with phase and amplitude control. IEEE Trans. Ind. Appl. 27:355–364.
Wu R, Dewan SB, Slemon GR(1988). A PWM ac to dc converter with fixed switching frequency, Int. Conf. Rec. IEEE-IAS Annu. Meet. pp. 706–711.
Ziogas PD, Kang YG, Stafanovic VR (1985). PWM control techniques for rectifier filter minimization. IEEE Trans. Ind. Appl. IA-21:1206–1214.

Optical observation of streamer propagation and breakdown in seed based insulating oil under impulse voltages

Essam A. Al-Ammar

Saudi Aramco Chair in Electrical Power, Department of Electrical Engineering, King Saud University, Riyadh, Saudi Arabia.

Increasing demand of electricity and high consumption of natural fuel resources has raised countless challenges for the electricity industry. For years, transformers were equipped with mineral oil for better and lasting performance but the recent uncertainty about the future of oil availability is motivating the researchers to find out the "Green Insulating Oils" for transformers. Thus, green insulating oils are being probed with great interest at research centers worldwide. In this paper, Canola oil was studied for applications in transformers as a sustainable solution. The paper also compared the proposed oil with mineral oil by studying the initiation and propagation of streamers in the oils when subjected to standard lightning impulse voltages. It also presents experimental results of a comparative study of initiation and propagation of streamers in canola seed based oil and mineral insulating oil when subjected under standard lightning impulse voltages. Moreover, streamer patterns, modes of propagation and their stopping lengths were investigated under both polarities of voltage in point-plane electrode gap. The paper concludes with six different findings, which states that canola oil is a better and sustainable choice of oil for transformer use.

Key words: Breakdown, impulse voltage, oil, seed based insulating oil, streamer propagation.

INTRODUCTION

Mineral oil based dielectric fluids are applied in a large variety of electrical power equipment. Worldwide estimates show that almost 30 to 40 million tons of mineral insulating oils are in use. The functions of these oils are to fill the air cavities of the porous solid insulation in order to improve its partial discharge behavior as well as to act as a heat transfer medium to dissipate the losses (Oommen, 2002). However, mineral oils are easily flammable, are poorly degradable and can contaminate soil and water, if spillover takes place. In addition, the increasing crisis of petroleum oil that is currently leading to uncertainty in its sustainable supply has forced the researchers worldwide to find suitable alternate sources (Claiborne et al., 1999; Amanullah et al., 2005; Perrier et al., 2004).

In this context, the vegetable seed oils are natural

Table 1. Measured properties of the investigated fluids.

Property		Unit	Canola Oil	Mineral Oil
Density at 23°C		kg/dm^3	0.98	0.89
Viscosity at:	30°C	Mm^2/s	28	23
	60°C		12	6.5
Acidity		mg/KOH/g	0.1315	0.056
Water Content		ppm	125	15
(60 Hz) AC Breakdown Voltage using IEC Mushroom Electrodes with 1.0 mm separation		kVrms	43	34
Dielectric Dissipation Factor (tan δ)	23°C		0.0027	0.0011
	80°C		0.019	0.01
Permittivity (ε_r) at	23°C		3.07	2.2
	80°C		2.85	2.05

products and have renewable sources with plenty of supply and present suitable "green products" that can replace the mineral based oils. Their most attractive features are the high biodegradability (95 to 100%) and higher fire point (~360°C) instead of ~160°C for most mineral based insulating oils. Researches on pre-breakdown and breakdown phenomena of mineral based insulating liquids have been progressing since 1950s (Sharbaugh et al., 1978; Beroual et al., 1998) respectively. Martin and Wang (2008); Rapp et al. (2009) and Yasuda et al. (2010) exemplify the use of vegetable oils for dielectric applications in power distribution transformers and other high voltage equipments. Nevertheless, recently transformer-grade vegetable oils are also commercially available. The first commercial product was BIOTEMP, patented in the U.S. in September 1999 by ABB. Another U.S. patent was also issued in September 1999 for transformer oil which uses regular soybean oil. Moreover, another U.S. patent was granted to Cooper Industries, Inc. under the trademark of Envirotemp FR3 (Moumine et al., 1995).

This fluid is based on standard-grade oleic oils, and is used commercially in some distribution transformers. Subsequent patents were also issued to the ABB inventors on the BIOTEMP fluid in August 2001 (Oommen and Claiborne, 1999; McShane et al., 2000).

In Badent et al. (2000) two types of oils that is, Vegetable Canola oil (VO) and Transformer grade mineral oil (MO) are investigated. Canola oil is kitchen grade pure oil and its main molecular composition is triglycerides and fatty acid containing both saturated as well unsaturated fatty components with up to 23 carbon chain lengths containing double bonds. Due to the presence of double bonds, it is prone to oxidation under thermal stress when in contact with copper or other metals. To overcome these problems, antioxidants such as Butylated Hydroxyanisole (BHA) and Butylated Hydroxytoluene are mixed in the bulk oil. Moreover, further reason to select the Canola oil is that several ester groups and other radicals are also present in Canola oil. The viscosity neutralization number and dissipation factor in Canola oil are higher than in mineral oil but these are within the accepted limits specified in international standards for mineral insulating oils. The higher values of permittivity at ε_r = 3.07 and hydrophilic character of Canola oil is expected to provide better edge to designers as compared to mineral oil, provided it is chemically synthesized and refined by stripping it off the radicals that make it prone toward oxidation and the ones that control its viscosity.

On the other hand, mineral oil molecules mainly consist of carbon and hydrogen atoms arranged in different structures such as paraffinic, naphthenic and aromatics. Their composition varies depending on the source, and the aromatic content plays a major role in the formulation of the streamer shapes (Badent et al., 2000; Devins et al., 1981). Some salient properties of these oils were measured in the laboratory and are summarized in Table 1.

EXPERIMENTAL SETUP

The experiment set up for detection of initiation and propagation of streamers in the investigated liquids is shown in Figure 1. A test cell comprising of point-plane electrode system was designed and used. Its main body was made of PTFE material and its top lid was of transparent PMMA ("Perspex") to facilitate the observation of inter electrode gap events. Tungsten and high carbon steel needles with tip radius (r_p) of 10 μm were used. The plane electrode was made of brass having a diameter of 50 mm with its edges rounded.

A Perspex sheet barrier of 3 mm thickness was embedded on its surface to protect the electrode and the attached detection equipment at the advent of oil breakdowns. The electrode gap was arranged in horizontal format and is shown in Figure 2. Haefely 10-

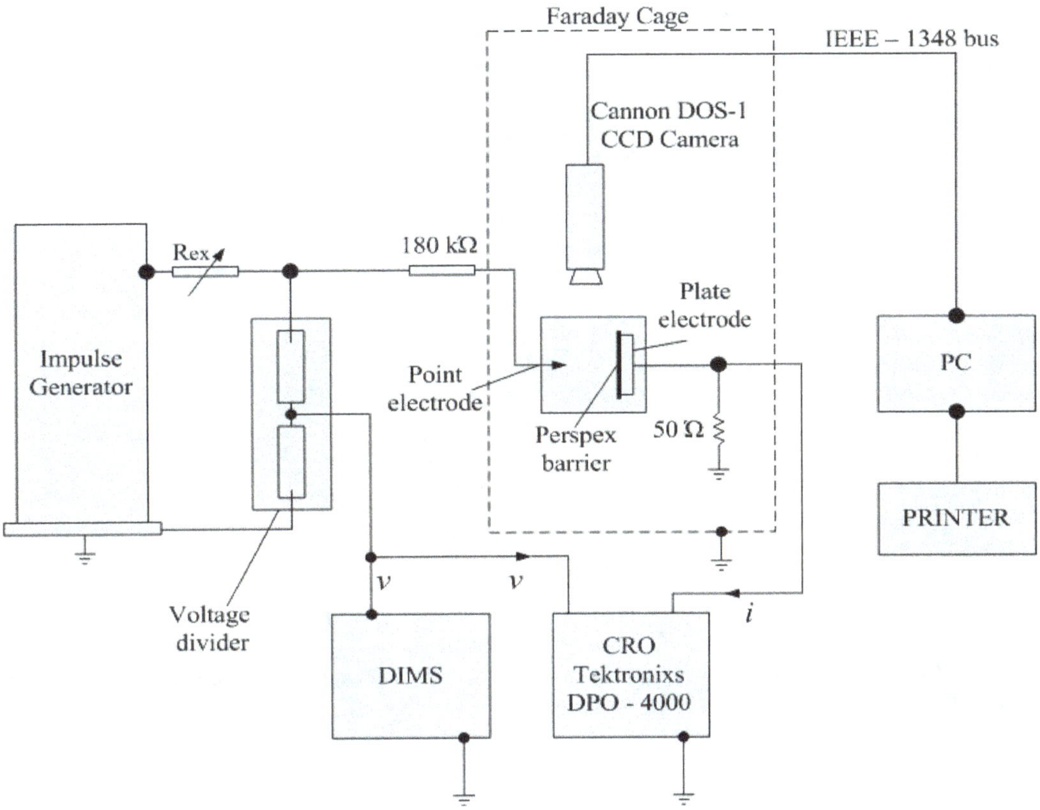

Figure 1. Sketch of experimental set up.

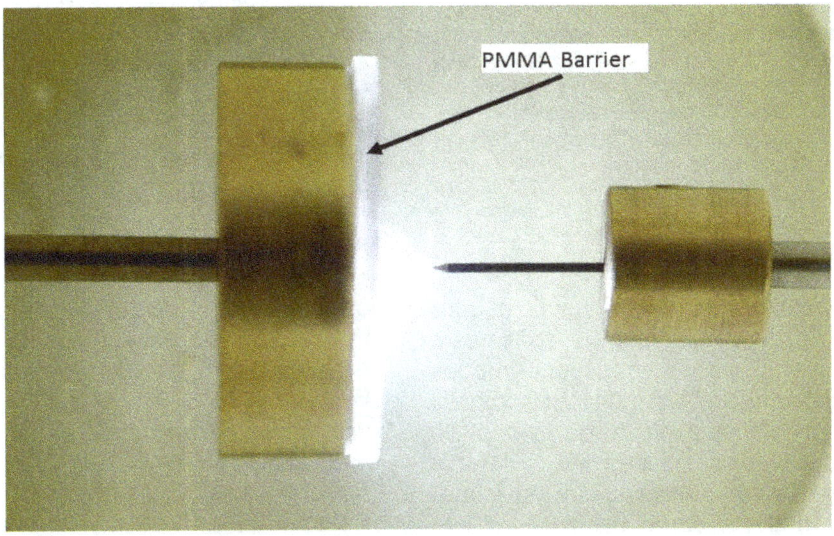

Figure 2. Electrode gap in the test cell.

stage impulse generator with 100 kV, 4 kJ output per stage was used to produce standard lightning impulse voltages which were measured through an RC voltage divider and Digital Impulse Measuring System (DIMS). The voltage signal and current signal coming from the precision non-inductive 50 Ω resistor were fed to a 400 MHz oscilloscope for data acquisition. A 180 kΩ resistor was connected at the output of the impulse generator to limit the breakdown current and injected energy into the test cell.

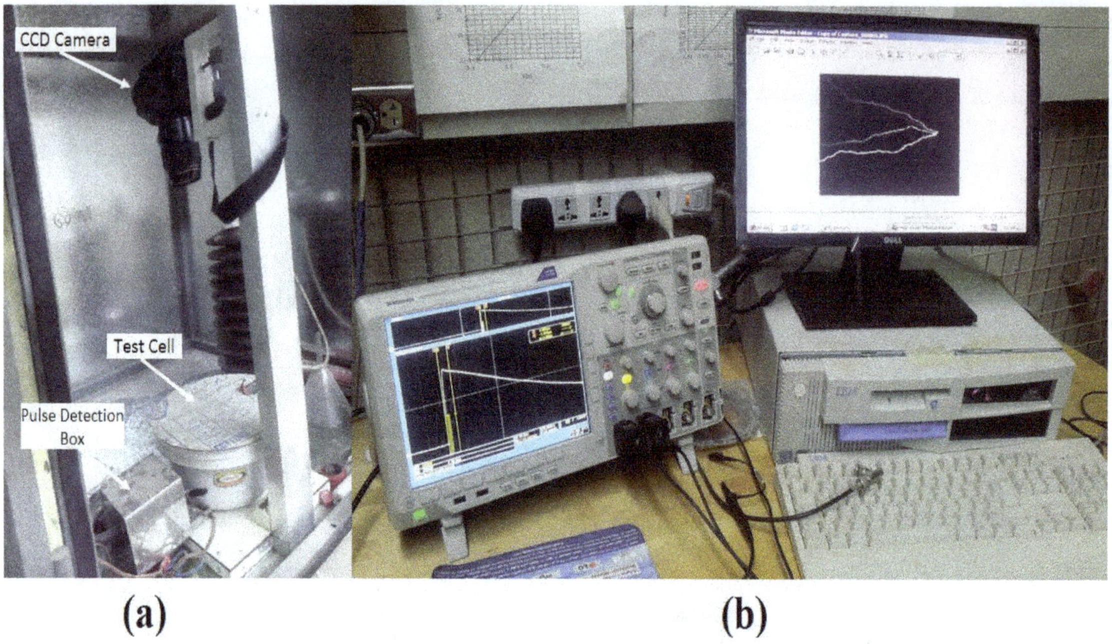

Figure 3. Experimental arrangement: (a) CCD camera, test cell and pulse detection box, (b) PC and oscilloscope connected with the incoming signal cables.

The shadowgraph system consisting of a 15 Mega pixel CCD Camera, which could be used with or without a flash was employed. Instead of conventional shadow graphic system in which the streamer channels are displayed in the background of a trigger light pulse, the self-irradiated light from filaments of streamers was used to capture the events. The camera was opened for 5 s synchronously with the operation of the trigger pulse applied to the impulse generator. This exposure time and any delay in the operation of trigger system were found optimum to capture the events associated with the applied voltage impulse. Moreover, Figure 3 displays photographically the set up with camera, test cell and other detection instruments connected together.

RESULTS AND DISCUSSION

Streamer characteristics of the two types of oils mentioned earlier were investigated. Comparisons were made for the shapes using propagation modes including their stopping lengths. This study was mostly carried out using a middle size electrode gap of 20 mm (from industrial application's point of view), with a point electrode tip of 10 μm radius. The streamer shapes in mineral oils had the same behavior as had already been reported in (Dang et al., 2012) and therefore these were not documented here to avoid duplication. Therefore, results presented here are for the streamers in Canola oil that were captured using both polarities of standard lightning impulse voltages (1.2/50 μs). However, the analysis and discussion are made for both types of the oils investigated that is mineral oil and canola oil.

Electrical characteristics of selected oils

The most important properties of insulating oils are the dielectric ones beside the usual physico-chemical characteristics. Vegetable oils possess high flash and fire point as compared to mineral based oil. This typical character lends strong support to adopt these oils in transformers located in hazardous locations. The combination of fire safety and high biodegradability (> 98%) can eliminate the traditional need for fire walls and deludge systems built around transformer banks (McShane et al., 1999; CIGRE WG A2-35 Brochure, 2010).

The A.C breakdown strength of vegetable oils was not affected with moisture intake up to 300 ppm, where as the moisture present in mineral oil (MO) has very deleterious impact on its properties (CIGRE WG A2-35 Brochure, 2010).The use of vegetable oil in combination with cellulose in transformer helps in drying out the later as it absorbs its moisture when in contact (Martin, 2010). Several groups of investigators (Perrier et al., 2004; Beroual et al., 1998; Rapp et al., 2009; Dang et al., 2012) have reported higher power frequency breakdown strength of conola oil as compared to mineral oil.

On the other hand, a number of studies on traditional mineral oil were published (Lesaint and Massala, 1998; Massala and Lesaint, 2001; Linhjell et al., 1994; Torshin, 2003, 2009; Lopatin et al., 1998) describing the streamer initiation and propagation in terms of velocity and shape, streamer mechanism and streamer modeling.

Figure 4. Streamer onset in Canola oil under negative lightening impulse voltage.

It is acknowledged that impulse pre-breakdown and breakdown characteristics are related to chemical composition of the liquid, so application of esters (with composition different to mineral oil) calls for detailed investigation under impulse voltage. Some papers on this topic were reported in Badent et al. (2000) and Duy et al. (2009). Streamer stopping length, propagation velocity, 50% breakdown voltage and acceleration voltage of natural ester (rape-seed oil) were also documented, which showed easier propagation of streamers in ester and consequently lower breakdown voltage than mineral oil. Hestac et al. (2004) reported the streamer inception voltages of rape-seed oil at 8 and 20 mm tip-plane gaps, were about 50% lower than that of mineral oil. Streamer initiation in a synthetic ester (Midel 7131) under strong non-uniform field was studied in Viet-Hung et al. (2012) which was found to be about 60% higher than that of mineral oil.

Streamer patterns in canola oil

Negative polarity streamer

At first the approximate breakdown level of the oil gap was determined and then starting from around 70% of this voltage level, the applied voltage three consecutive shots were applied at each set level. If no event was observed, the voltage was increased in steps of 2 kV_p. The streamer initiation voltage (V_i) was registered if it appeared under all three applied shots. At the onset, a faint minuscule light appears in the vicinity of the point electrode as shown in Figure 4a. With an increase in voltage, it expanded with instabilities appearing on its surface as shown in Figure 5b. One or two of the instabilities enlarged with increase in voltage and propagated toward the plane electrode. As the streamers increased in size, more filaments appeared as off-shoots from these enlarged filaments as shown in a sequence of events captured independently at different intervals as shown in Figure 5b-g. Figure 5g displays the breakdown event as captured at well above the breakdown voltage level which it impinged on the plane electrode.

It is clear that as the voltage is increased, more number of luminous branches appears and was propagated towards the plane electrode. The first one which came in contact with the plane electrode caused a flow of large current leading to the formation of a plasma channel which emanates strong light intensity light. This channel was also clearly seen in this image. Since the picture captured was in 2-dimensions while the streamer pattern propagated in 3-dimensions the filaments that were closer to the observation system appeared brighter while the ones that were away from the observation point exhibited less luminosity. It is to be noted here that the streamer propagating in mineral oil were less branched and more luminous than observed in Canola oil.

Positive polarity streamer

It was more difficult to capture the initiation event of streamers under positive polarity lightning impulses. The reason is that the streamers once initiated, propagate very swiftly with long branches. Figure 6a exhibits a white spot which indicated initiation of a positive streamer. It was much brighter than the initial initiation spot that was observed in the vicinity under negative impulses point electrode. As soon as the voltage was increased to the next step, large size filaments appeared and propagated much faster compared to streamers under negative polarity. Figure 7c-f show propagated streamer shapes captured with increasing voltage.

In case of the positive streamer, a single large streamer with bright luminosity resulted in the breakdown of the gap as shown in Figure 7g and it was observed to exert a much stronger shock-wave than the corresponding

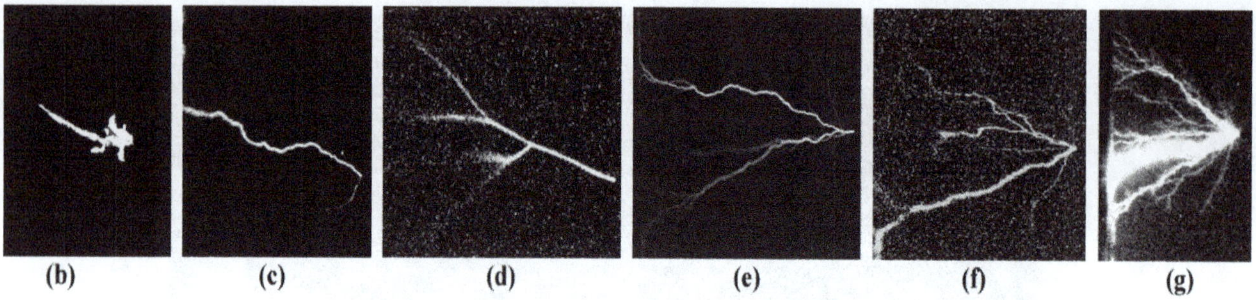

Figure 5. Sequence of initiation and propagation of negative streamers in vegetable oil. Electrode gap = 20 mm; Point tip radius = 10 µm (b) V = 65 kVp, (c) V = 73 kVp, (d) V = 75 kVp, (e) V = 85 kVp, (f) V = 90 kVp, (g) Breakdown event at -97 kVp.

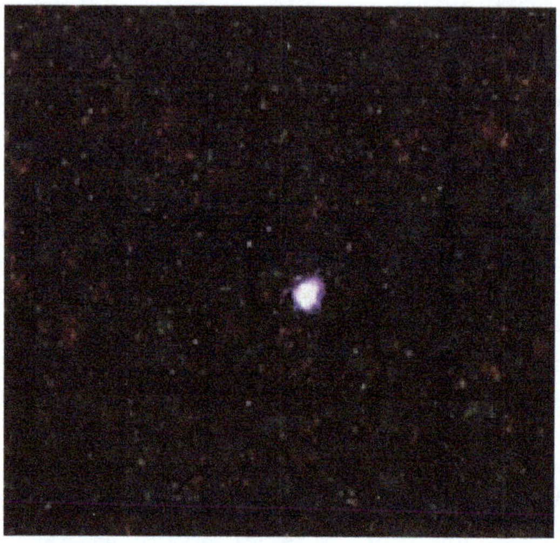

Figure 6. Positive streamer onset.

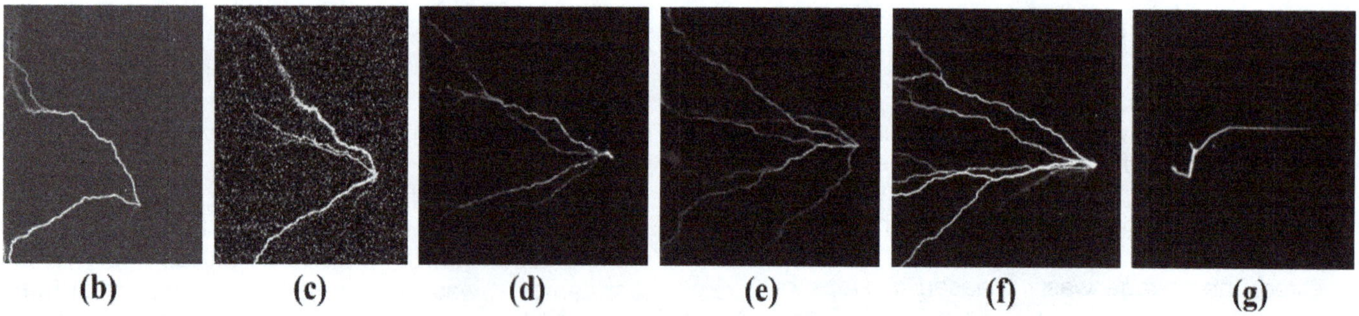

Figure 7. Sequence of initiation and propagation of positive streamers in vegetable oil. Electrode gap = 20 mm; point tip radius = 10 µm (b) V=57 kVp (c) V=62 kVp, (d) V=70 kVp, (e): V=81 kVp, (f): V=83 kVp, (g) Breakdown streamer at V=98 kVp.

negative streamer breakdown. It was also observed that the positive streamers, both in mineral oil as well as Canola oil were filamentary and less branched than the negative streamers.

Furthermore, in case of canola oil the streamers propagated to longer lengths and they were faster, especially once they cross the mid-gap spacing. These observations suggest that positive streamers in Canola

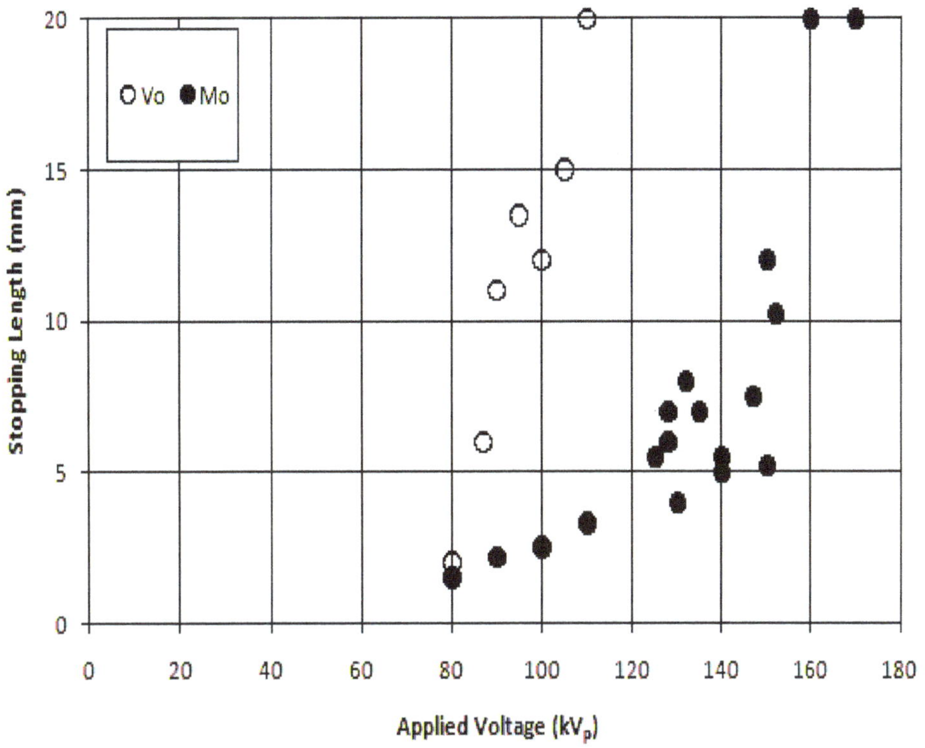

Figure 8. Final stopping lengths of negative streamers in Canola and mineral oils as a function of applied voltage. Gap spacing = 20 mm; point tip radius = 10 μm.

oil underwent different propagation modes as compared to these in mineral oil. However, this observation needs further experimental studies in order to isolate the occurrence of these differing modes and the shapes of associated streamers.

Stopping lengths of streamers

Figure 8 displays the comparison of final stopping lengths (L_f) of streamers under the application of preset voltage impulses of negative polarity lightning impulses for canola oil and mineral oil. These lengths were measured as the axial distance from the point electrode to the plane electrode. In case of mineral oil, the L_f increases almost linearly up to a voltage level of about 125 kV_p, where after, a large scatter was observed in the L_f values over the applied voltage range of 125 to 147 kV_p. Beyond this voltage level the L_f tends toward a very rapid rise up to breakdown voltage level of the gap. This was consistent with the results reported earlier by Dang et al. (2012) for mineral oil under similar gap configuration and by Liu et al. (2009) in 50 mm gap spacing. Moreover, from Figure 8, it was also observed that mineral oil had an exponential increase in applied voltage against the increase of stopping length.

This certainly explains that with the rise in voltage amplitude the negative streamer undergoes three different modes of propagation. The first mode in which the L_f propagates with lower velocity is generally confined to a gap range of about 0.3 d (where d = gap spacing). The second mode consists of medium velocity and displays large scatter in L_f values and occurs in the gap range of 0.3 to 0.5 d. Beyond the gap range of 0.5 d, the negative streamers propagate at much faster speed.

In Canola oil, too, the slower mode was confined in gap spacing of about 0.3 d. However, in this case, the second mode was not present, since the L_f values increased rapidly with further increase in the voltage. At a fixed voltage level, the L_f values in Canola oil were comparatively much longer than in mineral oil. This is why the negative polarity breakdown voltage of mineral oil was about 40% higher than in the Canola oil. This effect is opposite to AC breakdown voltage values reported in Table 1, where the AC breakdown voltage of Canola oil was almost 27% higher than in MO. However, its higher permittivity (ε_r = 3.1) which is comparable to transformer press board (ε_r = 3.7) and its typical hydrophilic character shall impose much beneficial effect

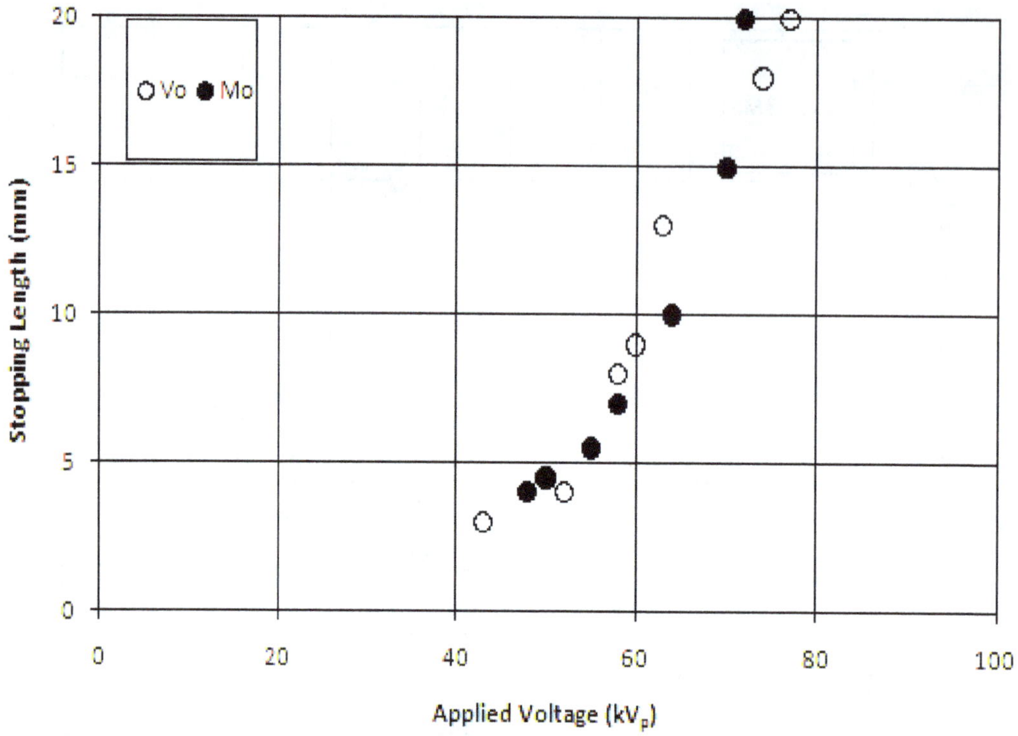

Figure 9. Final stopping lengths of positive streamers in Canola and mineral oils as a function of applied voltage. Gap spacing = 20 mm; point tip radius = 10 μm.

on the drying of cellulose and thus shall provide an offset toward its lower impulse breakdown strength. Therefore, by this experimental study, it could be concluded that Canola oil can be further refined to come at par with mineral insulating oil's properties and could be used as a good "Green Insulating Oil".

It has been reported earlier in their pioneer work by Devins et al. (1981) and later by others (Yasuda et al., 2010; Chadband and Sufian, 1985; Nelson, 1980; Kelley and Hebner, 1981) that the electron affinity of liquid molecules played a dominant role in the propagation of streamers. The molecule of Canola oil contains three esters (−C OOR) which have electro-negative oxygen atoms. These atoms may be the source behind swift propagation of negative streamers as well as for more branches that emanated from the initiated streamer trunk, as portrayed in Figure 5.

Figure 9 compared the L_f values of positive streamers as a function of applied voltage for both investigated oils. The streamer length versus transitions observed under negative polarity was not distinct under positive polarity. Moreover, the increase in L_f values with increase in the applied voltage was almost similar in both oils. Small scatter was noticed in L_f values but it was present in both oils. Moreover, it could also be observed that in positive streamer, both canola and mineral oil had a linear increase in applied voltage with respect to stopping length. The transition in this case occurred, around 56 kV_p. Under positive polarity lightning impulses, which streamers propagate very fast the streamers propagate much faster than under negative polarity, while their more luminous filaments indicated that they were more conducting than their negative counterparts. Whereas several negative streamer filaments impinge on plane electrode at the breakdown only one filament of the positive streamer that came in contact with the plane electrode became very luminous and caused a breakdown. Its stem remained initially bluish but this bluish color changed to bright white color in time as well as with the increase in the applied voltage. This aspect of positive streamer needs further investigations using a faster optical event capturing system.

As the negative or positive streamer propagates in the gap it traversed with certain velocity. Moreover, the voltage drop (ΔV) across the length of the streamer from the point electrode to the streamer head caused it to finally stop. Therefore, it depended on the local field at the point electrode as well as the mean electric field in the inter-electrode gap. The average field in the streamer could be evaluated from the 'L_f - V' characteristics of streamers in each oil and for each polarity of the applied voltage. Thus, average longitudinal electric field 'E_l' within the streamer channel followed the relation $E_l = \Delta V/\Delta L_f$. The deduced value of E_l for positive streamers for both oils was about 1.6 MV/m, whereas it was about 2.0 MV/m for the faster mode and around 6 MV/m for the slower negative streamer's mode observed near the inception in the mineral oil. Interestingly, E_l value for faster mode of negative streamers in Canola oil approached close to that of positive streamer values. The lower E_l values indicated that these streamers were more conducting. It is a well known fact that the more conducting the streamer, the more rapid it propagated through the electrode gap (Beroual et al., 1998; Badent et al., 2000; Devins et al., 1981; Nelson, 1980).

Conclusions

This paper has reported on streamer initiation and their propagation in seed based Canola oil and mineral insulating oil used in transformers. This study leads to the following main conclusions:

1. Negative polarity streamers in Canola oil were more branched than in mineral oil.
2. Propagation of negative steamers in mineral oil exhibit three distinct modes, whereas only two modes were detected in canola oil.
3. The streamer stopping length in Canola oil for the same voltage was much longer than in mineral oil. This leads to lowering of the impulse breakdown strength of Canola oil as compared to the mineral oil.
4. The conductivity of positive streamers was higher than those of negative streamers in both oils, while the stopping lengths of those streamers were almost close to each other in both liquids.
5. Positive streamers were less branched than negative streamers.
6. The AC breakdown strength of Canola oil was 27% higher than mineral oil but it was 40% lower under lightning impulse voltage.

However, it is tough to explicitly declare that Canola oil is better than mineral oil. But from the given set of experiment and results, it has been found that canola oil has high potential to be used in transformers. These results also support a statement and new area of research that slight alteration in canola oil can produce significant effect over its properties and could be perfectly suitable for transformers.

Conflict of Interest

The author(s) have not declared any conflict of interest.

ACKNOWLEDGEMENTS

The author would like to thank Saudi Aramco Chair in Electrical Power for technical guidance and financial funding. Special gratitude is due to Prof. Abdulrahman Arainy, Prof. Nazar Malik, Prof. Abdhulrhaman Beroual, and Dr. Mohammed I. Qureshi for always being around and providing invaluable comments. The author also wishes to thank Engr. Muhammad Babar and Eng. Nissar Wani for their help in preparing the experimental setup.

REFERENCES

Amanullah M, Islam S, Chami S, Ienco G (2005). "Analyses of electro-chemical characteristics of vegetable oils as an alternative source to mineral oil-based dielectric fluid," in IEEE Int. Conf. Dielectric Liquids, 2005, pp. 397–400.

Badent R, Hemmer M, Konekamp U, Julliard Y, Schwab A (2000). "Streamer inception field strengths in rape-seed oils," in Annual Report Conference on Electrical Insulation and Dielectric Phenomena, 2000, vol. 1. IEEE, 2000, pp. 272–275.

Beroual A, Zahn M, Badent A, Kist K, Schwabe A, Yamashita H, Yamazawa K, Danikas M, Chadband W, Torshin Y(1998). "Propagation and structure of streamers in liquid dielectrics," IEEE Elect. Insul. Magazine 14(2):6–17.

Chadband W, Sufian T (1985). "Experimental support for a model of positive streamer propagation in liquid insulation," IEEE Trans. Electr. Insul. 2:239–246.

CIGRE WG A2-35 Brochure, No. 436, October 2010.

Claiborne C, Walsh E, Oommen T (1999). "An agriculturally based biodegradable dielectric fluid," in IEEE Trans. Distribution Conf. 2:876–881.

Dang V, Beroual A, Perrier C (2012). "Investigations on streamers phenomena in mineral, synthetic and natural ester oils under lightning impulse voltage," IEEE Trans. Dielectrics Elect. Insul. 19(5):1521–1527.

Devins J, Rzad S, Schwabe R (1981). "Breakdown and prebreakdown phenomena in liquids," J. Appl. Phys. 52(7):4531–4545.

Duy CT, Lesaint O, Denat A, Bonifaci N (2009). "Streamer Propagation and Breakdown in Natural Ester at High Voltage", IEEE Trans. Dielectrics Electr. Insul. 16(6):1582-1594.

Hestac OL, Berg G, Ingebrigtsen S, Lundgaard LE (2004). "Streamer Injection and Growth under Impulse Voltage: A Comparison of Cyclohexane, Midel 7131 and Nytro 10X", IEEE. Conf. Electr. Insul. Dielectrics Phenomena (CEIDP), Colorado, USA, pp. 542-546.

Kelley E, Hebner R (1981). "The electric field distribution associated with pre-breakdown phenomena in nitrobenzene," J. Appl. Phys. 52(1):191–195.

Lesaint O, Massala G (1998). "Positive streamer propagation in large oil gaps: Experimental Characterization of Propagation Modes", IEEE Trans. Dielectrics Electr. Insul. (3):360-370.

Linhjell D, Lundgaard L, Berg G (1994). "Streamer propagation under impulse voltage in long point-plane oil gaps", IEEE Trans. Dielectrics Electr. Insul. I(3):447-458.

Liu Q, Wang Z, Perrot F (2009). "Impulse breakdown voltages of ester-based transformer oils determined by using different test methods," in IEEE Conf. Electr. Insul. Dielectric Phenomena, CEIDP'09, IEEE, 2009, pp. 608–612.

Lopatin V, Noskov MD, Badent R, Kist K, Schwab AJ (1998). "Positive discharge development in insulating oil: optical observation and simulation", IEEE Trans. Dielectrics Electr. Insul. 5(2):250-255.

Martin D, Wang Z (2008). "Statistical analysis of the ac breakdown voltages of ester based transformer oils," IEEE Trans. Dielectrics Elect. Insul. 15(4):1044–1050.

Martin MAG (2010). "Vegetable oils, and alternative to mineral oil for power transformers", IEEE Electrical Insulation Magazine, 26(6):7-13.

Massala G, Lesaint O (2001). "A comparison of negative and positive streamers in mineral oil at large gaps", J. Phys. D: Appl. Phys. 34(10):1525-1532.

McShane C, Corkran J, Harthun R, Gauger G, Rapp K, Howells E (2000). "Vegetable oil based dielectric coolant," Mar. 14 2000, US Patent 6,037,537.

McShane CP, Gauger GA, Luksich J (1999). "Fire Resistant Natural Ester Dielectric Fluid and Novel Insulation System for its Use. Proc. of IEEE/PES Transmission and Distribution Conference, 1999.

Moumine I, Gosse B, Gosse J, Clavreul R, Hantouche C (1995). "Vegetable oil as an impregnant in hv ac capacitors," in IEEE 5th Int. Conf. Conduction Breakdown Solid Dielectrics, ICSD'95, 1995, pp. 611–615.

Nelson J (1980). "Insulating liquids: Their uses, manufacture and properties," Electronics Power 26(6):481.

Oommen T (2002). "Vegetable oils for liquid-filled transformers," IEEE Electrical Insul. Magazine, 18(1):6–11.

Oommen T, Claiborne C (1999). "Electrical transformers containing electrical insulation fluids comprising high oleic acid oil compositions," Sep. 7 1999, US Patent 5,949,017.

Perrier C, Beroual A, Bessede J (2004). "Experimental investigations on different insulating liquids and mixtures for power transformers," in Conference Record of the IEEE Int. Symposium Electrical Insul. pp. 237–240.

Rapp K, Corkran J, Mcshane C, Prevost T (2009). "Lightning impulse testing of natural ester fluid gaps and insulation interfaces," IEEE Trans. Dielectrics Elect. Insul. 16(6):1595–1603.

Sharbaugh A, Devins J, Rzad S (1978). "Progress in the field of electric breakdown in dielectric liquids," IEEE Trans. Elect. Insul. 4:249–276.

Stockton DP, Bland JR, McClanahan T, Wilson J, Harris DL, McShane P (2007). "Natural Ester Transformer Fluids: Safety, Reliability & Environmental Performance," Petroleum and Chemical Industry Technical Conf. (PCIC '07): 1-7.

Torshin YV (2003). "Prediction of Breakdown Voltage of Transformer Oil from Predischarge Phenomena", IEEE Trans. Dielectrics Electr. Insul. 10(6):933-941.

Torshin YV (2009). "Initiation and propagation of the negative leader in transformer oil under impulse voltage", IEEE Trans. Dielectrics Electr. Insul. 16:1536-1542.

Viet-Hung D, Beroual A, Perrier C (2012). "Investigation on streamers phenomena in mineral, synthetic and natural ester oils under lightning impulse voltage", IEEE Trans. DEI. 19(5):1521-1527.

Yasuda K, Arazoe S, Igarashi T, Yanabu S, Ueta G, Okabe S (2010). "Comparison of the insulation characteristics of environmentally-friendly oils," IEEE Trans. Dielectrics Elect. Insul. 17(3):791–798.

On the nature of electric charge

Jafari Najafi, Mahdi

1730 N Lynn ST apt A35, Arlington, VA 22209 USA.

A few hundred years have passed since the discovery of electricity and electromagnetic fields, formulating them as Maxwell's equations, but the nature of an electric charge remains unknown. Why do particles with the same charge repel and opposing charges attract? Is the electric charge a primary intrinsic property of a particle? These questions cannot be answered until the nature of the electric charge is identified. The present study provides an explicit description of the gravitational constant G and the origin of electric charge will be inferred using generalized dimensional analysis.

Key words: Electric charge, gravitational constant, dimensional analysis, particle mass change.

INTRODUCTION

The universe is composed of three basic elements; mass-energy (M), length (L), and time (T). Intrinsic properties are assigned to particles, including mass, electric charge, and spin, and their effects are applied in the form of physical formulas that explicitly address physical phenomena. The meaning of some particle properties remains opaque. For example, there is no intelligible explanation of an electric charge. What is known of electric charge is its ability to generate force and an electromagnetic field and its attractive and repulsive reactions to other charged particles. It is known that an electric charge obeys the law of conservation and quantization. Constants have been identified, such as h (Planck constant), v_c (speed of light), and G (gravitational constant). Meaningful interpretations of mass-energy, length, and time in relation to these constants are also necessary. Attempts have been made to clarify the nature of electric charge, but none are comprehensive because they fail to provide explicit general formulas for the relation between electric charge and known physical parameters. The present study provides a new approach to this problem and an explicit formula that addresses the relation between electric charge and known physical parameters. This approach is of great generality and mathematical simplicity that simply and directly postulates a hypothesis for the nature of the electric charge. Although the final formula is a guesswork based on dimensional analysis of electric charges, it shows the existence of consistency between the final formula and proven physical facts.

The explanation begins with a brief introduction to dimensional analysis and how it is used. The new method is then applied to develop a formula to describe G in terms of known physical universal parameters. The new formula will be shown to be almost identical to the solution of the Friedmann equation. The guideline will be proposed to formulate an explicit definition of the nature of an electric charge. It will be shown that electric charge is equivalent to mass change in a particle. To overcome deficiencies in the new approach, the results will be compared to current proven knowledge. No contradictions or inconsistencies will be raised and excellent compatibility will be confirmed. Practically, any ambiguity in knowledge may unexpectedly produce multiple unknowns. The main purpose of this study is to eliminate obscurity about the origin of the electric charge.

DIMENSIONAL ANALYSIS

Almost all physical parameters (constants and variables) have a combination of the dimensions of mass, length, and time. The present study did not employ basic units for absolute temperature (Θ), amount of substance (N) and luminous intensity (J), and other parameters related to them. Because physical parameters are addressed, the gravitational constant and electric charge can be sufficiently expressed using a combination of M, L, and T. Constants without dimensions express values such as angle or proportions such as the ratio of particle speed to light speed. Some constants merely establish equality in the system of measurement. Some physical constants and variables do not have clear dimensions of a combination of M, L, and T. For example, the dimension of linear momentum is ML/T. This is an explicit combination of mass, length and time. The dimension of electric charge $\sqrt{4\pi\varepsilon_0 \frac{ML^3}{T^2}}$ is more complicated. The vacuum permittivity constant is ε_0, but it is not explicitly defined in terms of M, L, and T. One can easily decompose ML/T into, e.g., mass (M) and velocity of a particle (L/T), and find a formula for the linear momentum of a particle. This has not been accomplished for electric charge.

Dimensional analysis is a simple tool for understanding the relationship between physical parameters and equations based on their dimensions, but it is not used to explore and formulate unknown phenomena. It is often used to check the accuracy of calculations and equations and to express physical parameters based on their associations with other types of parameters.

The present study uses a slight generalization of this tool to find a meaningful and explicit description of the obscure physical parameter of electric charge. This method parses the dimension of an unintelligible parameter into separate parts and performs a simple mathematical manipulation on each part to convert it into a real physical parameter. Finally, the physical parameters derived for each part, keeping the final composition, will be used to present a precise definition of the relationship between the combination of M, L, and T and the original unknown parameter.

Crediting a physical parameter to a specific dimension must be based on logic and known physical facts. This method avoids any decomposition of a dimension or assigning a physical parameter to each part that promotes ambiguity in the meaning of a part. This constraint appears as a lemma that should be studied and developed. This method may not work in all cases, but gives good results for gravitational constant and electric charge.

It should be emphasized that this method and its related restrictions will not sufficiently support the final results. Each equation obtained from this method is guesswork, although dimensions for both sides of the equation are established. Experiments, measurement, and mathematical calculations based on pre-established relationships and equations confirm the conjecture; however, it is good practice to compare these results withproven physical facts.

First, the symbols [] and $[]^{-1}$ should be defined in relation to the dimensions of the physical parameters:

Definition 1: $[y] = M^{\alpha} L^{\beta} T^{\gamma}$

In this equation, $M^{\alpha} L^{\beta} T^{\gamma}$ is the dimension of physical quantity *y*. In fact, the units of any physical quantity can be expressed using a power law (Sedov, 1993).

Definition 2: $y = [M^{\alpha} L^{\beta} T^{\gamma}]^{-1}$

where a physical quantity y governs the dimension combination $M^{\alpha} L^{\beta} T^{\gamma}$.

NATURE OF THE GRAVITATIONAL CONSTANT G

Gravitational constant (G) is a constant parameter that appears in the equation of gravity:

$$F = G\,\frac{m_1 m_2}{r^2} \tag{1}$$

This equation itself does not provide a tangible interpretation of the nature of G. This formula does not define its source, amount, or relation to other actual parameters in the universe. To discover this relationship, the dimensions of both sides of Equation (1) for M, L and T are:

$$\frac{ML}{T^2} = [G]\frac{M^2}{L^2} \tag{2}$$

Therefore:

$$[G] = \frac{L^3}{MT^2}\,\frac{M}{M} \tag{3}$$

Equation (3) represents the dimension of G and states that physical parameters exist in the universe whose integration (as a formula with the above final dimension) gives the value of G. To fit this into an equation to determine the value of G in terms of physical parameters, the right side of Equation (3) is parsed into two parts and multiplied by dimensionless parameter β on the right side, obtaining:

$$G = \beta\,[\frac{L^3}{M}]^{-1}[\frac{1}{T^2}]^{-1} \tag{4}$$

The fraction *M/M* should not be omitted, but should

encompass in β for simplicity. This can be rewritten as:

$$G = \beta \left[\frac{4\pi/_{3}L^{3}}{4\pi/_{3}M}\right]^{-1}\left[\frac{1}{T^{2}}\right]^{-1} \quad (5)$$

If L is assumed to be the dimension of the radius of a sphere, the numerator of $\frac{4\pi/_{3}L^{3}}{4\pi/_{3}M}$ is the volume of that sphere:

$$G = \beta \left[\frac{V}{4\pi/_{3}M}\right]^{-1}\left[\frac{1}{T^{2}}\right]^{-1} \quad (6)$$

Obtaining:

$$G = \frac{3}{4\pi}\beta \left[\frac{1}{\left(\frac{M}{V}\right)}\right]^{-1}\left[\frac{1}{T^{2}}\right]^{-1} \quad (7)$$

The mass-to-volume ratio represents the mass density (ρ). Assuming that $\frac{1}{\rho}$ and $\frac{1}{T^2}$ are real physical parameters that govern dimensions $\frac{1}{\left(\frac{M}{V}\right)}$ and $\frac{1}{T^2}$, respectively, produces:

$$G = \beta \frac{3}{4\pi\rho T^{2}} \quad (8)$$

Because G is a universal constant independent of any particle and reference system, the parameters of Equation (8) should be relevant to the whole universe. Therefore, let the supposed sphere be the entire universe and M its mass; consequently, ρ is the mass density of the universe. Also let T be the age of the universe. Now, choosing constant β gives an equation that expresses the value and nature of gravitational constant G. Regardless of the value of constant β, Equation (8) states that G is inversely proportional to the squared age and matter density of the universe. That is, if G is constant, ρ must decline steadily because of the continual increase in the value of T. Therefore, V (the volume of the universe) increases steadily because of the law of mass-energy conservation. Now, we compare the derived formula for G (Equation (8)), with what has been obtained, using Einstein field equations. In the matter-dominant universe, solving the Friedmann equation leads to (Komissarov, 2012):

$$G = \frac{3\Omega_{m}H_{0}^{2}}{8\pi\rho_{m}} \quad (9)$$

In this equation, the Hubble constant (H_0) is approximately the inverse of the age of universe and Ω_m is the ratio of mass density of the universe to the critical mass density. This Ω_m may be related to *M/M* in Equation (3) that was canceled and temporarily included in β.

Therefore, if we substitute β with $\frac{\Omega_m}{2}$, there will be no difference between Equations (8) and (9).

Using dimensional analysis, plus guess and intuitive analysis based on physical realities, obtains an equation to determine the physical parameters defining G. Although, this process began with a good guesswork, evaluating it using the Friedmann equation confirmed the strength of new method and correctness of its result.

NATURE OF ELECTRIC CHARGE

The question remains about the nature of electric charge and why there is no comprehensible interpretation of electric charge based on M, L and T, or a combination of known parameters of particles, such as spin and velocity. It is not known why same-charged particles repel and oppositely-charged particles attract or if electric charge is really an intrinsic property of a particle. The complicated dimension of electric charge $\sqrt{4\pi\varepsilon_0\frac{ML^3}{T^2}}$ makes it difficult to accept it as an intrinsic property of particles, when compared with simple and explicit properties like mass and spin.

Electric charge seems to be an abstract of other basic properties and physical constants. If this is true, how can an electric charge and electromagnetic field be directly measured and formulated? The answer is that the style that mass-space-time exerts its characteristic of the primary intrinsic property (related to the electric charge) leads to an observable, measurable, and summarized physical parameter known as electric charge. The reason that the main intrinsic property cannot be identified is that value is beyond current accurate measurement. It must be emphasized that this study does not refute current knowledge on electric charges and electromagnetic fields. Electric charge is certainly a known, observable, and measurable parameter of the universe. Accepted knowledge, especially Maxwell's equations, that has been subjected to countless experimentation based on the current definition of electric charge is definitely correct; however, it is necessary to overcome the obscurity about the nature of electric charge. Physics currently face unresolved problems and opened line research on the subjects such as the unified field theory, baryon asymmetry, and magnetic monopoles (Christianto et al., 2007). Defining the nature of an electric charge may also shed light on these unknown areas in physics.

Previous attempts to find the origin of electric charge include work by Shpenkov and Kreidik (2004), who developed a hypothesis based on new definition of elementary particles and their exchange in the matter-space-time field. Their theory that an electric charge is a mass change rate seems to be consistent with the results of the present study. Shpenkov ("What the Electric Charge is") stated, "The erroneous form of the Coulomb law gave rise to a phenomenological system of notions with measures having fractional powers of base units that are really senseless. Cognition of the nature of electric charges has become impossible". This notion is the opposite of the conclusion of this study.

Olah (2009) proposed a hypothesis that is in contrast with the current proven model of electric charges and fields. It is difficult to accept such theories. Tiwari (2006) suggested the origin of an electric charge in terms of fractional spin. Sasso (On Primary Physical Transformations of Elementary Particles: the Origin of Electric Charge) proposed a hypothesis based on relation between electric charge and spin. Nguyen (2013) discussed the change in electron charge in an external magnetic field. He stated: "the variability (or the constancy) of the mass and the electric charge of the electron still remains as a foundational problem in modern physics, awaiting to be justified". Krasnoholovets (2003) and McArthur (1999) also discussed this.

None of these works provides an explicit formula for the exact value and the nature of electric charge based on proven facts. There are two remarkable points among these discussions: there is a possible relation between electric charge and mass change, and variability of electric charge is possible. The strength of the new approach is: (1) it utilizes current proven physical equations to develop a solution; (2) a simple novel method of dimension analysis is proposed; and (3) an explicit formula for an electric charge is provided. No previously mentioned research encompasses all three benefits. A new theory is not proposed, but a different way of looking at previous hypotheses is advanced.

Using the successful dimensional analysis from the previous section, the same procedure is followed to discover the nature of electric charge beginning with Coulomb's law:

$$F = \frac{q_1 q_2}{4\pi\varepsilon_0 r^2} \quad (10)$$

Replacing the dimensions of the parameters leads to:

$$[q^2] = 4\pi\varepsilon_0 \frac{ML^3}{T^2} \quad (11)$$

Assuming dimensionless parameter (γ) and multiplying it by the right side of Equation (11) produces the following:

$$q^2 = 4\gamma\pi\varepsilon_0 [\frac{ML^3}{T^2}]^{-1} \quad (12)$$

Up to this point, the equation represents the dimension of electric charge. It also states that there must be physical parameters and constants that are integrated as a formula (with this final dimension) that equals q. The next point is to assign meaningful parameters to the components of the q dimension, beginning by decomposing $\frac{ML^3}{T^2}$. Clearly, $\frac{ML^3}{T^2}$ can be categorized in different ways using parameters with proper and valid physical interpretations. One solution is:

$$[\frac{ML^3}{T^2}]^{-1} = [\frac{ML^2}{T}]^{-1} [\frac{L}{T}]^{-1} \quad (13)$$

In this equation, $\frac{ML^2}{T}$ is the dimension of angular momentum, and $\frac{L}{T}$ is the dimension of linear velocity. Since q is the intrinsic property of the particle, angular momentum should be interpreted as particle spin. It is known that particles with non-zero spin have zero electric charge (such as electron-neutrinos). This inconsistency is troubling. In addition, in Equation (12), the roots of the parameters of angular momentum and linear velocity describe electric charge, but the root of the parameters is senseless and physically meaningless. Guesswork in this method should not conflict with known principles, although the dimension relationship is true. Slight manipulation of Equation (12) produces:

$$q^2 = 3\gamma\varepsilon_0 [\frac{4\pi L^3}{3M}]^{-1} [\frac{M^2}{T^2}]^{-1} \quad (14)$$

As in Equations (5) and (6):

$$q^2 = 3\gamma\varepsilon_0 [\frac{1}{(\frac{M}{V})}]^{-1} [\frac{M^2}{T^2}]^{-1} \quad (15)$$

Producing:

$$q^2 = 3\gamma\varepsilon_0 [\frac{1}{\rho}]^{-1} [\frac{M^2}{T^2}]^{-1} \quad (16)$$

In this equation, ρ (mass density) should be interpreted as the ρ of the particle, or a known ρ. The former produces the same problem mentioned above, which is that the root of ρ is a meaningless physical parameter. If the ρ of space is considered to be where the particle is (e.g. mass density of the universe), the problem is resolved. Using Equation (8) in Equation (16) produces:

$$q^2 = (4\pi\varepsilon_0 \alpha G T^2) [\frac{M^2}{T^2}]^{-1} \quad (17)$$

By applying final guess that governs $\frac{dm}{dt}$ to $\frac{M}{T}$ dimension produces:

$$q = \sqrt{4\pi\varepsilon_0 \alpha G}\, T \frac{dm}{dt} \quad (18)$$

In this equation, α is a dimensionless parameter and is currently unknown; T is the age of universe; $\frac{dm}{dt}$ is the mass change in the particle over time (in the particle reference frame), and q is the electric charge. Here, $\frac{dm}{dt}$ was chosen to govern $\frac{M}{T}$ because it is more pertinent, simple and meaningful physical parameter.

Of the possible solutions for Equation (12), Equation (18) is unique and the most consistent with current proven facts and principles. When using this approach, it should be noted an independent formula for q was not proposed, but is concluded from existing formulas. Briefly, Equation (18) is an abstract of the law of gravity, Newton's 2nd law, Coulomb's law and the Freidmann equation.

Equation (18) says that the origin of an electric charge is the particle (rest-) mass change over time. In other words, mass creates a field of gravity and its change creates another field (electromagnetic). This is a surprising result because charged particles such as electrons and protons are inherently mass-variable; however, if $\left|\frac{dm}{dt}\right|$ (absolute value of $\frac{dm}{dt}$) is minuscule, it does not conflict with current proven facts. Note that Equation (18) establishes a two-way relationship between mass change and electric charge that any mass change (over time) in a particle (or physical system) will produce an electric charge (and electric field). The exact value of $\left|\frac{dm}{dt}\right|$ will be given based on experimentation to determine the exact value of α. It is possible to estimate values α and $\left|\frac{dm}{dt}\right|$ (for electrons), but should be kept in mind that they are not conclusive. Suppose there are two electrons separated by distance r (m). Equation (1) gives:

$$F_g = G\,\frac{m_e^2}{r^2} \tag{19}$$

where m_e is electron mass, and F_g is the gravitational force exerted on the two electrons. Equations (10) and (18) give:

$$F_e = \alpha G\,T^2\,\frac{\left|\frac{dm_e}{dt}\right|^2}{r^2} \tag{20}$$

Ratio $\frac{F_e}{F_g}$ is calculated as:

$$\frac{F_e}{F_g} = \alpha\,T^2\,\frac{\left|\frac{dm_e}{dt}\right|^2}{m_e^2} \tag{21}$$

If $\frac{F_e}{F_g}$ is alternately calculated using Equations (1) and (10), it produces:

$$\frac{F_e}{F_g} = \frac{1}{4\pi G\,\varepsilon_0} \times \frac{q_e^2}{m_e^2} \tag{22}$$

If the current values of Equation (22) are substituted:

$\varepsilon_0 = 8.854 \times 10^{-12}\ Fm^{-1}$
$G = 6.674 \times 10^{-11}\ m^3\,kg^{-1}sec^{-2}$
$q_e = -1.602 \times 10^{-19}$ coulomb
$m_e = 9.109 \times 10^{-31}\ kg$

This gives the following for the electron:

$$\frac{F_e}{F_g}(electron) = 4.165 \times 10^{42}$$

Equation (21) has two unknown parameters, α and $\left|\frac{dm_e}{dt}\right|$. Supposing $\frac{F_e}{F_g}(electron) = \alpha$, then $\left|\frac{dm_e}{dt}\right|$ will be:

$$\left|\frac{dm_e}{dt}\right| = \frac{m_e}{T} \tag{23}$$

Substituting the values in Equation (23) or Equation (18), gives:

$T = 4.350 \times 10^{17}\ sec$
$\left|\frac{dm_e}{dt}\right| = 2.094 \times 10^{-48}\ kg/sec$

With the proposed assumptions and estimations, the value of $\left|\frac{dm_e}{dt}\right|$ is outside of the apparatus accuracy range. The exact value of m_e is:

$m_e = 9.10938291(40) \times 10^{-31}\ kg$

Its relative standard uncertainty is 4.4×10^{-8}. Using the value calculated for $\left|\frac{dm_e}{dt}\right|$, it takes $10^9\ sec$ (~ 30 years) for a change to occur in the least significant digit of m_e.

Using the above assumptions and calculations that summarized in Equation (23), Equation (18) may be written as:

$$q = \left(\frac{e}{m_e}\right) T\,\frac{dm}{dt} \tag{24}$$

where $\frac{e}{m_e}$ is a well known physical constant (although e and m_e are both time variable) that can be directly measured by experimentation. Experimental or mathematical confirmation of Equation (18) produces the following results:

(i) A change in particle mass is the source of the electric charge and electric field. A charged particle keeps its structure during mass change. There is a definite threshold for mass change in each particle; thereafter, the particle decays or annihilates.

(ii) A change in particle mass increases or decreases, making the electric charge positive or negative.

(iii) It is necessary to investigate how the mass of a

charged particle changes and the role of photon particles in this process. Two overall scenarios exist: (1) mass (energy) exchange directly between charged particles; (2) mass (energy) exchange independently between each charged particle and space. In any case, the mass of negatively charged particle increases and, for positively-charged particles, it decreases, regarding Equation (24). A justification for attractive and repulsive characteristics of charged particles must be found based on their mass change.

(iv) If $\left|\frac{dm}{dt}\right|$ is constant, for each charged particle:

$$m_0(t) = \pm \left|\frac{dm}{dt}\right| \times t + m_0(t=0) \quad (25)$$

where t is the age of the particle, m_0 is the rest mass of the particle, $m_0(t=0)$ is the initial mass of the particle and $t=0$ is the time at which the particle began to act as a charged particle (continuous mass change).

(v) If $\left|\frac{dm}{dt}\right|$ is constant, then $\frac{d^2m}{dt^2}$ will be zero, since:

$$T = t + \tau \,;\, \tau \geq 0 \quad (26)$$

where T is the age of universe, t is the age of the charged particle; thus, τ is the difference between them. Equation (18) gives:

$$\frac{dq}{dt} = \pm\sqrt{4\pi\varepsilon_0 \alpha G}\left|\frac{dm}{dt}\right| \; C/sec \quad (27)$$

where $\frac{dq}{dt}$ is constant and non-zero for all charged particles. Because the magnetic field is proportional to $I = \frac{dq}{dt}$, this means that charged particles necessarily create the magnetic field (Biot-Savart law). Today, it is believed that there is inherent magnetism in charged particles such as electrons. This agrees with the proposed conclusion.

There is currently no approved experiment or mathematical calculation that directly assesses the conjecture about the nature of electric charge, but the present results do not create any contradiction with current knowledge about the electric charge and its effects. The validity of the results is supported by the law of gravity, Newton's 2nd law, Coulomb's law and the Freidmann equation.

DISCUSSION

Utilizing dimensional analysis to govern physical parameters to combinations of basic dimensions to identify unknown physical phenomena provides weaker mathematical support for the derived equation. This is the inherent deficiency of the approach. To overcome this deficiency, two further steps should be taken; comparing the final result with known and proved physical facts and conducting experiments to directly evaluate the results. This section compares the proposed hypothesis on the nature of electric charge with proven facts.

(a) The possibility of a charged particle without mass continues to be advanced by some parties. A similar question arises in the Reissner-Nordstrom solution of Einstein's field equations. The Reissner-Nordstrom metric (for a black hole with mass M and charge q) is:

$$ds^2 = \left(1 - \frac{r_s}{r} + \frac{r_q^2}{r^2}\right)dt^2 - \left(1 - \frac{r_s}{r} + \frac{r_q^2}{r^2}\right)^{-1} dr^2 - r^2 d\theta^2 - r^2 \sin^2\theta \, d\varphi^2 \quad (28)$$

where $r_s = \frac{2GM}{c^2}$ (Schwarzschild radius) and $r_q^2 = \frac{Gq^2}{4\pi\varepsilon_0 c^4}$.

It is possible to show that it is physically impossible to make the mass M in the Reissner-Nordstrom solution vanish, because the charge itself generates an electromagnetic mass that is part of M or constitutes all of mass M. The electromagnetic mass vanishes only when the charge vanishes (Pekeris, 1982). In agreement with this result, in Equation (18), the electric charge will be zero; if the mass of particle is constant (zero or nonzero value), and electric charge exists if and only if the mass of the particle is variable (and definitely exists). However, there is no restriction for the mass of a charged particle to become zero instantaneously. Ibohal and Kapil (Charged black holes in Vaidya backgrounds: Hawking's Radiation, Department of Mathematics, Manipur University, India) discussed a similar case for the Reissner-Nordstrom solution.

(b) It is more realistic to consider the Reissner-Nordstrom metric for a black hole in a non-flat background Friedman-Robertson-Walker universe. It can be shown that the mass and charge of the black hole both vary with the evolution of the universe (Chang and Shuang, 2004; Ibohal, 2002). The variability of mass and charge of charged particles is the pivotal result of Equation (18).

(c) The influence of cosmological expansion on local systems is still a subject of research. Some authors support the view that cosmic expansion affects only systems larger than a certain spatial scale and that there is no effect below that scale. Others believe that all systems are subject to the effect of cosmic expansion, although this effect is numerically negligible for small systems (like atoms) and stronger for larger objects. This expands the validity of the Friedmann-Lemaitre-Robertson-Walker metric down to small scale (Bochicchio et al., 2013; Jose J. Arenas: The effect of the cosmological expansion on local systems: Post-Newtonian approximation). In Equation (18), factor T is $\frac{1}{H_0}$ (where H_0 is the Hubble constant). This result strongly supports the latter idea and vice versa.

(d) Based on Equation (18), particles have an absolute equal electric charge if and only if they have an equal $\left|\frac{dm}{dt}\right|$. Thus, the amount of electric charge is independent of the amount of mass. This has been seen for charged particles, such as electron and proton, yet it continues to be expected that charge is dependent on mass (e.g. mass change).

(e) All stars, black holes, and planets experience eras during which they experience mass-energy exchange with space. According to Equation (18), all of them should be considered to be charged particles for this period. Thus, they have an electromagnetic field surrounding them. This was proven in nature of electric charge.

(f) Equation (18) states that charged particles with decreasing mass (e.g., positive charge) have finite life times because of their finite mass; thus, charged particles are not fundamentally stable. This provides a good explanation of proton decay as proposed by GUTs and is still a matter of subject and observation (Senjanovi´c, 2009).

(g) Experiments show that electric charge is quantized; the q of every charged particle is an integer multiple of elementary charge e. Equation (24) clearly shows that each charged particle is a multiple of e, although it cannot singly guarantees that the coefficient is integral.

This discussion shows the excellent consistency of Equation (18) with currently-accepted physical facts. It sufficiently supports the hypothesis offered herein to describe the nature of electric charge, and adequately eliminates the weakness of the method applied to formulate the electric charge in terms of known physical parameters.

CONCLUSION

This study introduced a generalization of dimensional analysis of the physical equations and parameters and utilized this method to identify an explicit relation between gravitational constant G, and two universe parameters (age and mass density). Also, It was found that the origin of an electric charge (and electromagnetic field) is mass change of particle(s) over time. Therefore, mass change should be considered as the primary intrinsic property of charged particle rather than electric charge. It is a surprising result that, if experimentally or mathematically proven, will significantly influences some areas of physics and our view of the universe. It appears that dimensional analysis is not only a reliable method for assessing the validity of equations, but also it can help to find a meaningful interpretation for a category of unknown physical parameters. This method effectively uses speculation and intuition that is founded on proven facts and logic.

Conflict of Interests

The author(s) have not declared any conflict of interests.

REFERENCES

Arenas JJ (2013). The effect of the cosmological expansion on local systems: Post-Newtonian approximation. arXiv:1309.3503 [gr-qc].

Bochicchio I, Faraoni V (2013). Cosmological expansion and local systems: A Lemaˆıtre-Tolman-Bondi model. arXiv:1111.5266v3 [gr-qc].

Chang JG, Shuang NZ (2004). Reissner-Nordstr¨om Metric in the Friedman- Robertson-Walker Universe. arXiv:gr-qc/0407045v2.

Christianto V, Smarandache F (2007). Thirty Unsolved Problems in the Physics of Elementary Particles. Progress Phys. 4:112-114.

Ibohal Ng (2002). On the variably-charged black holes in general relativity: Hawking's radiation and naked singularities. Class. Quantum Grav. 19 4327 doi:10.1088/0264-9381/19/16/308.

Komissarov SS (2012). Cosmology. Lecture, Room: 10.19 in Maths Satellite, email: S.S.Komissarov@leeds.ac.uk.

Krasnoholovets V (2003). On the nature of the electric charge, Hadronic J. Supplement 18(4):425-456.

McArthur W (1999). The Nature of Electric Charge, the general science journal.

Nguyen HV (2013). A Foundational Problem in Physics: Mass versus Electric Charge.

Olah S (2009), The Electric Charge, *Copyright © 2009, the general science journal.*

Pekeris CL (1982). Gravitational field of a charged mass point. Proc. NatL Acad. Sci. USA, 79:6404-6408.

Sedov LI (1993). Similarity and dimensional methods in mechanics. 10th edn. CRC Press, Boca Raton.

Senjanovi´c G (2009). Proton decay and grand unification. arXiv:0912.5375v1 [hep-ph].

Shpenkov GP, Kreidik LG (2004). Dynamic Model of Elementary Particles and Fundamental Interactions. *GED* Special Issues, GED-East, pp. 23-29.

Tiwari SC (2006). The Nature of Electronic Charge.Foundations of Physics Letters, 1(19):51-62.

Feasibility and technical studies of two water recirculating systems using two different power sources, solar photovoltaic and fuel generator

Oparaku N. F. and Nnaji C. E.

National Centre for Energy Research and Development, University of Nigeria, Nsukka, Nigeria.

Recirculating water systems are designed to minimize or reduce dependence on water exchange and flushing in fish culture units. Water is typically recirculated when there is a specific need to minimize water replacement, to maintain water quality conditions which differ from the supply water, or to compensate for an insufficient water supply. In this work, recirculating system of fish was used in rearing fish which comprised of fish ponds and treatment pond. Submersible pump was powered by solar energy while the electropome pump was powered by generator provided electricity. Simple annual costs analysis as well as net present value (NPV) method were used to compute the profitability. The total fixed cost of using recirculation system with solar powered pump was higher by ₦163, 500.00 while the total variable cost of using recirculation system with generator was higher by ₦244000.00. NPV's recorded were 299607, -66323, -40409 for generator powered system and 1336085, 575047, 626113 for solar powered system at r = 0.1, 0.2, 0.19, respectively. Results also indicated a shorter payback period for solar system. Solar as power source was more profitable than generator despite its high initial capital.

Key words: Recirculating system, solar photovoltaic, fish culture, net present value, financial feasibility, fuel generator.

INTRODUCTION

Recirculating system maximizes water re-use by employing comprehensive water treatment system. Water treatment processes typically are solid removal, infiltration, gas balancing, oxygenation, and disinfection. By addressing each of the key water concern through treatment rather than flushing as is used in flow-through and the partial reuse systems, ultimate control over culture conditions and water quality is provided.

There is growing interest in recirculation aquaculture system (RAS) technology in the world, as a result of perceived advantages over the conventional aquaculture (Emperor Aquatics, 2008; The Fish Site, 2010; Zhang et al., 2011; Food and Water Watch, 2008). Recirculating system can help in reduction of water and land usage. Recirculating system offers a high degree of control over the culture environment and fish biomass can be determined easily and accurately than in biomass. Even though it is capital intensive, claim of impressive yields with year-round production is attracting growing interest from prospective aquaculturists" (Losordo et al., 1998; Poulson, 2013; Rakacy, 2006). To evaluate the profitability of the venture, indicators of investment returns were determined such as net present value (NPV) and internal rate of return (IRR), payback period, (NAERL, 2000) and (Parin and Lupin, 1995). The operation of RAS which are mechanically sophisticated and biologically complex requires education, expertise and dedication (Duning et al., 1998). Many commercial

RAS have failed because of component failure due to poor design and inferior management (Masser et al., 1999; Sioux Indian Reservation, 2006). Good knowledge of the design of the system, specification of the technical components and operation of the system is therefore a prerequisite for a sustainable RAS farm. The water treatment process could increase operation costs and failure of the treatment system would result in huge economics losses (Summerfelt et al., 2001). Therefore, the aspect of economic feasibility has to be taken into consideration before embarking on the system. Generally, a feasibility study is conducted during the planning stage prior to obtaining approval for funds or financing of a project. The study analyzes and assesses feasibility of using solar photovoltaic and generator that uses fuel. Financial feasibility and other factors that could influence the sustainability of the project. It is important to critically evaluate the outcome or conclusions of a feasibility study. A good study may uncover alternatives and save significant time and money for the stakeholder of the project. The aims and objectives were to analyze the profitability of recirculating systems powered with generator and electricity and technical feasibility of the project.

MATERIALS AND METHODS

The project was carried out between January, 2009 and December, 2009 at the National Centre for Energy research and Development, University of Nigeria Nsukka. Nsukka is located at 6.9°N and 7.4°E and 445 m above sea level.

Treatment tank installation

Procurements of biofilters namely bioblocks, biobrush, Maifan stones, coral sand, ceramic ring, activated charcoal and Ultra Violet (UV) light were used for this study. They were arranged inside the treatment tank in the following order:

Biobrush ⟶ Bioblock ⟶ Maifan stones ⟶ Coral sands ⟶ Ceramic ring and Activated Charcoal ⟶ UV light (the arrows shows the order of arrangement of the compartments of the treatment tank)

The dimensions of treatment tank which was constructed with concrete are (3.4 × 1×1.5) m. There were four compartments in the water treatment tank each measuring (1× 0.6 ×1.25) m. The first compartment contains the biobrush, the second has bioblocks, the third contains maifan stones, coral sands, ceramic ring and activated charcoal, finally the last chamber houses the UV fluorescent tube which was placed at close proximity to the water surface but was not immersed in the water. Two pumps, Interdab electropome Jet 100 M 1horse power pump and Grundfos KPBasic 300A submersible pump were procured at Onitsha and Lagos respectively. Interdab electropome Jet 100 M uses electricity while Grundfos submerssible pump was powered by solar modules (photovoltaic) to ensure constant power supply and to serve as comparative between electric and solar energy. The quantity of water pumped by both pumps is 50 L/min at the depth of 1.25 m. Air stone aerator supply oxygen constantly to the ponds. Ceramic rings - surface area 1200 m^2/L and weighing 10 kg, bamboo carbon (activated carbon) - surface area 1200 m^2/L and weighing 10 kg were purchased at Kingdom Aquarium and fisheries Ltd. Lagos, Nigeria. Two overhead plastic tanks, volume 1000 L each were procured at Onitsha for water storage.

Treatment process

Water from the overhead tank (Inlet water) entered the pond where fishes are kept and then flowed into the treatment tank as waste water. As waste water flowed through biobrush, bioblocks, maifan stone, coral sand, ceramic ring and activated carbon it is filtered. Solar powered pump water and electric powered pump water were then collected. Water lastly flowed into the UV light compartment where it was disinfected (UV treated water). After the waste water had passed through the treatment tank, the treated water was air lifted into the culture tank for use by the fish and recirculated back again into the filter again for purification.

Methods of estimating profitability of recirculating systems

The methods used for evaluating profitability were the following: Rate of return on the original investment (i_{ROI}), Present-worth (PW), Net Present Value and Pay out time (n_R) (Parin and Lupin, 1995).

Rate of return on the original investment (i_{ROI})

The annual net profit divided by total initial investment represents the fraction which, when multiplied by 100, is known as the percentage return on investment. The procedure used was to find the return on total original investment, with the value of the average net profit being the numerator and thus, the rate of return on the original investment, i_{ROI} =

$$NP_a = \frac{1}{n} \times \sum_{j=1}^{n} NP_j = \frac{NPa}{It}$$

Np_a = annual net profit , I_t = total initial investment.

Present-worth (PW)

This method compared the present-worth (PW) of all the cash flows with the original investment. It assumed equal opportunities for re-investment of the cash flows at a pre-assigned interest rate.

$$PW = \sum_{j=1}^{n} \frac{CF_j}{(1+i)^j} - I_T \quad , \quad PW' = \frac{\sum_{j=1}^{n} \frac{CF_j}{(1+i)^j}}{I_T}$$

Where, CF = cash flow; I_T = initial Investment; i = interest rate.

Net present value

The net present value (NPV) of a project is the difference between the sum of the discounted cash flows which are expected from the investment and the amount which is initially invested. A trial and error method was used to establish the interest rate to be applied to the cash flow each year, such that the original investment would be reduced to zero (or salvage value, plus land, plus working capital) during the useful life of the project. Internal rate of return, r, is calculated by trial and error:

DCFRR = IRR = r,

Table 1. Description of fixed and variable investment of using recirculating system with solar photovoltaics.

S/N	Description of fixed Investment	Unit cost	Price	Variable investment	Cost
1	Pond Construction	150,000	N150,000		
2.	Treatment Tank Construction	N80,000.00	N80,000.00	Cost of paying a labourer every month (N10000.00 a month)	N120000.00
3.	Plumping materials and connections cost	N58,000	N58,000	3500 Fingerlings at N200 each	N70,000.00
4.	Electric wiring of the pond	-	N5,000		
5.	Cost of Roofing for air pump mounting		N50,000.00	Cost of rearing a fish for 1 year- N200 x 3500	N700,000.00
6.	Grundfos water pump	N36,000.00	N36,000.00		
7.	4 Panels (100 Amps)	N55,000.00	N210,000		
8.	Charge Controller	N18,000.00	N18,000.00		
9.	Stand for the Panels	N7,000.00	N7,000.00	Annual servicing cost	N5000.00
10.	Inverter	N55,000.00	N55,000.00		
11.	UV Flourescent Tube	N36,000.00	N36,000.00		
12.	Oxygen Pump with (air stones)	N49,000.00	N49,000.00		
13.	Biobrush (4)	N1,500.00	N6,000.00	Miscellanous	20,000.00
14	Bioblocks	N27,000.00	N27,000.00		
15.	Hand net for scoping the fish out of the pond	N4,000.00	N8,000.00	**Total variable**	**N915,000.00**
16	Booth & Polythene Trouser	N10,000.00	N10,000.00		
17.	Water analysis kit	N40,00.00	N40,000.00		
18.	2 Battery (12v)	N25,000	N50,000.00		
19.	Ground Artermia (one tin)	N9,000.00	N9,000.00		
20.	Grinding mill(3Horse power)		N55,000.00		
21.	20 packets of Coral Sands (1000 g)	N1,000.00	N20,000		
22.	20 packets of activated carbon(500 g)	N1,000.00	N20,000		
	Total fixed Cost		N1059,000.00		
	Total Cost		N1974000.00		

$$\sum_{j=1}^{n}\frac{CF_j}{(1+r)^j} - I_T = 0$$

where

NPV typically is calculated over a specific time period of interest, e.g., 3 or 5 years. If the project NPV is greater than zero, the project is considered to be profitable over that time period. If the project NPV is less than zero, the project is considered to be not profitable over that time period.

Pay out time/Payback period

This method focus on recovering the cost of investment. Pay out time represents the amount of time that it takes for a capital budgeting project to recover its initial costs pay out time, in years = Fixed depreciable investment / (average profit/year) +(average depreciation/year).

$$Average \cdot Cash \cdot Flow = CF_a = \frac{1}{n} \times \sum_{j=1}^{n} CF_j \quad Pay \cdot out \cdot time, n_R = \frac{I_F}{CF_a}$$

$I_F =$ Fixed depreciable Investment; C_F = average profit/year; a= average depreciation/year.

RESULTS

Methods of estimating profitability of recirculating systems

Total Cost =Total Fixed Cost (TFC) + Total Variable Cost (TVC)
Total Cost = N1059, 000.00 + N915,000.00 = N1974 000.00 (Table 1)
Total revenue= (price of 1 mature fish=N400 × 3500) = N1, 400,000.00
Annual- profit= TR-TVC = N I, 400,000 – N915, 000 = N485, 000.00

Annual cost analysis

Total Cost =Total Fixed Cost (TFC) + Total Variable Cost (TVC).
Total Cost = N895, 500.00 + N1,159, 000.00 = N2054,500 (Table 2).
Total revenue= (price of 1 mature fish=N400 × 3500) = N1, 400,000.00.

Table 2. Description of Fixed and variable Investment of using recirculating system with generator.

1.	Pond Construction (3)	N150,000	Variable Investment	Costs
2.	Treatment Tank Construction	N80,000.00	3500 Fingerlings at N20 each	N70,000.00
3.	Plumbing materials and Connection Cost	N58,000.00	Cost of rearing a fish for 1 year N200 × 3500	N700,000.00
4.	Wiring of the pond	N5,000.00	Cost of Paying a labourer per month(N10000.00) for 1 year	N120000.00
5	Cost of Roofing for air pump	N50,000.00		
6	Electric Pump	N26,000.00	Cost of fuel for 1 month N19500 (30 × 650).	N234000.00
7.	2 Generators (model 2700)	N 70,000.00		
8	UV Flourescebt Tube	N36,000.00	Cost of oil filter, oil, fuel filter, after every 600 hrs of operation (7 times a year) for a sduty cycle of 12 h / day	N 10, 000.00
9	Bioblocks (1 cubic metre)	N27,000.00		
10	4 Biobru	N6,000	Annual inspection and servicing cost	N 5000. 00
11.	Cost of 10 Brood Stocks	N10,000.00		
12.	Cost of Ovaprim	N3,500.00	Miscellanous	N20,000.00
13	Oxygen Pump with Air stones	N49,000.00	**Total variable cost**	**N1,159,000.00**
14.	Hand Net	N 8000.00		
15.	Bo oth and Polythene Trouser	N10,000.00		
16.	Pelleting machine	N104,000.00		
17.	Water kit Analysis	N40,000.00		
18.	Cost of ground Artermia (one tin)	N8,000.00		
19.	Grinding machine	N55,000.00	**Total Cost**	**N2,054,500.00**
	Filter Media	N20,000		
20.	20 packets of Maifan Stones	N20,000		
	20 Ceramic Rings	N20,000.00		
21	20 packets of Coral sands	N 20,000.00		
22	20 packets 0f bamboo (activated charcoal)	N20,000.00		
	Total fixed cost	**N 895500.00**		

Annual profit = TR-TVC =N 1, 400,000 – N1,159,000 = N241, 000.00

(i) The total fixed cost of using recirculation system with solar powered pump is higher by (N1059, 000.00 – N895, 500.00) =N163, 500.00

(ii) The total variable cost of using recirculation system with generator is higher by (N1159000.00 – N915, 000.00) = N244000.00 (cost of fuel for 1 year) (Table 2).

Decision

Adopting any of the recirculation system is profitable. However, it is more profitable to adopt recirculation system with solar powered pump since the 1 year variable cost (raw materials + labour) was lowered by N244, 000.00. The cost of a generator (model 2700) was N35000 (Table 3), salvage value of generator was N10, 000 while the useful life was put at 5 years, depreciation

Table 3. Statement of sources and application of funds for a for a recirculating system using generator.

Activity	2005	2006	2007	2008	2009	2010	2011	2012	2013	2014
Capital	N2,054,500									
Working capital	N1159,000									
Applications										
Fixed investment	N895500.00									
variable cost	1159000.00	1159000	1159000	1159000.00	1159000	1159000	1159000.00	1159000	1159000	1159000
Total revenue	1,400,000	1400,000	1400,000	1400,000	1400,000	1,400,000	1400,000	1,400,000	1400,000	1400,000
Costs of production	1159000	1159000	1159000	1159000	1159000	1159000	1159000	1159000	1159000	1159000
Annual profit	241,000	241000	241000	241000	241000	241,000	241000	241000	241000	241000
Minus 10% tax	48200	48200	48200	48200	48200	48200	48200	48200	48200	48200
Net profit	192800	192800	192800	192800	192800	192800	192800	192800	192800	192800
Plus depreciation	5000	5000	5000	5000	5000	5000	5000	5000	5000	5000
Cash flow	197800	197800	197800	197800	197800	197800	197800	197800	197800	197800

value was N5000.00. The salvage value of solar panels was N15000.00 and useful life was 25 years, depreciation was put at N6, 400.00. A tax assumed to be levied on the fish produced was 20% per annum and was deducted from the annual gross profit. Operating costs include fixed costs and variable costs. Fixed costs are associated with the long-term operation of a catfish farm. Examples include: taxes (on property), insurance, depreciation, interest, amortization payments (for repayment of borrowed money). These costs are often overlooked but must be considered in assessing the financial situation of a catfish farm. Variable costs are the costs that vary with the size of the catfish farm or the number of ponds being stocked. Larger farms (or stocking more ponds) have much greater total variable production costs than smaller farms. Examples include: feeds, seed/fingerlings, fuel and/or power, chemicals, fertilizers, harvesting costs, and labour. Expected returns include the money that the catfish farmer receives from the sale of catfish. Profit is the most important return and is determined by subtracting the costs of production from the amount received when the catfish are sold. (Note: start-up costs, annual fixed costs, and variable production costs must all be used in calculating production costs). Returns from catfish farming may be reported as "gross" or "net" returns –the distinction between the two is important.

Gross return refers to the total amount of money received for the catfish that are sold. Not much consideration is given to how much it cost to produce the crop. Gross return is calculated by multiplying the total number of kilograms sold by the price received per kilogram for the fish. Net return refers to the total amount of money remaining after all costs of production have been subtracted from gross returns. Net return is also known as profit.

It is a more important measure of a catfish farm than gross return. Net return also reflects on the efficiency of the catfish farm. These costs and returns were summarized in table form (Tables 3 and 4).

Rate of return on the original investment

The percentage return on original investment for recirculating system that uses solar photovoltaic was 36.6% while that of generator was 21.5% the time value of money was not considered, since only the average profit was used, not its timing. Recirculating system with solar is the best in terms of profitability because the value of rate of return on original investment was greater than values in the generator. The profits from years 1 through 10 could be reversed and the return on original investment would be the same.

The present-worth

The present-worth and the PW' relationship for the recirculating system was calculated by applying a rate of i = 15% per year in equation, the following results were obtained for generator powered recirculating system: The result for the present worth was N3713 and photovoltaic was

Table 4. Statement of sources and application of funds for a recirculating system using photovoltaics as power source.

Activity	2005	2006	2007	2008	2009	2010	2011	2012	2013	2014
Capital	N1,974,000									
Working capital	N915,000									
Applications										
Fixed investment	N1,059,000									
Variable costs	915,000	915,000	915,000	915000	915,000	915,000	915,000	915,000	915,000	915,000
New battery procurement after three years				50,000						
Total Revenue	1,400,000	1400,000	1400,000	1400,000	1400,000	1,400,000	1400,000	1,400,000	1400,000	1400000
Costs of production	915,000	915,000	915,000	965,000	915,000	915,000	915000	915,000	915000	915000
Annual Gross profit	485,000	485000	485000	435000	485000	485000	485000	485000	485000	485000
Minus 10% tax	97000	97000	97000	97000	97000	97000	97000	97000	97000	97000
Net profit	388,000	388,000	388,000	348,000	388,000	388,000	388,000	388,000	388,000	388,000
Plus depreciation	6400	6400	6400	64000	6400	6400	6400	6400	6400	6400
Cash flow	394400	394400	394400	354400	394400	394400	394400	394400	394400	394400

N8146296. At the end of ten years, the cash flow to the project, compounded on the basis of end-of-year income, will be: for generator empowered recirculating system, F = N93303 For photovoltaics empowered recirculating system; F = N8146296 . The relationship between the present-worth of the annual cash flow and the total capital investment for generator was PW' = 988803/ 895500 = 1.1042, for photovoltaics was PW' = 8146296 / 1059000 = 7.692.

Net present value (NPV)

Net present values recorded were 299607, -663232, -40409 at r = 0.1, 0.2, 0.19, respectively for the generator powered recirculating system. NPV values for photovoltaics powered recirculating system were as follows 1336085, 575047, 626113 at r = 0.1, 0.2 and 0.19, respectively. Net present values recorded were positive for photovoltaic systems while it is positive at r =0.1 in generator powered system (Tables 5 and 6).

However, solar as power source was more profitable than generator. It is the present value of future net cash inflows minus the initial capital cost. Each year's net cash flows can be reduced by the present value by multiplying it by $\frac{1}{(1+r)}$ where r = interest and n is the year considered. This process is known as discounting. The present values of all the annual net cash flows can then be summed up to give the total present value. If the initial investment is subtracted from the total present value, the result is called the net present value (NPV).

Discounted cash flow rate of return

The values calculated for r = 0.15 and 0.2, respectively for photovoltaic and generator, the resulting rate of return calculated (Figures 1 and 2) was equivalent to the maximum interest rate that could be paid to obtain the necessary funds to finance the investment and completely paid back by the end of the useful life of the project. The interpolation to determine the correct value of r was done by plotting the relationship between the original investment and the total present-worth as a function of r, as is shown in Figures 3 and 4.

Planning farm operations

The profitably model was used to plan the cash flows over the 10 year planning horizon. The investment and finance schedule indicated how much finance the farmer needed (equity plus loan), interests, repayment and depreciation (depreciation needed for tax calculation). The operations statement showed the net profits after

Table 5. Calculation of internal rate of return for the recirculating system of fish pond powered by generator.

Year (m)	Cash flow (Naira) d_m	Trial for r = 0.1		Trial for r = 0.2		r = 0. 19	
		Factor	Present-worth (Naira)	Factor d_m	Present-worth (Naira)	Factor d_m	Present-worth (Naira)
0	(895500)						
1	197800	0.909	178020	0.833	164767	0.840	166152
2	197800	0.826	163383	0.694	137273	0.705	139449
3	197800	0.751	148548	0.579	114526	0.592	117098
4	197800	0.683	135097	0.482	95340	0.497	98307
5	197800	0.621	122834	0.402	79515	0.417	82483
6	197800	0.564	111559	0.335	66263	0.350	69230
7	197800	0.513	101471	0.279	55186	0.294	58153
8	197800	0.466	92175	0.232	45890	0.247	48857
9	197800	0.424	83667	0.194	38373	0.207	40945
10	197800	0.385	76153	0.162	32044	0.174	34417
Total			1195107		829177		855091
Relationship = Total present-worth / Original investment			$\frac{1195107}{895500}$ =1.3346		$\frac{829177}{895500} = 0.9260$		$\frac{855091}{895500}$ =0.9549
NPV			299607		-66323		-40409

Table 6. Calculation of internal rate of return for the recirculating system powered by photovoltaic solar system.

Year (m)	Cash flow (Naira) d_m	Trial for r = 0.1		Trial for r = 0.2		r = 0. 19	
		Factor	Present-worth (Naira)	Factor d_m	Present-worth (Naira)	Factor d_m	Present-worth (Naira)
0	(1059000)						
1	394400	0.909	358510	0.833	328535	0.840	331296
2	394400	0.826	325774	0.694	273714	0.705	278052
3	394400	0.751	296194	0.579	228358	0.592	233485
4	354400	0.683	242055	0.482	170821	0.497	176137
5	394400	0.621	244922	0.402	158549	0.417	164465
6	394400	0.564	222442	0.335	132124	0.350	138040
7	394400	0.513	202237	0.279	110038	0.294	115954
8	394400	0.466	183791	0.232	91501	0.247	97417
9	394400	0.424	167226	0.194	76514	0.207	81641
10	394400	0.385	151844	0.162	63893	0.174	68626
Total			2395085		1634047		1685113
Relationship = Total present-worth / Original investment			$\frac{2395085}{1059000}$ = 2.262		$\frac{1634047}{1059000}$ = = 1.543		$\frac{1685113}{1059000} = 1.591$
NPV			1336085		575047		626113

subtracting the costs from the revenue. The cash flow statement indicated the surplus (losses and /or gains) over the 10 year period. Also, the cash flow indicated how much of the loan could be repaid and during what period in the years of production. The balance sheet was used to keep track of the accounting of the farm. The profitability measurements showed how the cash flows could be used in the calculations of NPV and the IRR.

It should be noted that besides serving as a decision support tool for investment analysis, the profitability model can be used during operations as a planning tool year by year. The balance sheet reflected the assets and liabilities during the operations. Profitability measurements, IRR and financial ratios indicated the

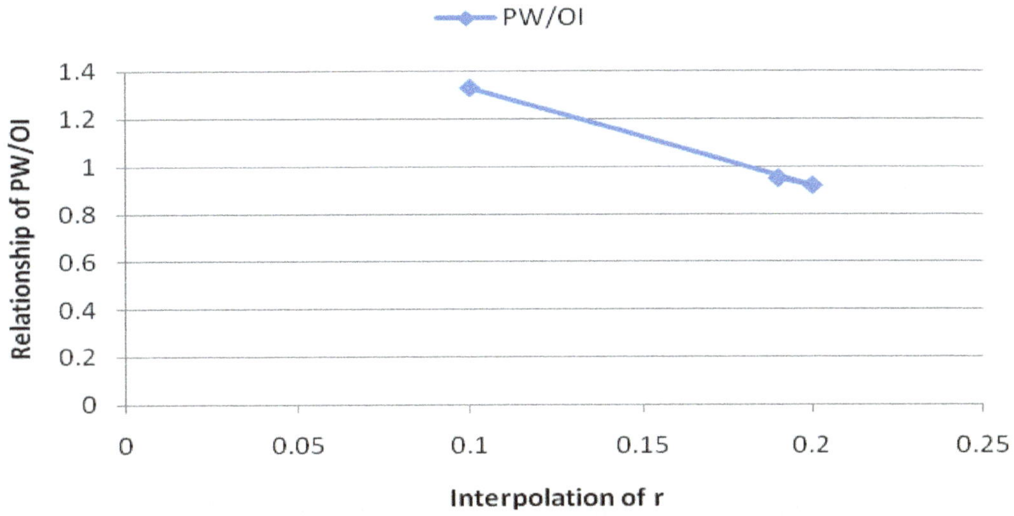

Figure 1. The relationship of PW/ O I and r in recirculating system powered with generator.

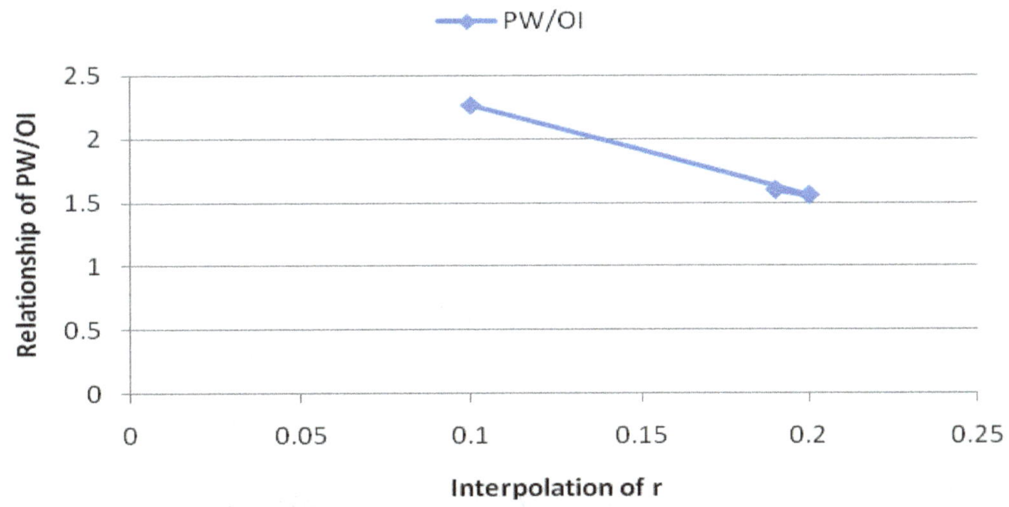

Figure 2. The relationship of PW / OI and r in recirculating system powered with photovoltaic.

feasibility of the venture over the years.

Pay time

Pay time for generator and solar photovoltaic were 4.53 and 2.69 years as calculated from the equation and can be determined by plotting the graph of accumulated cash against years (Tables 7 and 8). Tables 7 and 8 show accumulated cash flows for the recirculating system that was powered by generator and photovoltaic as power source. The cash flow accumulated, moving from negative to positive, and when the project ends, the capital invested in current assets and land would be recovered, resulting in a positive final cash flow. The cash flow was negative for 0 to 4th year for the recirculating system that was powered by generator and was only negative in 0-2nd year for the recirculating system that was powered by photovoltaic. This is an indication of the success of the venture since the accumulated cash flow was consistently positive after the 2nd year and 4th year in photovoltaic and generator system respectively.

DISCUSSIONS

To evaluate the profitability of the venture, indicators of investment returns were determined such as NPV, IRR and payback period (NAERL, 2000; Parin and Lupin,

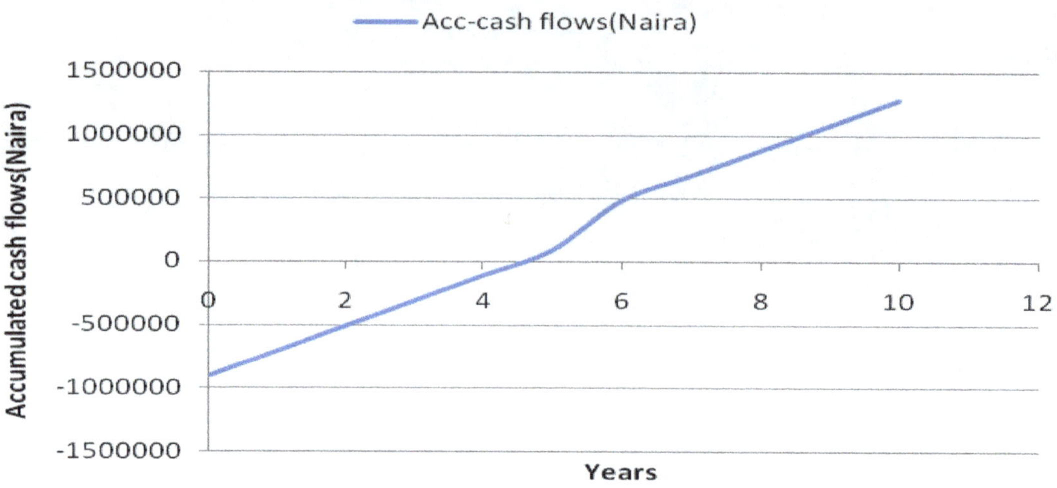

Figure 3. Accumulated Cash flow in generator powered recirculating system.

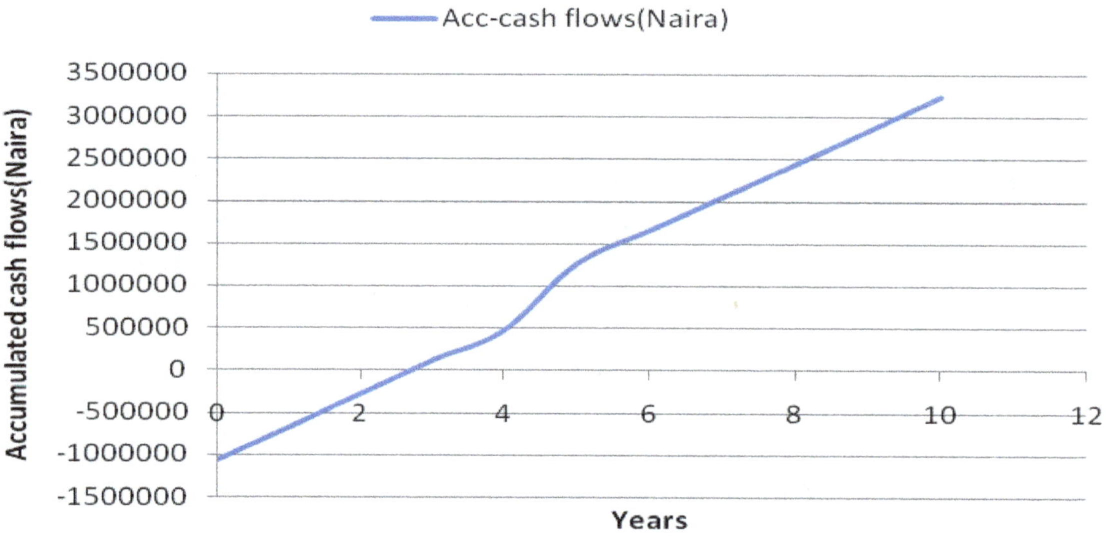

Figure 4. Accumulated Cash flow in recirculating system power with photovoltaic solar system.

Table 7. Accumulated cash flows for the recirculating system that uses generator as power source.

Years	Cash flow (Naira)	Accumulated cash flow (Naira)
0	-895500	-895500
1	197800	-697700
2	197800	-499900
3	197800	-302100
4	197800	-104300
5	197800	93500
6	197800	489100
7	197800	686900
8	197800	884700
9	197800	1082500
10	197800	1280300

Table 8. Accumulated cash flows for the recirculating system powered with photovoltaic.

Years	Cash flow (Naira)	Accumulated cash flow (Naira)
0	-1059000	-1059000
1	394400	-664600
2	394400	-270200
3	394400	124200
4	354400	478600
5	394400	1267400
6	394400	1661800
7	394400	2056200
8	394400	2450600
9	394400	2845000
10	399400	3244400

1995). The results obtained indicated positive NPV's in photovoltaic powered recirculating system and positive NPV in generator powered system where r = 0.1 while r = 0.2 and 0.19 recorded negative values of NPV. Key factors which affect profitability of operations in fish plants are generally cost and quality of raw material and the yield from processing, as long as the raw material is available and the market for the resulting products is stable (Parin and Lupin, 1995). The result of IRR and a payback period of 2.69 and 4.53 years obtained for photovoltaic and generator respectively were within the range that would be acceptable and profitable. Reduction in payback period is better in photovoltaic system because the project was able to recoup the original investment within a shorter period. Positive values of NPV as well as higher values of IRR in recirculating system powered with solar and reduced payback period are all indications that solar is a better option than generator despite its high initial capital investment. The methodology developed here can easily be adapted to evaluate any type of investment for instance fish farming enterprises of other species or fishery operations.

The challenge to designers of recirculating systems is to maximize production capacity of capital invested through employing the use of efficient energy sources to power the systems. Components should be designed and integrated into the complete system or existing fish ponds to reduce cost while maintaining or even improving reliability. There are many alternative technologies for each process and operation. The selection of a particular technology depends upon the species being reared, production site infrastructure, production management expertise, and other factors. Prospective users of recirculating aquaculture production systems need to know about the required water treatment processes, the components available for each process, and the technology behind each component. A recirculating system maintains an excellent cultural environment while providing adequate feed for optimal growth.

Conclusion

The result of IRR and a payback period of 2.69 and 4.53 years obtained for photovoltaic and generator, respectively were within the range that would be acceptable and profitable. Reduction in payback period is good because the project was able to recoup the original investment within a shorter period. Positive values of NPV as well as higher values of IRR in recirculating system powered with solar and reduced payback period are all indications that solar is a better option than generator despite its high initial capital. Further, it is anticipated that a successful and vibrant small scale recirculating system powered with solar could trigger a commercial recirculating system in the country. In addition, the small scale farmers might grow in capital and knowledge and transform themselves into medium and eventually large scale farmers.

REFERENCES

Duning RD, Losordo TM, Hobbs AO (1998). The Economics of Recirculating Tanks Systems, A Spreadsheet for Individual Analysis, Southern Regional Aquaculture Center, SRAC Publication No. 456.

Emperor Aquatics (2013). "Recirculating Systems in Aquculture". Available at http://www.emperoraquatics.com/aquaculture-recirculationsystems.php#.Um AMx6B59kg.

Food and Water Watch Report (2008). "Land Based Recirculating Aquacultural Systems". Available at www.foodandwaterwatch.org › Reports › Fish.

Losordo TM, Masser MP, Rakosy J (1998). Recirculating Aquaculture Tanks Production System, An Overview of Critical Considerations. Southern Regional Aquaculture Center, SRAC Publication No. 451.

Masser MP, Rakosy J, Losordo TM (1999). Recirculating Aquaculture Tanks Production System, Management of Recirculating System. Southern Regional Aquaculture Center, SRAC Publication No. 452.

NAERLS (2000). Econonomics of Aquacultural production. Extension Bulletin No107, Fisheries series No.5 published by National Agricultural Extension and Research Liaison services Ahmadu Bello University, Zaria.

Parin AZ, Lupin MH (1995). Food and Agriculture Organization of the United Nations. Fisheries Technical Paper, P. 351.

Poulson TM (2013). "Recirculating Aquaculture". Available at http://www.eurofish.dk /~efweb/ images/stories/files/Turkey/8-TMP.pdf.

Rakacy JE (2006). "Recirculating Aquaculture Tank Production System". Available at *www.ca.uky.edu/wkrec/454fs.PDF.*

Sioux Indian Reservation (2006). "Design Guide for Recirculating Aquaculture System", Available at http://www.rowan.edu/colleges/engineering/clinics/engwoborders /Rep orts/EWB_fish_hatchery_final.pdf.

The Fish Site (2010). "Recirculating Water Systems used for Fish Production", Available at http://www.thefishsite.com/fishnews/13400/recirculating-water-systems-used-for-fish-production.

Zhang SY, Gu L, Hui BW, Xing GL, Yan HY, Ling T, Hung L (2011). "An Integrated RecirculatingAquaculture System for Land-Based Fish Farming. The Effects of Water Quality and Fish Production. Aquacultural Eng. 45(3):93-102.

Crustal and upper mantle electrical conductivity structure in north central Nigeria

Obiora, Daniel N. and Okeke, Francisca N.

Department of Physics and Astronomy, University of Nigeria, Nsukka, Enugu State, Nigeria.

Separated spherical harmonic analysis coefficients of the external and internal parts of the observed quiet-day geomagnetic field variations (Sq) for the North Central Nigeria were used to determine the conductivity profile to depths of about 873 km by Schmucker equivalent substitute conductor method. Within the crust, the conductivity increased from 0.027 S/m at a depth of 7.4 km to 0.074 S/m at 15.5 km and 0.098 S/m at 24 km depth. It suddenly rose to 0.181 S/m at 26.5 km and then decreased to 0.131 S/m at 37.1 km depth. The conductivity within the upper mantle rose gradually from 0.043 S/m at 60.4 km to 0.045 S/m at 100.7 km and reached 0.071 S/m at 220.6 km. It fluctuated from 0.092 S/m at 273.6 km to 0.105 S/m at 457.5 km and got to 0.118 S/m at 523.1 km. Finally, it reached 0.163 S/m at 601.9 km and 0.271 S/m at 727.3 km depth. There seemed to be some evidence of discontinuities near 71-165 km, 165-221 km, 221-405 km and 405-666 km. The region showed a roughly exponential increase of conductivity with depth. The profile gave evidence of a less steep increase in conductivity with depth to about 405 km and very steep increase in conductivity thereafter.

Key words: North Central Nigeria, crust, upper mantle, spherical harmonic analysis coefficient (SHA), geomagnetic field variation, quiet day, electrical conductivity-depth structure.

INTRODUCTION

The solar quiet daily field variations (Sq) provide a natural signal source with frequencies appropriate to upper mantle conductivity studies. Field variation measurements at the observatories are sensitive indicators of a number of physical changes that transpire between the sun and the Earth's surface. The principal cause of the geomagnetic quiet day field variations is the ionospheric dynamo current created when there is a force on the ionized region of the atmosphere in the presence of the Earth's main field. Selective characteristics of the collision frequencies of the ionized atmospheric particles make the E-region electrons near 100 km the most suitable current carriers (Campbell, 1987). A force on these electrons is created by the day-to-night thermotidal changes in the atmosphere as the sun rises and falls daily through the year and by some upper atmospheric winds of global scale.

The quiet condition ionospheric source currents induce secondary currents in the conducting Earth. The fields from the Sq system penetrate beneath the crustal levels to a depth dependent upon the effective wavelength of the source and the conducting properties of the deep Earth. At the surface observatories, a summation of the source and secondary fields are recorded. A Gaussian spherical harmonic analysis (SHA) method allows the separation of the source and induced fields representing their potential functions as two converging series of terms whose coefficients are indentified by order m and degree n indices. It is assumed that the ionospheric current system can be considered as fixed with respect to the sun as the earth rotates under the system. In this view, an observatory samples the field of this current through

360° of longitude in 24 h. With this assumption, mathematical hemisphere has been established on whose surface the field variations are responding to an external vortex current source fixed in longitude and to a conducting Earth structure that is symmetric about the axis of the sphere (Campbell and Schiffmacher, 1988).

Determination of electrical conductivity as a function of depth helps to provide knowledge of physical state and chemical composition at different depths of the Earth's interior. Anomalies of electrical conductivity are very useful in identifying the zones of melting and dehydration. Hence, delineation of these zones is very important in understanding the mobile areas of the Earth's crust and upper mantle, where tectonic movements and regional metamorphism lead to distinct patterns of subsurface conductivity (Chandrasekhar, 2011). The study of the physics of the Earth's interior, particularly in terms of the variation of the electrical conductivity with depth is very essential.

The primary purpose of this study is to estimate the general conductivity profile of the crust-upper mantle portion of the Earth in the North Central Nigeria. The resulting conductivity profile will be compared with conductivity profiles of other researchers.

METHODOLOGY

Source of data

The average hourly geomagnetic data used in this study were obtained from geomagnetic stations established in parts of the region (Abuja; 9° 40'N, 7° 29'E and Ilorin; 8°30'N, 4°33'E) by magnetic data acquisition set (MAGDAS), Japan for the years 2008, 2009 and 2010.

Method of analysis

The spherical harmonic analysis coefficients are determined from the global distribution of Fourier coefficients. With the order m (values 1- 4) and degree n (values 1-12), the external cosine and sine coefficients are computed from (Campbell, 2003):

$$a_n^{me} = \frac{(n+1)a_n^m + c_n^m}{2n+1} \quad (1a)$$

$$b_n^{me} = \frac{(n+1)b_n^m + d_n^m}{2n+1} \quad (1b)$$

and the internal cosine and sine coefficients from

$$a_n^{mi} = \frac{na_n^m - c_n^m}{2n+1} \quad (1c)$$

$$b_n^{mi} = \frac{nb_n^m - d_n^m}{2n+1} \quad (1d)$$

a_n^m, b_n^m, c_n^m and d_n^m are called intermediate coefficients and they are computed from

$$a_n^m = \frac{2n+1}{4n(n+1)} \int_0^{180} \left[X_c^m \frac{dP_n^m}{d\theta} \sin(\theta) + Y_s^m m P_n^m \right] d\theta \quad (2a)$$

$$b_n^m = \frac{2n+1}{4n(n+1)} \int_0^{180} \left[X_s^m \frac{dP_n^m}{d\theta} \sin(\theta) - Y_c^m m P_n^m \right] d\theta \quad (2b)$$

$$c_n^m = \frac{2n+1}{4} \int_0^{180} Z_c^m P_n^m \sin(\theta) d\theta \quad (2c)$$

$$d_n^m = \frac{2n+1}{4} \int_0^{180} Z_s^m P_n^m \sin(\theta) d\theta \quad (2d)$$

The integral sign in equations 2a-d means a summation over a θ range of 0 to 180°. The angle θ is the geomagnetic colatitudes and dθ is the step increment (2.5°) of the analysis; P_n^m is the Schmidt normalized associated Legendre function (Campbell, 1997). The size of these steps is selected to be appropriate to the wavelength resolution that is to be accomplished by the SHA fitting.

The conductivity determination depends upon the separated external and internal SHA coefficients computed for the analysis area. Schmucker (1970) first introduced the method of profiling the Earth's conductivity with a transfer function using the external and internal spherical harmonic coefficients at a given site. This function gave the depth to equivalent substitute conductors that would produce the observed fields at the Earth's surface. Schmucker's complex transfer function C_n^m, has real z and imaginary –p parts which Campbell and Anderssen (1983) wrote in terms of external and internal SHA coefficients:

$$z = \frac{R}{n(n+1)} \left\{ \frac{A_n^m \left[na_n^{me} - (n+1)a_n^{mi} \right] + B_n^m \left[nb_n^{me} - (n+1)b_n^{mi} \right]}{(A_n^m)^2 + (B_n^m)^2} \right\}, \quad (3)$$

and

$$p = \frac{R}{n(n+1)} \left\{ \frac{A_n^m \left[nb_n^{me} - (n+1)b_n^{mi} \right] - B_n^m \left[na_n^{me} - (n+1)a_n^{mi} \right]}{(A_n^m)^2 + (B_n^m)^2} \right\} \quad (4)$$

where R in kilometer is the Earth's radius, z and p are in kilometers and the coefficient sums are given by

$$a_n^{me} + a_n^{mi} = A_n^m \text{ and } b_n^{me} + b_n^{mi} = B_n^m \quad (5)$$

For each n, m set of coefficients, the depth (km) to the uniform substitute layer is given by

$$d_n^m = z - p, \quad (6)$$

with a substitute layer conductivity (S/m) of

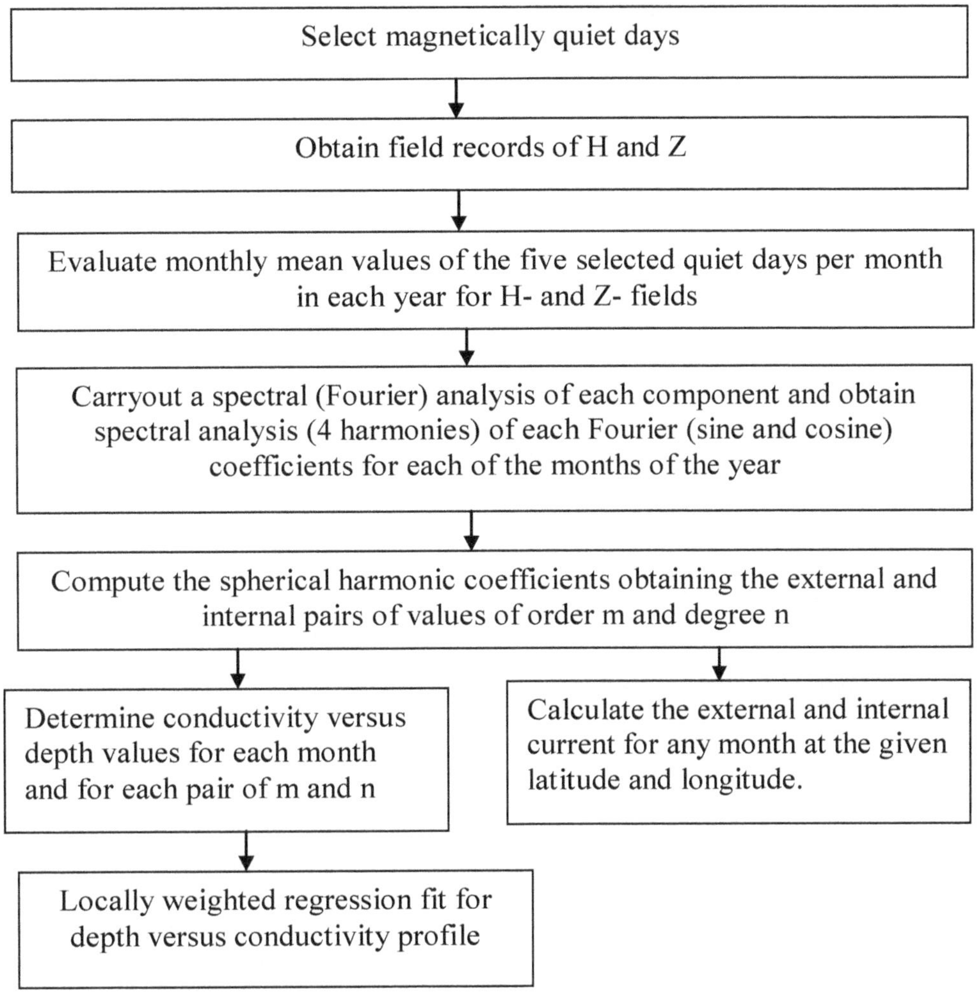

Figure 1. Data processing flow chart.

$$\sigma_n^m = \frac{5.4 \times 10^4}{m(\pi p)^2} \tag{7}$$

The ratio S_n^m of the internal to external components of the geomagnetic surface field is

$$S_n^m = u + iv \tag{8}$$

Where

$$u = \frac{\left(a_n^{me}\right)\left(a_n^{mi}\right) + \left(b_n^{me}\right)\left(b_n^{mi}\right)}{\left(a_n^{me}\right)^2 + \left(b_n^{me}\right)^2} \tag{9}$$

and

$$v = \frac{\left(b_n^{me}\right)\left(a_n^{mi}\right) - \left(a_n^{me}\right)\left(b_n^{mi}\right)}{\left(a_n^{me}\right)^2 + \left(b_n^{me}\right)^2} \tag{10}$$

The validity of Equations 6 and 7 is limited by the conditions that:

$$0^0 \geq \arg\left(C_n^m\right) \geq -45^0 \tag{11a}$$

and

$$80^0 \geq \arg\left(S_n^m\right) \geq 10 \cdot 5^0 \tag{11b}$$

Also the SHA coefficient amplitudes must not be too small because the relative errors inherent in the SHA coefficients increase as the amplitudes of the coefficients decrease.

The procedure for method of analysis is summarized in Figure 1. The original data set for this study comprised the Sq variations of five quietest days for the stations for each month of the year; 2008, 2009 and 2010. The analysis started with the selection of five magnetically quietest days from international quiet days (IQDs) in each month for the years in which the data were obtained. The hourly values for the five quietest days were summed for each month hour by hour and the average value calculated. This monthly average helps to eliminate the daily variability in the data. The Fourier coefficients were appropriately smoothed with respect to geomagnetic latitude and then, the SHA method was used on 2.5°

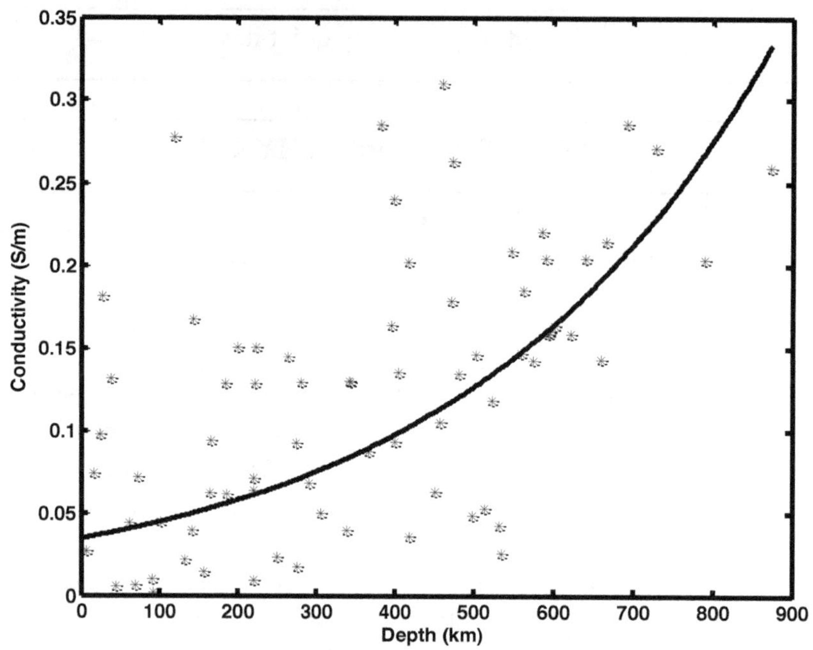

Figure 2. Crust-mantle electrical conductivity profile of North Central Nigeria.

latitude samples to obtain the Gauss coefficients for order 4 and degree 12. The local time change of field was made equivalent to longitude position. The depths to conductive layers were then computed from Equations 3, 4 and 6, while the associated conductivities were calculated from Equation 7.

RESULTS AND DISCUSSION

Figure 2 shows the resulting conductivity profile. The conductivity values clustered more between 60 and 666 km depths. The small blocks represent the conductivity-depth computation values while the solid line is the exponential regression fitted curve represented by:

$$\sigma = 0.0344e^{0.0026d} \text{ (S/m)} \quad (12)$$

where d is depth in kilometers. Within the crust, the conductivity increased from 0.027 S/m at a depth of 7.4 km to 0.074 S/m at 15.5 km and 0.098 S/m at 24 km depth. It suddenly rose to 0.181 S/m at 26.5 km and then decreased to 0.131 S/m at 37.1 km depth. The conductivity within the upper mantle rose gradually from 0.043 S/m at 60.4 km to 0.045 S/m at 100.7 km and reached 0.071 S/m at 220.6 km. It fluctuated from 0.092 S/m at 273.6 km to 0.105 S/m at 457.5 km and got to 0.118 S/m at 523.1 km. Finally, it reached 0.163 S/m at 601.9 km and 0.271 S/m at 727.3 km depth. There seemed to be some evidence of discontinuities near 71-165 km, 165-221 km, 221-405 km and 405-666 km. These locations are near phase change depths identified on seismic records (Dziewonski and Anderson, 1981). Equation 12 indicates the general trend of the conductivity with depth in the studied range. Although the function is drawn from the surface to 873 km, the reliable section lies between about 15 to 666 km. The notable high conductivity value seen at a depth of 462 km may be due to the same effect noted in global studies and may correspond to the olivine-spinel phase transition (Garland, 1981; Lilley et al., 1981).

The profile indicates the existence of high conductivity zone at depths between 15.5 and 37 km in the crust which is close to the result obtained by Ritz (1984) in Kedougou site, Senegal. Two interpretations may be proposed to explain this high conductivity values in the crust: (i) the existence of conductive graphite associated with extensive shear zones in the Precambrian basement (Gough, 1983); (ii) the incorporation of hydrated conductive oceanic materials in the continental crust (Drury and Niblett, 1980). The North Central Nigeria is located within the Precambrian basement in the geology of Nigeria (Obaje, 2009).

Velocity-depth profiles obtained from seismic waves by Dziewonski and Anderson (1981) averaged for the full Earth show that an abrupt rise in velocity bounds the crust and upper mantle at about 6 to 75 km. Between about 100 and 220 km deep, a low velocity zone is encountered beyond which the velocity increases gradually with steps near 400 and 670 km and high conductivity values are noted at 462 and 666 km depths in this work. The noted correlation between our conductivity profile and seismic zones might be an indirect manifestation of common process acting differently on both parameters. For a particular composition and phase of Earth material, the electrical

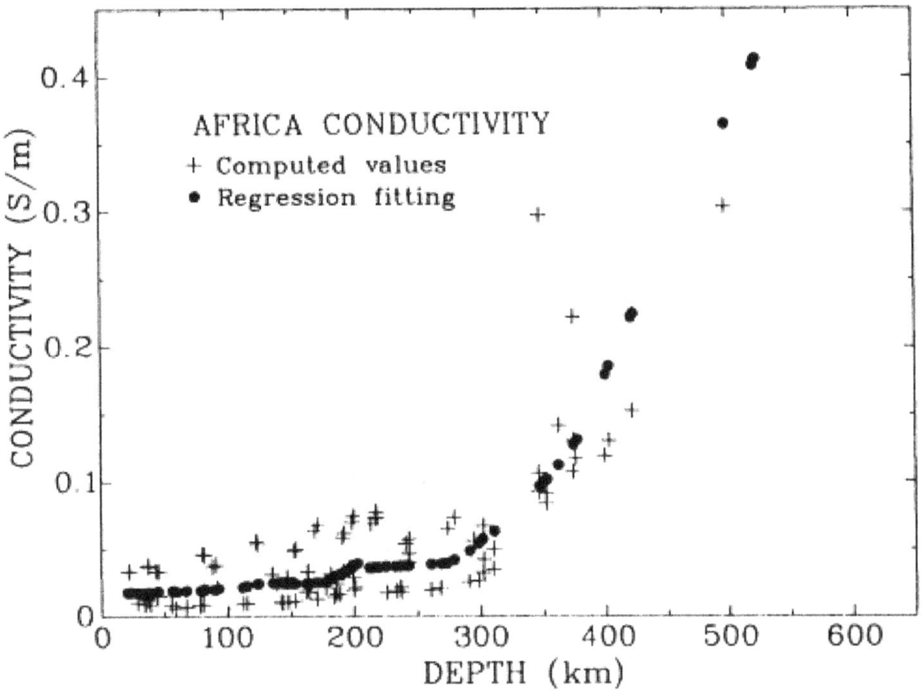

Figure 3. African region conductivity profile (Campbell and Schiffmacher, 1988).

conductivity rises almost exponentially with the negative reciprocal of the temperature (Tozer, 1970). The Earth's temperature increases with depth; therefore, for a homogeneous region the electrical conductivity increases with depth. Phase transition steps in seismic velocity occur at depths in the Earth where enhanced temperatures and pressures cause a readjustment of the mineral structure or a major composition change.

The results of this study could be compared with the results of Campbell and Schiffmacher (1988) on upper mantle electrical conductivity for seven sub-continental regions of the Earth. Their result for North American profile showed elevated conductivity from about 20 to 60 km depth; South America has an extremely high conductivity at depths less than 100 km; African and East Asian regions show no highly conducting regions at shallow depths (which corresponds with the result of this study); the central Asian region indicates increased values between about 150 to 200 km and at more than 400 km depth, Africa and central Asia have high conductivity values which agrees with the results of this study. Figure 3 is the African region conductivity profile by Campbell and Schiffmacher (1988) which is compared with the profile of the present study.

We also compared our work with that of Campbell and Anderssen (1983) on conductivity of the sub-continental upper mantle: an analysis using quiet-day geomagnetic records of North America. Their result showed that from depths of about 140 to 540 km, the conductivity in $(\Omega\text{-m})^{-1}$, may be represented by:

$$\sigma = 0.0067e^{0.0070d}$$

where d is depth in km. They also had small perturbations of conductivity indicating some layering at 140 to 220, 220 to 400 and 400 to 600 km which are close to our result. Figure 4 shows the conductivity profile of Campbell and Anderssen (1983).

These three profiles (present study, Campbell and Schiffmacher, 1988 and Campbell and Anderssen, 1983) show similar trend. Figure 5 shows the three profiles. The maximum depths for Campbell and Schiffmacher (1988) was not up to 600 km while that of Campbell and Anderssen (1983) was around 600 km, but the depth for this study exceeded 700 km. Our results have higher conductivity values above about 350 km depth than that of Campbell and Schiffmacher (1988) and above about 380 km for Campbell and Anderssen (1983). The conductivity values for the three profiles agreed between about 350 and 400 km. All the profiles show high conductivity values from about 380 km and below. Our conductivity values are smaller from about 380 km and below. The difference in the conductivity values may be due to lateral inhomogeneities.

Electrical conductivity in the Earth depends on the amount of free particle charges and their mobility. It is usually considered that the solid rock forming minerals are almost insulators and that any conductivity above about 10^{-4} S/m is due to interstitial water. The conductivity of this water varies depending on the amount and nature of dissolved salts, but the main controlling parameter is

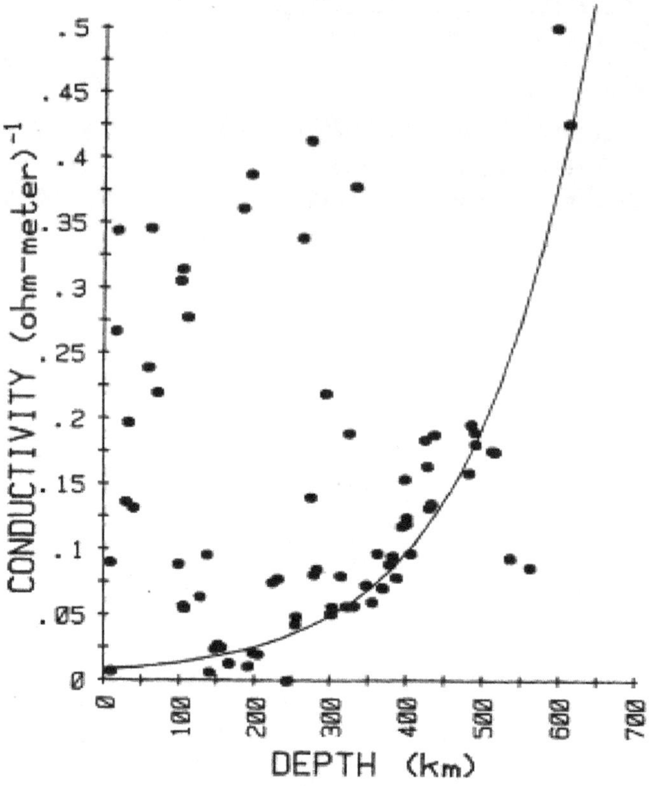

Figure 4. Conductivity profile of the Sub-continental upper mantle (Campbell and Anderssen, 1983).

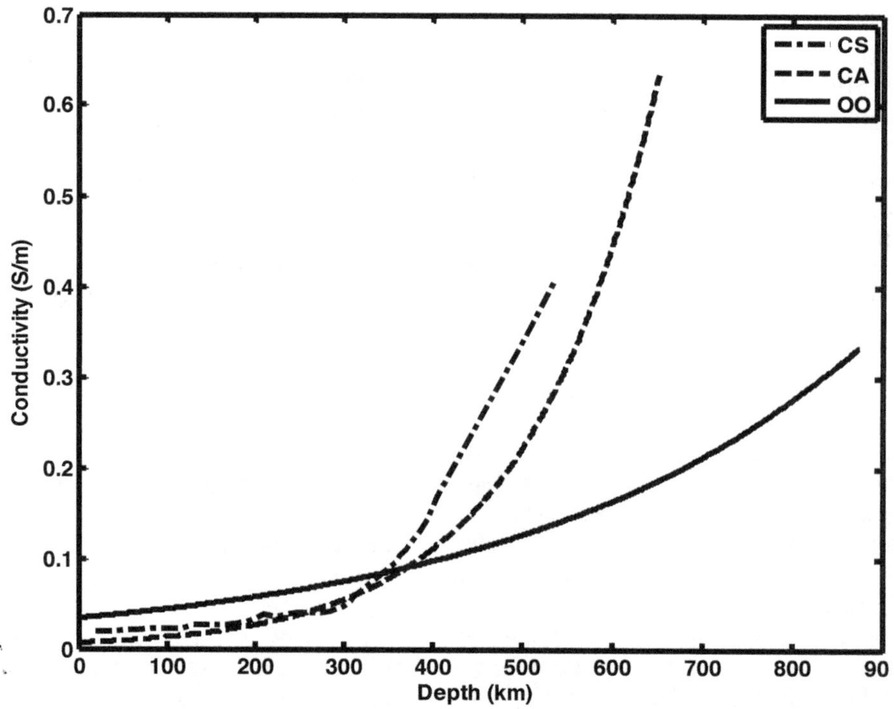

Figure 5. Comparison of the conductivity profile of the present work with other models: OO (Present study); CS (Campbell and Schiffmacher, 1988); CA(Campbell and Anderssen, 1983).

porosity. In many parts of the world, conductivities of about 0.2 S/m are found in the middle or lower crust (Parkinson and Hutton, 1989). The amphibolites-granulite transition causes the release of water. The presence of this water may be sufficient to account for the conductivity anomalies in the crust (Hyndman and Hyndman, 1968). Alternatively, the water released may depress the solidus sufficiently to cause partial melting (Adam, 1978). Hermance and Pedersen (1980) favour partial melting to explain the crustal (20 km) conductivity anomaly beneath the Rio Grande rift zone, which is an area of heat flow. Factors for increase in electrical conductivities of the continental crust include: free water with a high ionic content (fluids); free carbon (graphite); other conducting minerals (such as magnetic oxides or sulphur) and rock melts (Schwarz, 1990).

At depths of about 80 km under continents, seismic S waves are found to have a lower velocity and higher attenuation than immediately above and below this level. Conductivity is found to increase in the same depth range, giving rise to the intermediate conducting layer (ICL). This region was suggested for low velocity layer, known as the asthenosphere hypothesized to account for the replacement of crustal material on the release of stress (Jeffreys, 1952). The concept that explained these three physical parameters (conductivity, seismic velocity, and viscosity) is the presence of a small percentage of partial melt (Shankland and Waff, 1977; Adam, 1980). The best estimates of geotherms and solidus of mantle material as a function of depth indicate that, except where heat flow is unusually low, the geotherm crosses the solidus near this depth. Tozer (1981) pointed out the similarity between conductivity and seismic parameters and considered viscosity to be the common link between the two variables. He emphasized the importance of convection in the mantle and on this basis, believed that temperatures are never high enough for partial melt to be important and that water released by the instability of amphibole is the main agent controlling conductivity in the mantle.

Olivine is considered to be the chief constituent of mantle rocks. It is not stable at high temperature and pressure conditions (Katsura and Ito, 1989), implying that the mantle mineralogy changes with depth. It is believed that the transformation of α-phase of olivine into β - and γ- phases is primarily responsible for the discontinuous changes in the electrical conductivity at greater depths and that the much reported seismic discontinuity at 410 km is attributed to be α - β transformation at high pressure and temperature (Ito and Katsura, 1989). Both the β - and γ- phases have much higher elastic wave velocities than appropriate for mantle below 410 km (Anderson, 1989). The dissociation of γ-phase is believed to be related to the well-known seismic discontinuity at 660 km (Ito and Takahashi, 1988).

Magneto-telluric (MT) soundings in continental areas frequently showed a general reduction in mantle resistivity between depths of 80 and 190 km (Schmucker and Jankowski, 1972; Drury, 1978) and these depths might correspond to the lithosphere-asthenosphere boundary. This implies increase in conductivity which corresponds with our result. Patton (1980) in an analysis of data from the 'Northern Platforms and Shields', which included the East European platform, the Baltic shield, Greenland and part of the Canadian shield, concluded that there must exist a shear wave low-velocity-layer, LVsL, between 80-250 km depth underlying the whole region. Low velocity zones are identified with high conductivities.

Conclusion

The electrical conductivity profile of the crust and upper mantle of the North Central Nigeria at depths of about 7 to 873 km was determined using the quiet ionospheric current variations observed within the region. The conductivity increased exponentially with depth. There appeared to be distinct discontinuities near 71-165 km, 165-221 km, 221-405 km and 405-666 km. Although the function is drawn from the surface to 873 km, the reliable section lies within about 15 to 666 km. Our conductivity profile compares favourably with those of other regions (Campbell and Schiffmacher, 1988; Campbell and Anderssen, 1983). The trend of our conductivity profile is generally similar in values to that found for Africa and East Asia. Below 350 to 380 km, our conductivity values were smaller, and higher above 380 km depth. The high conductivity values observed in the crust may be due to the existence of conductive graphite associated with extensive shear zones in the Precambrian basement and the incorporation of hydrated conductive oceanic materials in the continental crust. The rapid increase in conductivity observed below 400 km depth is in conformity with the works of Campbell and Schiffmacher (1988) that the upper mantle under Africa and Asian regions is highly conductive.

ACKNOWLEDGEMENTS

We wish to thank International Center for Space Weather Science and Education, Kyushu University, Fukuoka, Japan, for providing us with data and Dan Okoh of Centre for Basic Space Science, University of Nigeria, Nsukka for helping us with matlab programme during the analysis of this work.

REFERENCES

Adam A (1978). Geothermal effects in the formation of electrically conducting zones and temperature distribution in the Earth. Phys. Earth Planet. Int. 17:21-28.

Adam A (1980). Relation of mantle conductivity to physical conditions in the asthenosphere. Geophys. Surv. 4:43-55.
Anderson DL (1989). Theory of the Earth. Blackwell, London. pp. 75-120
Campbell WH (1987). The upper mantle conductivity analysis method using observatory records of the geomagnetic field. Pure Appl. Geophys. 125:427-457.
Campbell WH (1997). Introduction to Geomagnetic Fields. Cambridge University Press, New York. pp. 22-24.
Campbell WH (2003). Introduction to Geomagnetic Fields, 2nd edition. Cambridge University Press, New York. pp. 26-27.
Campbell WH, Anderssen RS (1983). Conductivity of the subcontinental upper mantle: an analysis using quiet-day records of North America. J. Geomag. Geoelectr. 35:367-382.
Campbell WH, Schiffmacher ER (1988). Upper mantle electrical conductivity for seven subcontinental regions of the earth. J. Geomag Geoelectr. 40:1387–1406.
Chandrasekhar E (2011). Regional electromagnetic induction studies using long period geomagnetic variations. In: The Earth's Magnetic Interior, eds. Petrovský E et al. IAGA Special Sopron Book Series 1, 31 DOI 10.1007/978-94-007-0323-0. Springer Dordrecht Heidelberg London, New York. P. 32.
Drury MJ (1978). Partial melt in the asthenosphere: evidence from electrical conductivity data. Phys. Earth Planet. Int. 17:16-20.
Drury MJ, Niblett ER (1980). Buried ocean crust and continental crust geomagnetic induction anomalies: a possible association. Can. J. Earth Sci. 17:961-967.
Dziewonski AM, Anderson DL (1981). Preliminary reference Earth model. Phys. Earth Planet Int. 25:297-356.
Garland GD (1981). The significance of terrestrial electrical conductivity variations. Ann. Rev. Earth Planet. Sci. 9:147-174.
Gough DI (1983). Electromagnetic geophysics and global tectonics. J. Geophys. Res. 88:3367-3377.
Hermance J, Pedersen J (1980). Deep structure of the Rio Grande rift: A magnetotelluric interpretation. J. Geophys. Res. 85(B7):3899-3912.
Hyndman RD, Hyndman DW (1968). Water saturation and high electrical conductivity in the lower continental crust. Earth Planet. Sci. Lett. 4:427-432.
Ito E, Katsura T (1989). A temperature profile of the mantle transition zone. Geophys. Res. Lett. 16:425-428.
Ito E, Takahashi E (1988). Post-spinel transformations in the system $MgSiO_4$-Fe_2SiO_4 and some geophysical implications. J. Geophys. Res. 94:10637–10646.
Jeffreys H (1952). The Earth, 3rd edition. Cambridge University Press. P. 169.
Katsura T, Ito E (1989). The system of Mg_2SiO_4-Fe_2SiO_4 at high pressures and temperatures: precise determination of stabilities of olivine, modified spinel and spinel. J. Geophys. Res. 94:15663–15670.
Lilley FEM, Woods DV, Sloane MN (1981). Electrical conductivity profiles and implications for the absence or presence of partial melting beneath central and southeast Australia. Phys. Earth Planet. Int. 25:419-428.
Obaje NG (2009). Geology and Mineral Resources of Nigeria. Springer Dordrecht Heidelberg, London. pp. 2-15.
Parkinson WD, Hutton VRS (1989). The electrical conductivity of the earth. In: Geomagnetism, Vol.3 ed. Jacobs JA. Academic Press, London. pp. 261-321.
Patton H (1980). Crust and upper mantle structure of the Eurasian continent from the phase velocity and Q of surface waves. Rev. Geophys. Space Phys. 18:605-625.
Ritz M (1984). Inhomogeneous structure of the Senegal lithosphere from deep magnetotelluric soundings. J. Geophys. Res. 89:11317-11331.
Schmucker U (1970). An introduction to induction anomalies. J. Geomag. Geoelectr. 2:9–33.
Schmucker U, Jankowski J (1972). Geomagnetic induction studies and the electrical state of the upper mantle. Tectonophysics. 13:233-256.
Schwarz G (1990). Electrical conductivity of the Earth's crust and upper mantle. Surveys in Geophysics. 11:133-161.
Shankland TJ, Waff HS (1977). Partial melting and electrical conductivity anomalies in the upper mantle. J. Geophys. Res. 82: 5409-5417.
Tozer DC (1970). Temperature, conductivity, composition and heat flow. J. Geomag. Geoelectr. 22:35-51.
Tozer DC (1981). The mechanical and electrical properties of Earth's asthenosphere. Phys. Earth Planet. Int. 25:280-296.

25

Photoluminescence from GaAs nanostructures

Alemu Gurmessa[1], Getnet Melese[2], Lingamaneni Veerayya Choudary[3] and Sisay Shewamare[4]

Department of Physics, Jimma University, Ethiopia.

The confinement properties of semiconductor nanostructures have promising potential in technological application. The main objective of this study is to describe the dependence of Photoluminescence (PL) intensity on different parameters like temperature, excitation wavelength, time and photon energy of GaAs quantum dots (QDs). The model equations are numerically analyzed and simulated with matlab and FORTRAN codes. The experimental fitted values and physical properties of materials are used as data source for our simulation. The result shows that at low temperature the peak is quite sharp, as temperature increases the PL intensity decreases and get quenched at particular thermal energy.

Key words: Photoluminescence (PL) intensity, GaAs quantum dots, nanostructures, quantum confinement, thermal quenching energy.

INTRODUCTION

Nanomaterials are the cornerstones of nanoscience and nanotechnology and are anticipated to play an important role in future economy, technology, and human life in general. The strong interests in nanomaterials stem from their unique physical and chemical properties and functionalities that often differ significantly from their corresponding bulk counterparts. Exceptionally large surface area to volume ratios relative to the bulk produces variations in surface state populations that have numerous consequences on material properties (Jin and Christian, 2007).

The small size of nanostructures permits the infamous electronic device scaling for faster operation, lower cost and reduced power consumption. These unique properties enable the variety of electronic, photonic and optoelectronic information storage, communication, energy conversion, catalysis, environmental protection, and space exploration applications based on semiconductor nanostructures (Alivisatos, 1996). Semiconductor nanoparticles, generally considered to be particles of material with diameters in the range of 1 to 10 nm (Pan and Feng, 2008).

GaAs has advantages in electronic properties which are superior to those of silicon. It has a higher saturated electron velocity and higher electron mobility, allowing transistors made from it to function at frequencies in excess of 250 GHz. Unlike silicon junctions, GaAs devices are relatively insensitive to heat owing to their wider band gap. It tends to have less noise than silicon devices especially at high frequencies. Because of its wide direct band gap transition, GaAs is an excellent material for space electronics and optical windows in high power applications. Combined with the high dielectric constant, this property makes GaAs a very good electrical substrate and unlike Si provides natural isolation between devices and circuits (Blakemore, 1982).

Photoluminescence (PL) is the spontaneous emission of light from a material under optical excitation. The excitation energy and intensity are chosen to probe different regions and excitation concentrations in the sample. PL investigations can be used to characterize a variety of material parameters. PL spectroscopy provides electrical characterization; it is selective and extremely sensitive probe of discrete electronic states.

Intensity of the PL signal has received the most attention in the analysis of interfaces. This interest is due to the fact that, although several important mechanisms affect the PL response, it is generally found that large PL signals correlate with good interface properties (Timothy, 2000).

Dependence of PL intensity on different parameters

Temperature dependence of PL emission

The overall shift of the QD emission to lower energies is caused by band gap narrowing. At low temperature (T < 100K) the carriers are captured by the QDs randomly. Once captured, the carriers in the QDs cannot be thermally excited (Teo et al., 1998). So the emission reflects the normal distribution of the QD. Irrespective of the specific quenching mechanism, the temperature dependence PL intensity for GaAs nanostructures with intensity near absolute zero I_0, rate parameter C and activation energy E_T is given by Kittel (2005):

$$I(T) = \frac{I_o}{1 + C\exp\left(\frac{-E_T}{k_B T}\right)} \qquad (1)$$

This equation is used to simulate data on the dependence of PL intensity on temperature.

Optical absorption

The absorption coefficient α describes how the light intensity is attenuated on passing through the material. The intensity transmitted through the sample of thickness z with incident light intensity I_0 is given by:

$$I(z) = I_0 \exp(-\alpha z). \qquad (2)$$

The quantum confinement model

The low-temperature PL spectrum of the GaAs QD ensembles shows a Gaussian profile with a line width broadening of 30nm (Jung et al., 2004).

The photon energy emitted would be slightly larger than the band gap energy (Sze and Kwok, 2007):

$$h\upsilon = \hbar\omega = E = E_c + \frac{\hbar^2 k^2}{2m_e^*} - \left(E_v - \frac{\hbar^2 k^2}{2m_h^*}\right) = E_g + \frac{\hbar^2 k^2}{2m_r^*} \qquad (3)$$

According to the Quantum Confinement (QC) model, the emission wavelength and intensity depend on nanocrystal diameter, size distribution and concentration. This model can explain the general tendency of most experimental results such as the blue shift of the luminescence spectrum with decrease of the GaAs-nanocluster size.

Weak confinement

The electrons and holes can now be thought of as independent particles; excitons are not formed. Separate quantization of motion of the electron and hole is now important factor. The optical spectra should consist of a series of lines due to transitions between sub bands. The shift in energy is now:

$$\Delta E = \frac{\hbar^2 \pi^2}{2\mu R^2}. \qquad (4)$$

Strong confinement

When the excitonic mass is replaced by the reduced mass μ in the weak confinement regime the dominant energy term is the Coulomb term, and quantization of the motion of the exciton occurs. The shift in energy of the lowest energy state is:

$$\Delta E = \frac{\hbar^2 \pi^2}{2m_e^* R^2} \qquad (5)$$

Where, $\frac{1}{M} = \frac{1}{m_e^*} + \frac{1}{m_h^*}$

Assuming that a Gaussian size distribution about the mean diameter d_0 for the nanocrystallites (Mic′ic′ et al., 1997):

$$I(d) = N\frac{1}{\sqrt{2\pi\sigma}}\exp - \frac{(d - d_0)^2}{2\sigma^2} \qquad (6)$$

The number of electrons N in a column diameter d participating in the PL process is proportional to d^2.

Dependence of PL intensity on time

Photoluminescence experimental samples are excited with ultra-short light pulses and the change of the emitted light as a function of time is observed (Lingmin et al., 2009).

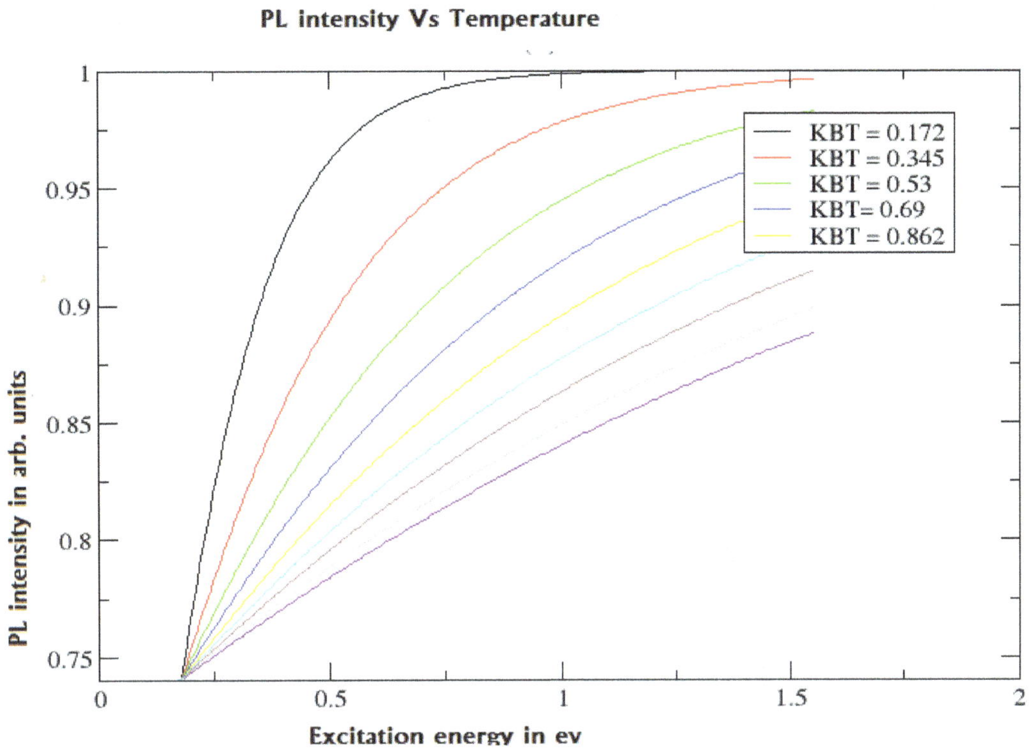

Figure 1. PL intensity versus thermal energy for GaAs nanostructures.

To gain information about the carrier dynamics is necessary to find the relationship between the photoluminescence decay and the carrier lifetimes. The photoluminescence intensity, that is, the light signal emitted by the material due to radiative recombination, is proportional to the product of electron and hole concentrations (GhodsiNahri et al., 2010):

$$I(t) = I_0 exp\left(\frac{-t}{\tau}\right)^{\beta} \quad (7)$$

Where I(t) and I_0 are the PL intensity as a function of time and at t = 0, respectively, τ is the PL decay time constant, and β is a dispersion factor ≤ 1. Thus, the capture rates can be extracted from PL transients if either the radiative recombination rate is known or at least much smaller than the capture rate. However it is not possible to distinguish between the influences of electron and hole capture (GhodsiNahri et al., 2010).

METHODOLOGY

In this work problems are solved analytically and numerical techniques had been used to determine the most important optical parameter for PL intensity. In order to obtain the desired result for this work we have simulated data using matalab codes for temperature and time dependent intensity. We assumed Gaussian model equation that describes the dependence of PL intensity on the size of the nanocluster and matlab Fortran 90 program have been developed for the model equation that simulates the data to compare with experimental results. The results obtained with this simulated data with the model equation agree with experimental results done by other researchers.

RESULT AND DISCUSSION

PL intensity versus temperature

Figure 1 depicts the PL intensity as a function of thermal energy for temperatures ($T = 20k, 40k, 60k, 80k, 100k, 120k, 140k, 160k, 180k$ and $E_a = 0.08\ eV$) fitting parameters are involved. This simulation result is in good agreement with experimental results; the emission spectra for GaAs show that PL intensity very high at low thermal energy. The activation energy is about 0.25 eV. Due to the increasing temperature the thermal energy increase at which 1.5 eV PL gets quenched, thus 1.5 eV is quenching thermal energy.

PL intensity versus wavelength

Figure 2 shows the PL intensity as a function of wavelength for GaAs nanostructures. For both values of

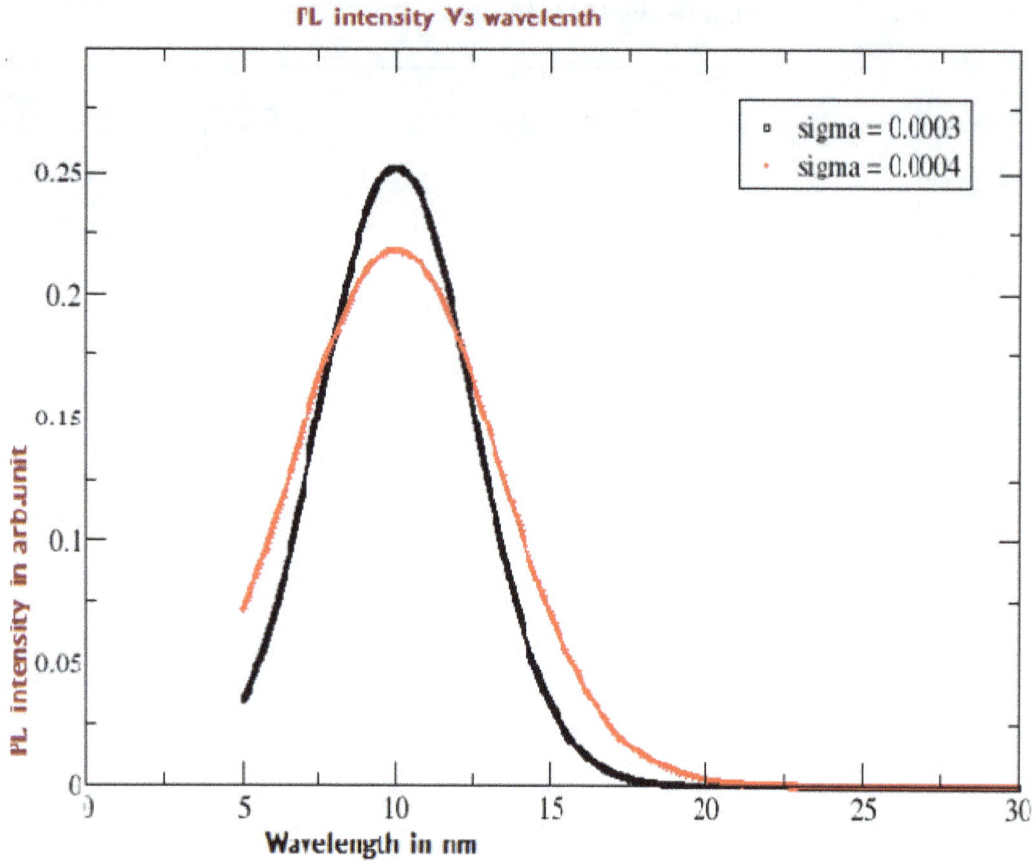

Figure 2. PL intensity versus wavelength.

the standard deviation sigma, the PL intensity has sharper peak when near the center of the Gaussian curve. Semiconductors are transparent to photons whose energies lie below their band gap and are strongly absorbing for photons whose energies exceed the band gap energy (Ardyanian and Ketabi, 2011). The simulation result shows that the PL intensity decreases as the wavelength increases to a certain peak and then decreases continuously.

PL intensity versus photon energy

Figure 3 depicts the simulation result of PL intensity versus Photon energy for GaAS nanostructures. The photo flux absorbed by semiconductor nanostructures enforces the material's property got to be changed from one phase to other. The excitation of electron- hole changed to tunneling of electron from valance band to conduction band.

PL intensity versus time

Figure 4 shows the simulated result of PL intensity versus time (a) and the experimental result carried out by other researchers (b) from literature (Jong, 2009) for comparison. It is observed that as decay life time increases the PL intensity decreases.

The PL decay time of GaAs nanostructures decreases monotonically with increasing time. A shorter decay time constant allows a higher modulation frequency, but reduces the efficiency. This is due to the decrease in oscillation period and resulting non radiative recombination of the particles in the nanocrystal.

Conclusion

The emission spectra of GaAs QD show that there is high PL intensity at low temperature. As the temperature increases, the thermal energy increases and the PL gets quenched. The result also shows that PL intensity decrease as wavelength increases. Sharp peak PL intensity is observed near the mean diameter of GaAs QD. PL intensity observed between limited visible photon energy. As photon energy increased exceeding energy gap the peak PL intensity become lowered. The PL intensity decays with time; this is because of the contributions of radiative and non-radiative transitions.

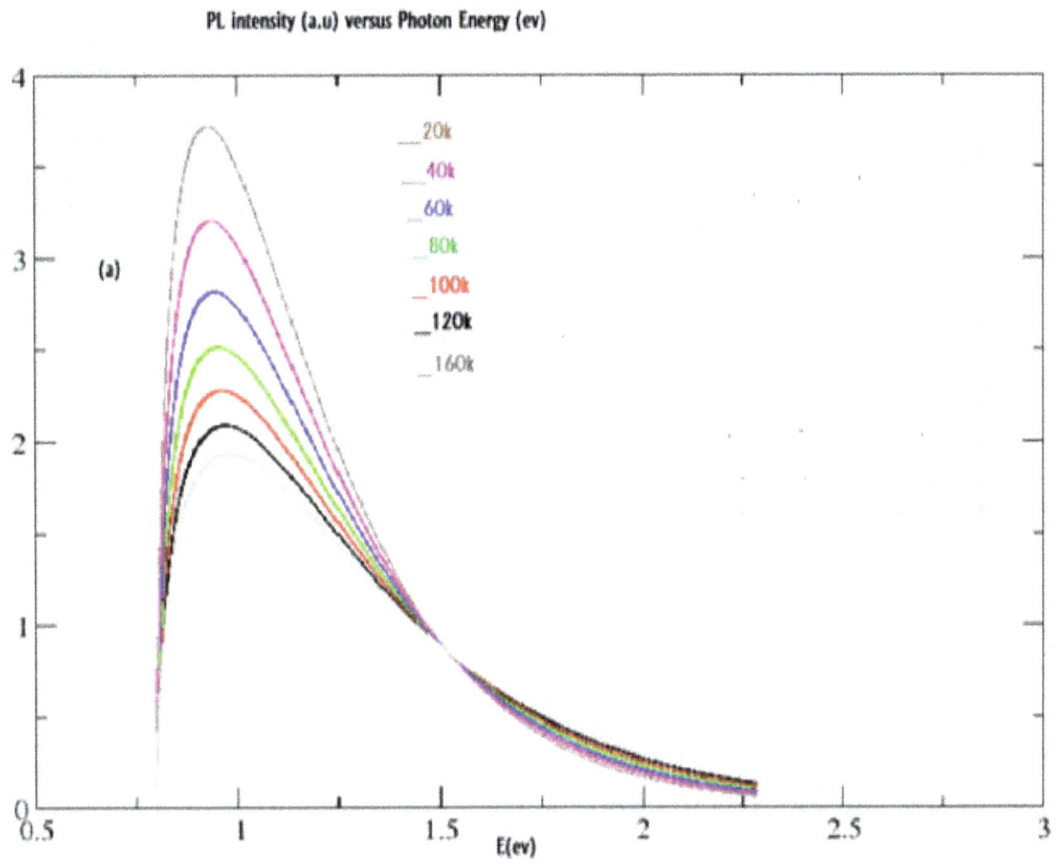

Figure 3. PL intensity Vs. photon energy.

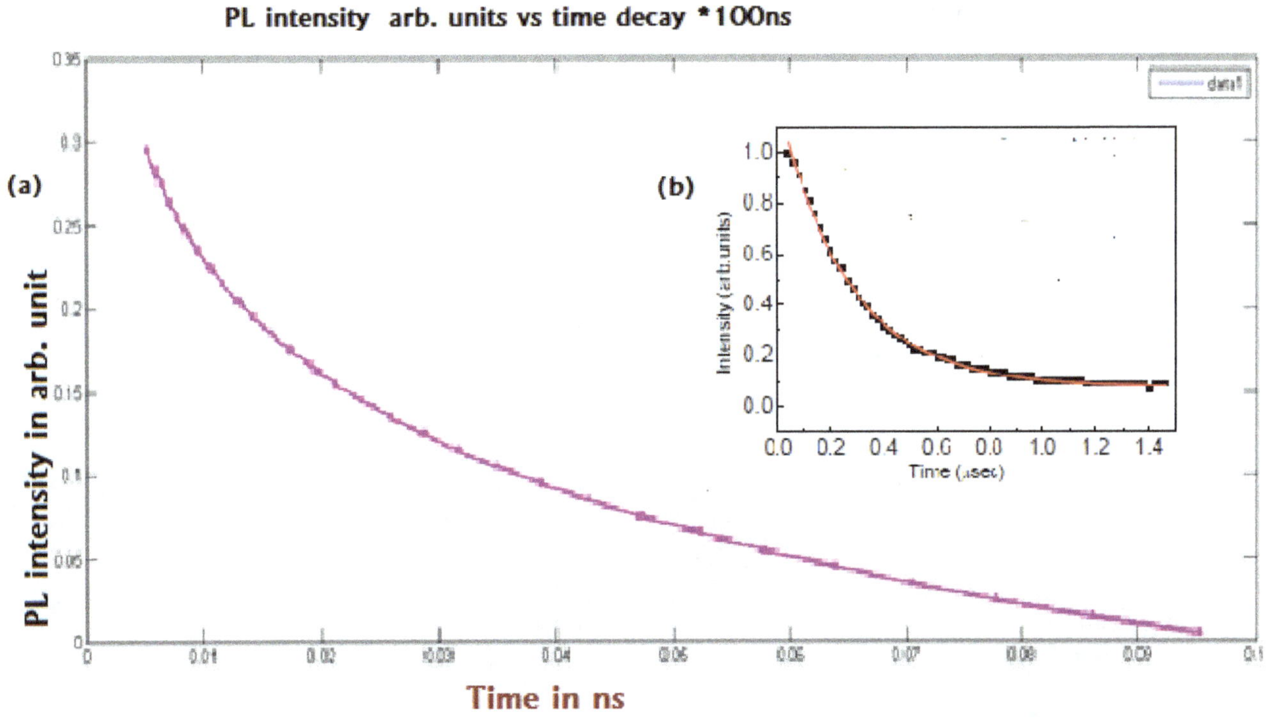

Figure 4. PL intensity vs. decay time.

Conflict of Interest

The authors have not declared any conflict of interest.

ACKNOWLEDGEMENT

The authors expresses their gratitude to Jimma University and Jimma College of Teachers Education for their financial support.

REFERENCES

Alivisatos AP (1996). Semi conductor clusters, nanocrystals and quantum dots science. Nano lett. 271(5251):933-937.

Ardyanian M, Ketabi SA (2011). Time-Resolved Photoluminescence and Photovoltaics, National Renewable Energy Laboratory (NREL), The University of Leeds, UK. 11(3).

Blakemore JS (1982). Semiconducting and other major properties of gallium arsenide, American Institute of Physics.

GhodsiNahri D, Arabshahi H, RezaeeRokn-Abadi M (2010). Analysis of Dynamic and Static Characteristics OF InGaAs/GaAs Self assembled quantum dot lasers. Armenian J. Phys. 3(2):138-149.

Jin ZZ, Christian DG (2007). Optical and dynamic properties of undoped and doped semiconductor nanostructures, University of California, Santa Cruz, CA 95064 USA (2007). https://e-reports-ext.llnl.gov/pdf/353172.pdf

Jong SK (2009). Size Dependence of the Photoluminescence Decay Time in Unstrained GaAs Quantum Dots. J. Korean Phys. Soc. 55(3):10511055.

Jung JH, Im HC, Kim JH, Kim TW, Kwack KD (2004). Optical properties and electronic structures in InAs/GaAs Quantum Dots. J. Korean Phys. Soc. 45:S622-S625.

Kittel C (2005). 8th ed. Introduction to Solid State Physics, eighth ed, John Wiley and Sons, USA.

Lingmin K, ZheChuan F, Zhengyun W, Weijie L (2009). Temperature dependent and time-resolved photoluminescence studie so fInAs self-assembled quantum dots with InGaAs strain reducing layer structure. J. Appl. Phys. 106:01351.

Mic´ic OI, Cheong HM, Fu H, Zunger A, Sprague JR, Mascarenhas A, Nozik AJ (1997). Size-Dependent Spectroscopy of InP Quantum Dots, National Renewable Energy Laboratory, 1617 Cole BouleVard, Colorado 8040(101):4904-4912.

Pan H, Feng YP (2008). Semiconductor nanowire and nanotubes: effects of size and surface to the volume ratio. J. Appl. Phys. 2(11):2410-2414.

Sze SM, Kwok KNg (2007). Physics of Semiconductor Devices. 3rd edition. A John Wiley and Sons, JNC. 3(2):602-607.

Teo KL, Colton JS, Yu PY (1998). An analysis of temperature dependent photoluminescence line shapes in InGaN. Appl. Phys. Lett. USA. 73(12).

Timothy HG (2000). Photoluminescence in Analysis of Surfaces and Interfaces, John Wiley and Sons Ltd, Chichester, pp. 9209-9231.

Permissions

All chapters in this book were first published in IJPS, by Academic Journals; hereby published with permission under the Creative Commons Attribution License or equivalent. Every chapter published in this book has been scrutinized by our experts. Their significance has been extensively debated. The topics covered herein carry significant findings which will fuel the growth of the discipline. They may even be implemented as practical applications or may be referred to as a beginning point for another development.

The contributors of this book come from diverse backgrounds, making this book a truly international effort. This book will bring forth new frontiers with its revolutionizing research information and detailed analysis of the nascent developments around the world.

We would like to thank all the contributing authors for lending their expertise to make the book truly unique. They have played a crucial role in the development of this book. Without their invaluable contributions this book wouldn't have been possible. They have made vital efforts to compile up to date information on the varied aspects of this subject to make this book a valuable addition to the collection of many professionals and students.

This book was conceptualized with the vision of imparting up-to-date information and advanced data in this field. To ensure the same, a matchless editorial board was set up. Every individual on the board went through rigorous rounds of assessment to prove their worth. After which they invested a large part of their time researching and compiling the most relevant data for our readers.

The editorial board has been involved in producing this book since its inception. They have spent rigorous hours researching and exploring the diverse topics which have resulted in the successful publishing of this book. They have passed on their knowledge of decades through this book. To expedite this challenging task, the publisher supported the team at every step. A small team of assistant editors was also appointed to further simplify the editing procedure and attain best results for the readers.

Apart from the editorial board, the designing team has also invested a significant amount of their time in understanding the subject and creating the most relevant covers. They scrutinized every image to scout for the most suitable representation of the subject and create an appropriate cover for the book.

The publishing team has been an ardent support to the editorial, designing and production team. Their endless efforts to recruit the best for this project, has resulted in the accomplishment of this book. They are a veteran in the field of academics and their pool of knowledge is as vast as their experience in printing. Their expertise and guidance has proved useful at every step. Their uncompromising quality standards have made this book an exceptional effort. Their encouragement from time to time has been an inspiration for everyone.

The publisher and the editorial board hope that this book will prove to be a valuable piece of knowledge for researchers, students, practitioners and scholars across the globe.

List of Contributors

Widad M. Faisal
Department of Electronic and Communication, Faculty of Engineering and Petroleum, Hadramout University of Science and Technology, Yemen

Salwan K. J. Al-Ani
Department of Physics, College of Science, University of Baghdad, Baghdad, Iraq

D. S. Yadav
Department of Physics, Ch. Charan Singh P G College, Heonra, Etawah-206001 (U.P.) India

Chakresh Kumar
Department of Electronics and Communication Engineering Tezpur University, Napam-784001, India

Seyed Hossein Hosseini
Department of Chemistry, Faculty of Science, IslamShahr Branch Islamic Azad University, Tehran-Iran

Ghasem Asadi and
Department of Chemistry, Faculty of Science and Engineering, Shar-e-Rey Branch, Islamic Azad University, Tehran- Ghom Express Way, Tehran-Iran

S. Jamal Gohari
Department of Chemistry, Faculty of Science, Imam Hossein University, Babaee Express Way, Tehran - Iran

Sagadevan Suresh
Crystal Growth Centre, Anna University, Chennai-600 025, India

D. N. Obiora
Department of Physics and Astronomy, University of Nigeria, Nsukka, Nigeria

F. N. Okeke
Department of Physics and Astronomy, University of Nigeria, Nsukka, Nigeria

K. Yumoto
Space Environment Research Centre, Kyushu University, Fukuoka, Japan

K. H. Kamarudin
Advanced Materials Research Group, Renewable Energy Research Interest Group, Department of Physical Sciences, Faculty of Science and Technology, Universiti Malaysia Terengganu, 21030 Kuala Terengganu, Terengganu, Malaysia

M. I. N. Isa
Advanced Materials Research Group, Renewable Energy Research Interest Group, Department of Physical Sciences, Faculty of Science and Technology, Universiti Malaysia Terengganu 21030 Kuala Terengganu, Terengganu, Malaysia

M. A. Al-Eshaikh
Research Center, College of Engineering, King Saud University, P. O. Box 800, Riyadh 11421, Saudi Arabia

M. Iqbal Qureshi
Research Center, College of Engineering, King Saud University, P. O. Box 800, Riyadh 11421, Saudi Arabia

Jassim M. Najim
Department of Physics, KHAWLAN Faculty of Education Arts and Science, Sana'a University, Yemen

S. Barış
Department of Mechanical Engineering, Faculty of Engineering, Istanbul University, 34320 Avcilar-Istanbul, Turkey

M. Ş. Demir
Department of Mechanical Engineering, Faculty of Engineering, Istanbul University, 34320 Avcilar-Istanbul, Turkey

N. Ahmed
Department of Mathematics, Gauhati University, Guwahati - 781014, Assam, India

M. Dutta
Department of Mathematics, Gauhati University, Guwahati - 781014, Assam, India

A. O. Awodugba
Department of Pure and Applied Physics, Ladoke Akintola University of Technology, Ogbomoso, Nigeria

Y. K. Sanusi
Department of Pure and Applied Physics, Ladoke Akintola University of Technology, Ogbomoso, Nigeria

J. O. Ajayi
Department of Pure and Applied Physics, Ladoke Akintola University of Technology, Ogbomoso, Nigeria

Sunil Yadav
Department of Applied Science and Humanities, Faculty of Mathematics, Alwar Institute of Engineering and Technology, Alwar North Ext. Alwar-301030, Rajasthan. India

D. L. Suthar
Department of Applied Science and Humanities, Faculty of Mathematics, Alwar Institute of Engineering and Technology, Alwar North Ext. Alwar-301030, Rajasthan. India

Hatice Asil
Faculty of Education, Kilis 7 Aralik University, 79000 Kilis/Turkey

Kübra Çinar
Department of Physics, Faculty of Sciences, Atatürk University, 25240 Erzurum/Turkey

Emre Gür
Department of Physics, Faculty of Sciences, Atatürk University, 25240 Erzurum/Turkey

Cevdet Co skun
Department of Physics, Faculty of Arts and Sciences, Giresun University, 28100 Giresun/Turkey

Sebahattin Tüzemen
Department of Physics, Faculty of Sciences, Atatürk University, 25240 Erzurum/Turkey

Muzahim I. Azawe
Department of Physics, College of Education, University of Mosul, Mosul, Iraq

Cliff Orori Mosiori
Department of Physics, School of Pure and Applied Sciences, Kenyatta University, Kenya

Amin Mahmoudi
Department of Electrical Engineering, University of Malaya, Kuala Lumpur, Malaysia

Seyed Mahdi Moosavian
Department of Engineering, Shahrood Branch, Islamic Azad University, Shahrood, Iran

Solmaz Kahourzade
Department of Electrical Engineering, University of Malaya, Kuala Lumpur, Malaysia

Seyed Nabi Hashemi Ghiri
Department of Electrical Engineering, Shiraz University, Shiraz, Iran

B. Ardjani
Laboratory of Theoretical Physics, Science Faculty, Tlemcen University, P. O. Box 119, 13000 Tlemcen, Algeria

B. Liani
Laboratory of Theoretical Physics, Science Faculty, Tlemcen University, P. O. Box 119, 13000 Tlemcen, Algeria

I. S. El-Hallag
Department of Chemistry, Faculty of Science, Tanta University, Tanta 31527, Egypt

E.H. El-Mossalamy
Department of Chemistry, Faculty of Science, Benha University, Benha, Egypt

M. K. El-Mansy
Department of Physics, Faculty of Science, Benha University, Benha, Egypt

L. M. Al-Harbi
Department of Chemistry, Faculty of Science King Abdulaziz University, P. O. Box 42805, Jeddah 21551, Saudi Arabia

Hamdi Abdi
Electrical Engineering Department, Faculty of Engineering, Razi University, Kermanshah, Iran

Ramtin Rasoulinezhad
Department of Electrical Engineering, Science and Research Branch, Islamic Azad University, Kermanshah, Iran

Md. Nazmul Hasan
International Islamic University Chittagong (Dhaka Campus), 147, Green road, Dhaka-1205, Bangladesh

Md. Shamimul Haque Choudhury
International Islamic University Chittagong (Dhaka Campus), 147, Green road, Dhaka-1205, Bangladesh
Bangladesh University of Engineering and Technology, Bangladesh

M. Shafiul Alam
International Islamic University Chittagong (Dhaka Campus), 147, Green road, Dhaka-1205, Bangladesh
Bangladesh University of Engineering and Technology, Bangladesh

Muhammad Athar Uddin
International Islamic University Chittagong (Dhaka Campus), 147, Green road, Dhaka-1205, Bangladesh
Bangladesh University of Engineering and Technology, Bangladesh

Essam A. Al-Ammar
Saudi Aramco Chair in Electrical Power, Department of Electrical Engineering, King Saud University, Riyadh, Saudi Arabia

Jafari Najafi, Mahdi
1730 N Lynn ST apt A35, Arlington, VA 22209 USA

N. F. Oparaku
National Centre for Energy Research and Development, University of Nigeria, Nsukka, Nigeria

C. E. Nnaji
National Centre for Energy Research and Development, University of Nigeria, Nsukka, Nigeria

Daniel N Obiora
Department of Physics and Astronomy, University of Nigeria, Nsukka, Enugu State, Nigeria

Francisca N. Okeke
Department of Physics and Astronomy, University of Nigeria, Nsukka, Enugu State, Nigeria

Alemu Gurmessa
Department of Physics, Jimma University, Ethiopia

Getnet Melese
Department of Physics, Jimma University, Ethiopia

Lingamaneni Veerayya Choudary
Department of Physics, Jimma University, Ethiopia

Sisay Shewamare
Department of Physics, Jimma University, Ethiopia

www.ingramcontent.com/pod-product-compliance
Lightning Source LLC
Chambersburg PA
CBHW080704200326
41458CB00013B/4959
* 9 7 8 1 6 8 2 8 5 0 5 8 9 *